Electronic Transitions in Organometalloids

ORGANOMETALLIC CHEMISTRY

A Series of Monographs

BRIAN G. RAMSEY: Electronic Transitions in Organometalloids, 1969.

Other volumes in preparation.

Electronic Transitions in Organometalloids

BRIAN G. RAMSEY

DEPARTMENT OF CHEMISTRY
THE UNIVERSITY OF AKRON
AKRON, OHIO

1969

ACADEMIC PRESS New York and London

ACADEMIC PRESS, INC.
111 Fifth Avenue, New York, New York 10003

United Kingdom Edition published by
ACADEMIC PRESS, INC. (LONDON) LTD.
Berkeley Square House, London W.1

LIBRARY OF CONGRESS CATALOG CARD NUMBER: 68-59168

PRINTED IN THE UNITED STATES OF AMERICA

Once upon a time
When rain made little people in the street
and more elvin things under toadstools hid
I think I remember that
with a new pair of tennis shoes
I could run so fast that I
would
reach
immortal
speeds
outspeeding
breath itself
to breathe
Then stopping just short of the speed of light
would turn to look at the where I was and thus
examine the changeling Me

Book and poem for Margaret, Michael and Sean

Foreword

Organometallic chemistry continues to be one of the most vigorously developing fields in all of chemical science. During the last few years alone, organometallic systems have been central in the development of new theoretical insights, new laboratory syntheses, and even new industrial processes. It is now apparent that great areas of organometallic chemistry have remained underdeveloped for decades, in part because they fell between conventional inorganic and organic chemistry. Indeed, it is now possible to visualize the development of organometallic chemistry as a third major branch of chemistry, overlapping with organic chemistry on one side and inorganic on the other.

With this volume, we initiate a series of monographs devoted to special topics in organometallic chemistry. This book by Professor Brian Ramsey represents the first attempt to systematize the electronic spectra of organometallic compounds of the nontransition metals, and to apply modern theory to their explanation. The subject seems particularly timely, for neither the data nor the theory were available for such a treatment even a few years ago.

Succeeding volumes in this series will treat various aspects of organometallic chemistry — theoretical, structural, synthetic, and practical. We hope that these books will be useful, not only to specialists in the field, but to all who need information on rapidly developing topics in organometallic chemistry.

May 1969

P. M. MAITLIS
F. G. A. STONE
ROBERT WEST

Preface

Five years ago there would have been little justification for writing even a review article on the subject of this book, electronic transitions in the non-transition metal organometallic compounds. Now, however, the author finds new activity and publications appearing faster than they can be incorporated into the manuscript, even with the presence of an Addendum. The delicate fact is that much of the interpretive opinion expressed here is open to revision or total change with the now rapid appearance of better data and theoretical calculations. The matter of better data is particularly important since many molecules discussed here are notoriously difficult to purify, handle, or even synthesize, and details about such matters are often unhappily missing in the literature. The reader is therefore urged to exercise, according to his own knowledge, some judgment about, for example, the appearance of weak or moderate intensity transitions in the reported spectra of some of the less well-characterized molecules discussed here, for we have had to assume in the absence of special knowledge that all that is reported is truth.

This book is written from the personal viewpoint of the author. In order to avoid the criticism of writing a book which was in fact a "nonbook" book in being either only a collection of reprints or a bibliographic collection of numbers and abstracts, I have tried to approach the majority of papers, whose results comprise this book, as if I were in fact the referee of an unpublished paper. We often set our students a similar task. The reader must take into account the fact that where in the text I have been severely critical and attacked the results of some workers for what I believe in my guise as a referee to be good reason, the authors of the work have not had the opportunity to defend themselves. If the point at issue is an important one to the reader, he may certainly want to read the original paper to decide for himself whether the objections raised are valid. In addition, in most instances, I have usually enjoyed not only the advantage of hindsight, but also that advantage which accrues to anyone who is able to view a very large amount of previously scattered work which has all been brought together for the first time.

It is sincerely hoped that this monograph will stimulate the organometallic chemist to obtain reliable ultraviolet visible spectra of the compounds which he synthesizes and that the theoretician will become aware of and feel challenged by the increasing number of new and interesting molecules whose spectra are now known.

I would like to acknowledge the assistance of those who have been kind enough to provide me with helpful discussion and the results of their own work, often before publication: H. Boch (University of Münich), A. Cowley (University of Texas), G. Fraenkel (Ohio State University), H. Gilman (Iowa State University), J. P. Oliver (Wayne State University), C. G. Pitt (Triangle Research Institute), J. R. Spielman (California State College at Los Angeles), Robert West (University of Wisconsin), and K. Yates (University of Toronto). Some special acknowledgment is due Mrs. Wanda Abruzzino who was able to translate my own illegible handwriting into beautiful typewritten copy.

March 1969 BRIAN G. RAMSEY

Contents

CHAPTER I

Introduction to Electronic Transitions

The purpose of this monograph is to collect and present to the average organometallic chemist in some cohesive fashion the considerable but scattered reports of the visible and ultraviolet spectra of nontransition metal organometallics.

Because most of the data in the literature are given in units of millimicrons or angstroms, units with which chemists are most familiar, we have reported all data in units of millimicrons (mμ) rather than reciprocal centimeters (cm^{-1}) or electron volts (eV). We would like to simply remind our reader that millimicron and angstrom units are not linear in energy and that for example a difference of 5 mμ near 200 mμ which represents an energy of 3.5 kcal/mole is at 650 mμ a difference of only about one-tenth of that. For this reason occasionally at points in the discussion where correlation of absorption maxima with energy was required, we have used some linear energy unit such as kilocalories per mole or electron volts for the transition maxima.

A more important arbitrary usage has been adopted. The sign $\rightarrow$, used for example in $\pi \rightarrow \pi^*$, indicates absorption of light and not emission. The opposite usage is that adopted by many spectroscopists. Beyond this point, this chapter consists of a brief elementary treatment of the molecular orbital and valence bond (principally charge transfer) descriptions of molecules and their electronic transitions. This may be passed over by any reader who feels well grounded in this area already.

A. Molecular Orbital Transition Energies

The most popular, most recent, and even in some respects best approach to the understanding of electronic transitions of large molecules has been by the molecular orbital description of molecules. In its simplest form the molecular orbital method consists of deriving n number of approximate one-electron wavefunctions as linear combinations of n atomic wavefunctions. The total wavefunction is then taken as the product of molecular orbitals over occupied orbitals. In the symmetrical case of M–M bonding, we can begin by simply taking the sum and the difference of two atomic orbitals with the same symmetry characteristics about the bond axis, to give two molecular orbitals one lower in energy than the starting M atomic wavefunction and one higher in energy. These new molecular orbitals are termed bonding and antibonding, respectively.

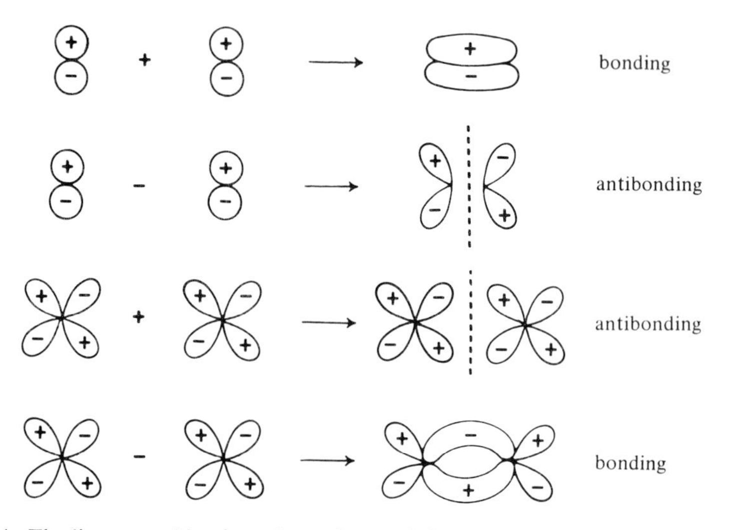

FIG. 1. The linear combination of atomic p and d orbitals into molecular orbitals.

This process is depicted in Fig. 1 for two atomic p orbitals, and for two atomic d orbitals. By inspection we can see from Fig. 1 that the plus combination of p orbitals gives a bonding π molecular orbital whereas it is the minus combination of d orbitals which is bonding.

Once we have obtained a description of a molecule in terms of molecular orbitals, these molecular orbitals are filled with the appropriate number of electrons according to the well-known Aufbau principle, two electrons of opposite spin to each molecular orbital in order of increasing energy. The electronic transitions, which result in the absorption maxima in the visible or ultraviolet spectra of a molecule, are now regarded in a first approximation as one-electron excitations from occupied molecular orbitals to vacant or half-filled orbitals. Two-electron excitation probabilities are normally regarded as vanishingly small. The energy of the transition is taken to be the difference in energy between the two molecular orbitals involved. In this fashion the 165 and 217 mμ transitions of ethylene and butadiene, respectively, are often described as $\pi \rightarrow \pi^*$ transitions promoting an electron from the highest filled to the lowest unfilled π molecular orbitals. Similarly the first weak, long wavelength transition, near 280 mμ, of a ketone which involves the excitation of one of the oxygen nonbonding electrons to the π^* orbital becomes in this notation an $n \rightarrow \pi^*$ transition.

The construction of molecular orbitals is often depicted graphically in figures known as correlation diagrams. Consider in Fig. 2 the formation of

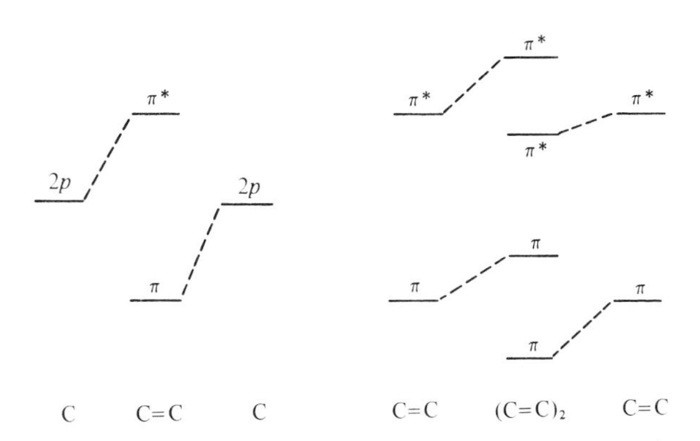

FIG. 2. A correlation diagram for simple molecular orbital energies of $CH_2{=}CH_2$ and $(CH_2{=}CH)_2$.

ethylene from two atomic *p* orbitals. Localized molecular orbitals may be used as basis functions for generating new molecular orbitals in the same fashion as atomic orbitals as long as certain symmetry rules are followed. For example, the molecular orbitals of ethylene may be used to construct those of butadiene in Fig. 2.

From a correlation diagram such as Fig. 2, we can attempt for instance to predict what would happen to the UV maximum of ethylene if we were to substitute a vinyl group for hydrogen. We would in this case correctly predict a shift of the first absorption maximum to the red. The maximum interaction between wavefunctions is obtained when they are of the same energy and decreases as the energy difference increases.

In many cases the description of the ground state, excited state, and electronic transition energy by the one-electron wavefunctions described above is sufficiently accurate to enable good qualitative conclusions and predictions to be made. Often the description of an electronic transition in terms of these simple one-electron wave functions, for which the term *configuration* is used, is completely inadequate. To illustrate this point the simple molecular orbital description of benzene and the configurations of the excited states derived only from transitions between the two highest filled and lowest unfilled molecular orbitals are given in Fig. 3a–d.

If only these simple configurations (Fig. 3a–d) were considered, we would expect that the first absorption maximum of benzene would occur at the same wavelength as ethylene but with about four times the intensity whereas, of course, benzene has not one, but three absorption maxima of varying intensity at 180, 203, and 256 mμ. In order to adequately describe ground states, excited states, and the electronic transitions between them, it is often necessary, as in the benzene example, to introduce what is called *configuration interaction.* Configuration interaction in molecular orbital theory plays the role of resonance in the valence bond approach.

What this means is that better wavefunctions for the electronic states of a molecule are obtained if new wavefunctions are obtained as linear combinations of the one-electron configurations of the same symmetry. As with atomic or molecular orbitals, the amount of interaction between configurations of the same symmetry is greatest when they have the same energy, and decreases as the energy difference between them increases. For this reason, configuration interaction is especially important for higher energy states where the configurations tend to lie closer together in energy. The term *state* is often reserved for the total wavefunction obtained after configuration interaction between the simple one-electron molecular orbital wavefunctions, or configurations. Returning to the benzene question, it can be shown that configuration interaction (or resonance) between the energetically degenerate configurations (Fig. 3) will give rise to three excited states, one of which is doubly degenerate. The observed absorption maxima in the benzene spectrum can therefore be regarded as transitions from the ground state to each of these three possible excited states. Figure 4 is a term level diagram for benzene in which the energies of configurations and of states appear rather than those of the separate molecular orbitals.

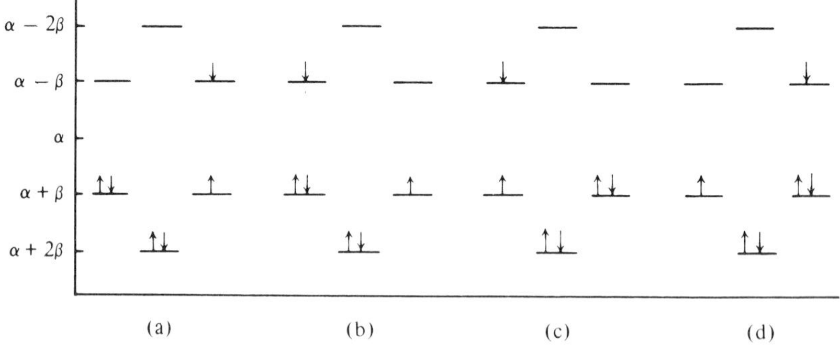

Fig. 3. The lowest energy benzene excited singlet configurations.

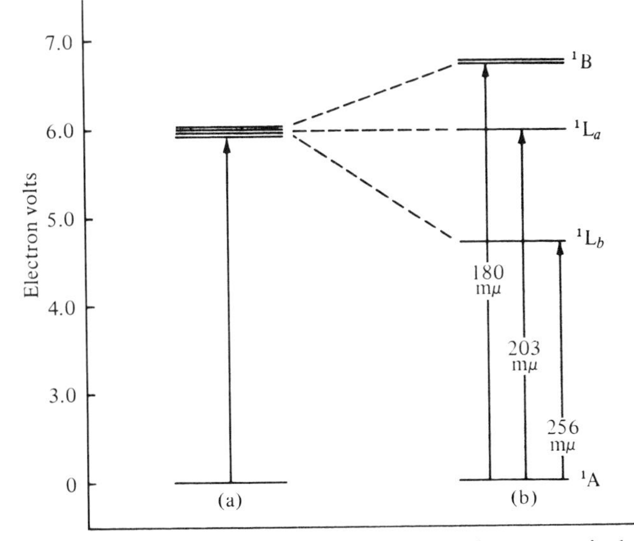

FIG. 4. A term level diagram for the three lowest energy benzene excited singlet states (a) before and (b) after configuration interaction.

The incorrect assumption that the "energy" of a molecular orbital is independent of the number of electrons occupying the orbital is still another potential pitfall in the use of simple molecular orbital diagrams to predict relative transition energies [2]. This assumption, inherent in simple molecular orbital treatments such as the Hückel method, ignores electron repulsion and exchange integrals and therefore predicts the same energy for both singlet–singlet and singlet–triplet transitions. In fact, because electron repulsion terms are less for electrons of the same spin, the triplet state is lower in energy than the singlet excited state by a difference of twice the molecular orbital exchange integral,

$$\delta_{12} = \iint \phi_1(1)\phi_2(2)e^2/r_{12}\phi_1(2)\phi_2(1)\, d\tau_1\, d\tau_2.$$

What is particularly important is that the exchange integral is a function of the spacial overlap of the molecular orbitals ϕ_1 and ϕ_2 involved, and it is necessary to keep this factor in mind when comparing energies of different types of transitions.

For example, the simple molecular orbital energy diagram of Fig. 5a for the R—CO_2 chromophore predicts that the first $\pi \rightarrow \pi^*$ transition will be lower in energy than the $n \rightarrow \pi^*$ transition. Experimentally, however, the $n \rightarrow \pi^*$ transition of carboxylic acids and esters is the lower energy transition. Consideration of the relative singlet–triplet energy splittings (Fig. 5b) offers a rationale for this observation. The atomic orbitals containing the nonbonding electrons are in a plane perpendicular to the antibonding π^*

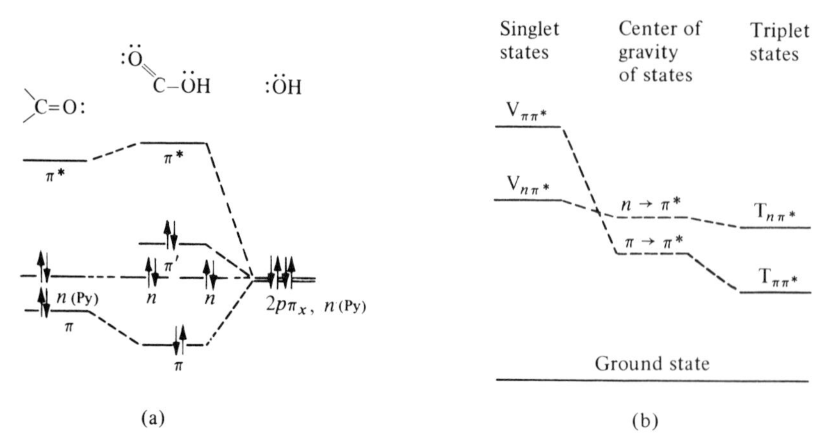

FIG. 5. (a) Qualitative molecular orbital energy correlation diagram for CO_2H. (b) Relative energies of $\pi \rightarrow \pi^*$ and $n \rightarrow \pi^*$ singlet and triplet states [2].

orbital, which results in poor overlap and therefore a small splitting of singlet and triplet excited states. On the other hand, the π and π^* orbitals occupy the same region in space, resulting in a larger singlet–triplet separation. The net result, as indicated in Fig. 5b, is to place the calculated $n \rightarrow \pi^*$ singlet transition in energy below the $\pi \rightarrow \pi^*$ transition in agreement with experiment.

B. INTENSITIES

Roughly speaking, an allowed transition is expected to have a molar extinction coefficient or molar absorptivity (abbreviated ϵ) of the order of 10^4–10^5. A nonallowed or forbidden transition should by the semantics of the title have an ϵ of zero; however, by various mechanisms ϵ's up to 10^2 are obtained. In actual fact the terms *allowed* and *nonallowed* usually need to be examined in context, although generalizations of the following sort can be made.

Transitions involving changes of multiplicity are not allowed; for instance, transitions between singlet ground states and triplet excited states are forbidden and have ϵ's less than one. Double excitations are also highly forbidden.

Because the observation of a physically observable quantity cannot depend on arbitrary choice of space coordinates, the mathematics of group theory can be applied to the prediction of whether or not a transition is allowed or forbidden, depending on the symmetry characteristics of the ground and excited states. Transitions which are symmetry forbidden such as the first transition $A \rightarrow L_b$ of benzene may attain respectable intensity because of vibrations which distort the basic symmetry of the molecule. Weakly allowed transitions may gain intensity by "intensity borrowing" from coupling with a strongly allowed transition of the same symmetry. Similarly, although we might expect a symmetry allowed transition to have a large ϵ, this also will not be the case if the molecular orbitals or wavefunctions do not have sufficient spatial overlap. For example, it is obviously ridiculous to expect to observe a transition between the molecular orbitals of two chromophores which are infinitely separated in space, even if the transition is symmetry allowed.

The integrated intensity of a transition is proportional to the square of the integral $\mathbf{M} = \int \Psi \sum e\mathbf{r} \, \Psi^* \, d\tau$, where Ψ and Ψ^* are the total wavefunctions of the ground and excited states. The transition moment **M** is a dipole moment vector which represents the charge displacement during the transition. (This is not, however, the same thing as the dipole moment of the excited state.) The increase in ϵ with chain length of the conjugated olefins of the first $\pi \rightarrow \pi^*$ transition reflects an increase in this transition dipole moment vector **M**.

The *Franck-Condon principle*, another important tool for interpreting electronic transitions in molecules, simply states that within the time of an electronic transition the nuclei of the molecule remain unchanged in position. This principle is the result of the slow time period (10^{-10} to 10^{-12} sec) of a nuclear vibration compared to the "motion" of an electron, which is of the order of 10^{-15} sec. One consequence of the Franck-Condon effect is that the molar extinction coefficient, ϵ, of the absorption maximum of a transition depends on the population distribution of molecules in rotational and vibrational excited states, although the absolute or integrated intensity is independent of such effects. Other consequences of this such as solvent effects on transition energies are considered later.

C. Charge Transfer States

Valence bond or resonance theory still provides in many instances useful means of accounting for and predicting effects on electronic transitions in molecules. In this approach to the problem, good guesses are made as to the structure or contributing structures (i.e., wavefunctions) of ground and excited states, which are then described as resonance hybrids of these resonance structures. It is true that an unwieldy number of resonance structures often are required, especially for excited states, and that the choice of these resonance structures springs largely from chemical intuition. Nevertheless, an intelligent estimate of the "structure" of an excited state makes possible many excellent qualitative descriptions of electronic transitions, and substituent and solvent effects on these transitions, which are not readily apparent from the molecular orbital approach or which require considerable configuration interaction and tedious calculations to self-consistency by the molecular orbital method. For example, the origin of the large, red, solvent shift of the $\pi \rightarrow \pi^*$ transition of α,β-unsaturated ketones $R_2C{=}CR{-}C(O)R$ with increasing solvent polarity is immediately apparent in terms of the contributing excited state structure $R_2C^+{-}CR{=}C(O^-)R$, since chemists are already familiar with the concept of increasing stabilization of ionic species with increasing solvent polarity.

The description of many electronic transitions as charge transfer transitions is fundamentally a valence bond approach. This method describes an intramolecular electronic transition in terms of a donor portion of a molecule, D, and an acceptor portion, A. The electronic transition is then regarded as an excitation which removes the electron from D and places it on A. An excellent example of such a transition is the intense long-wavelength absorption of triarylboranes [3] near 300 mμ in which the transition is nicely described as removing an electron from the aromatic ring, D, and placing it in the vacant acceptor boron $2p$ orbital. In the simplest sort of approximation the energy of a charge transfer transition, $D{-}A \rightarrow D^+A^-$, may be derived by Eqs. (1)–(5) in which (1) D and A are separated to an infinite distance, (2) an electron is removed from D and (3) placed on A, and finally (4) D^+ and A^- are brought back together within bonding distance and the covalent bond reformed. The terms I_p, E_a, and (e^2/r) represent then the ionization potential of D, the electron affinity of A, and the electrostatic work, in that order.

$$D{-}A \rightarrow D\cdot + A\cdot + Dd \tag{1}$$

$$D\cdot \rightarrow D^+ + e + I_p \tag{2}$$

$$A + e \rightarrow A^{(-)} \qquad - (E_a) \tag{3}$$

$$D^+ + A^{(-)} \rightarrow D^{(+)}{-}A^{(-)} - (e^2/r) - Dd \tag{4}$$

$$D{-}A \rightarrow D^+{-}A^- \tag{5a}$$

$$\Delta E = h\nu = I_p - E_a - (e^2/r) \tag{5b}$$

To the approximation of Eq. (5b) the ground state of the molecule or complex is represented by the resonance structure DA and the excited state by the charge transfer structure D^+A^-. If DA and $D^+A^{(-)}$ have the same symmetry, however, resonance interaction will "mix" some of the excited state structure into the ground state stabilizing it, and some of the ground state structure into the excited state increasing its energy. The wavefunctions of the ground state ψ_G and the excited state ψ^* are then expressed by Eqs. (6) and (7) where $\lambda \ll 1$.

$$\psi_G = \psi_{\mathrm{DA}} + \lambda\psi_{\mathrm{D^+A^-}} \tag{6}$$

$$\psi^* = \psi_{\mathrm{D^+A^-}} - \lambda\psi_{\mathrm{DA}} \tag{7}$$

If resonance between the structures DA and D^+A^- is possible, i.e., they are of the same symmetry, a correction term E_a which is the sum of the ground state stabilization and the excited state destabilization energies should be added to Eq. (5b) to give Eq. (8),

$$E_{\mathrm{CT}} = I_p - E_a - C + E_R \tag{8}$$

where $C \equiv e^2/r$.

The successful interpretation of an electronic transition in terms of a simple charge transfer transition depends on the mixing coefficient λ being small, that is, the resonance interaction between the excited state and the ground state must be small.

The description of an electronic transition in a molecule as intramolecular charge transfer is most appropriate when the transition in molecular orbital terms corresponds to electron promotion from a molecular orbital essentially localized in one region of the molecule to a higher energy molecular orbital localized in a different region of the molecule. Such a molecular orbital description is commonly obtained where there are large differences in atomic or group orbital electronegativities between atoms or groups in the molecule. It is for this reason that we will find many of the electronic transitions of organometallics are particularly well described as intramolecular charge transfer transitions.

D. Solvent Effects on Spectra

The loss of considerable rotational and vibrational structure in a spectrum is one consequence of a change from gas phase to hydrocarbon solution spectra. A further loss in vibrational band structure is often observed in the envelope of a spectrum as the solvent is further changed from a nonpolar solvent such as hexane to more polar solvents such as ethanol or water.

Because relaxation from the first electronic–rotational–vibrational excited state in solution takes place in the time scale of a few molecular vibrations of the solvent, a shorter excited state lifetime leads to "uncertainty" broadening of the transition. In addition, these solvent molecules engaged in specific solvation of the solute act as loosely bound solute substituents, providing numerous additional rotational–vibrational ground state levels from which the soluted solvent complex may be excited. Corresponding excited states may be dissociative. These factors, especially important in the case of polar solutes, account for the additional loss of structure in absorption bands in polar solvents.

A practical problem of considerable concern to those trying to compare spectra of the same or related compounds in different solvents is the effect of solvent on an absorption wavelength maximum, and one of the best descriptive papers [1] dealing with the effects of solvent on the spectra of organic molecules is that of Bayliss and McRae.

In changing from gas phase to solution spectra there is a general solvent polarizability effect which tends to move an absorption maximum towards lower energy. This spectral red shift is the consequence of stabilization of the excited state by the induced dipole interaction between the transition moment and the solvent molecule. This effect is not Franck-Condon restricted since polarization of solvent electrons can take place in the same time period as the solute electron excitation. For example, the absorption maxima of a number of intermolecular charge transfer complexes are correlated by the solvent index of refraction, which is of course directly related to molecular polarizability [4].

In polar or hydrogen bonding solvents with permanent dipole moments, particularly if the solute is also polar, additional effects may play a more important role than the polarizability term and in some cases lead to a blue shift of absorption maximum with increasing solvent polarity. As a result of the inability of solvent molecules to reorient themselves or their permanent dipoles during a solute electronic excitation, two types of strain may be introduced into the excited state, *packing strain* and *orientation strain*. If the excited state is bigger than the ground state, as is often the case for small molecules, then the solvent cage is likely to be too small to comfortably contain the excited state which will then be *destabilized* by the amount of this *packing strain*. Packing strain is thought to be negligible in the excitation of large molecules. A larger term for most polar molecules is the *orientation strain* resulting from the interaction of the permanent solvent dipole moment with the excited state dipole moment. The permanent solvent dipoles arrange themselves to stabilize any permanent dipole moment of the ground state. Thus, if the excited state dipole moment is less (i.e., either of smaller magnitude or in the opposite direction) than that of the ground state, a blue shift of the absorption maximum to higher energy with increasing solvent polarity is found because of greater ground-state than excited-state solvent stabilization. A special case of this is the blue shift of $n \rightarrow \pi^*$ transitions in hydrogen bonding relative to hydrocarbon solvents. If the solute dipole increases on excitation, then an unexpectedly large red shift is observed. These two cases of *orientation strain* are illustrated in Fig. 6.

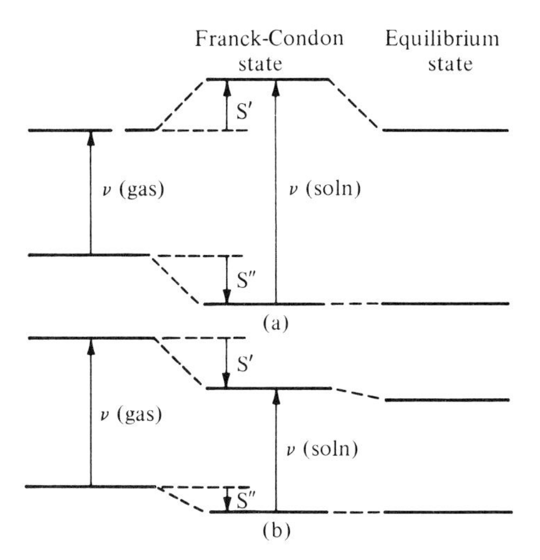

FIG. 6. Franck-Condon orientation strain with (a) increase in dipole moment on excitation, (b) decrease in dipole moment on excitation [1].

REFERENCES

1. N. S. Bayliss and E. G. McRae, *J. Phys. Chem.* **58**, 1002 (1954).
2. H. H. Jaffe, D. L. Beveridge, and M. Orchin, *J. Chem. Ed.* **44**, 383 (1967).
3. B. G. Ramsey, *J. Phys. Chem.* **70**, 611 (1966).
4. H. M. Rosenberg, *Chem. Commun.* **1965**, 312.

CHAPTER II

Organometallic Derivatives of Groups I and II

ORGANOMETALLIC DERIVATIVES OF GROUP I

A. General Comments on Structure

Of the organometallic derivatives of the Group I metals, only the structure of organolithium compounds has received more than passing attention. A review [6] of organolithium structure by Brown has recently appeared in "Advances in Organometallic Chemistry." The simple alkyllithiums form tetramers or hexamers by electron deficient four-center bonding in which the lithium atoms occupy the apexes and the alkyl groups occupy the faces of a tetrahedron or octahedron. In any given case the tetramer–hexamer equilibrium is found to be a function of both steric hindrance and solvent polarity. Normal alkyllithiums up to *n*-butyl are hexameric in hydrocarbon solvents [32]; more highly branched and long chain alkyllithiums are frequently tetrameric [40, 50]. Normal alkyllithiums are also reportedly [42] tetrameric in ether below $-50°C$.

The bonding of methyllithium tetramer has been treated [51] in terms of simple molecular orbital calculations by Weiss and Lucken. Under T_d symmetry the bonding orbitals are found at $T_2 \approx 10.5$ eV, $A_1 \approx 11.5$ eV, with the lowest corresponding antibonding orbitals near 4.5 (T_2) and 5.5 (A_1) eV.

In anticipation of possible solvent effects on electronic spectra, it should be pointed out that the hexameric octahedron has two vacant faces, i.e., orbitals, which may coordinate with Lewis bases without dissociation of the hexamer [7, 8], although if the base is sufficiently strong, dissociation may also occur [7].

Conjugated alkyl metals in which the incipient carbanion is resonance stabilized are probably dimeric, and possibly monomeric at low concentrations, in hydrocarbon and other solvents of low coordinating ability. This conclusion is based, however, on the limited number of cases of polystyryllithium [25, 34, 38, 46, 52], polyisoprenyllithium, benzyllithium, and 1,1-diphenylalkyllithium, in which steric factors are also important. It has recently become apparent however, due to work notably by Hogen-Esch and Smid [25], and Waack, and co-workers [46], that the major species present in solutions of conjugated organometallics in more polar (and more basic) solvents are intimate or contact ion pairs ($R^-MS_n{}^+$) and solvent separated ion pairs ($R^-/ /M^+S_m$) in which the metal ion becomes specifically solvated, i.e., coordinated by solvent molecules, S. In the case where the metal ion is lithium and the solvent is tetrahydrofuran (THF), $n = 1$–2 and $m = 3$–4 for 9-fluorenyl- and 1,1-diphenyl-*n*-hexyllithium [16, 25, 46].

Nuclear magnetic resonance studies of 9-fluorenyllithium·$(THF)_3$ indicate that when $m = 3$, the ion pair may still be an intimate ion pair in hydrocarbon solvent but that the solvated lithium cation is no longer bonded to a specific carbon, but is above the plane of the ring in the π cloud [16]. In general it would be anticipated that the specific number of solvent ligands coordinated by the metal ion would depend on the relative basicity of the "carbanion."

In any discussion then of the ultraviolet spectra of conjugated organometallic derivatives, or in discussing solvent effects on such spectra, it becomes imperative to identify which species or structure is under consideration. It seems more fruitful and logical for the purposes of this monograph to consider the dimer, contact ion pair, and solvent separated ion pair as different molecular species in a discussion of their spectra. Specific results will be considered in later sections.

Rodionov and co-workers [41] have reported that phenyl- and mesityllithium are dimeric in ether. More surprising is the observation of an intense ESR signal from freshly prepared solutions of lithium aryls.

B. Spectra of Simple Alkyllithiums

The absence of more than rising end absorption above 215 mμ in the spectra [47] of ethyl- and butyllithium compounds in tetrahydrofuran (THF) is in only qualitative agreement with the previously mentioned (Section II-A) Weiss and Lucken [51] calculations on the tetramer, which predict as the first UV absorption an allowed $A_1 \rightarrow T_2$ transition near 250 mμ (5 eV). Similar calculations for the hexamer have not been reported.

The ultraviolet spectra of *n*-butyl and but-3-enyllithium in hydrocarbon solvent has recently been reported to 180 mμ [36]. The hexameric *n*-butyllithium spectrum exhibits absorption maxima at 221 and 225 mμ. A new band appears at 251 mμ, in addition to several maxima in the 220 mμ region band, in the spectrum of but-3-enyllithium. These results and the supporting NMR data suggest an electron donor interaction between the π olefin and the alkyllithium hexamer framework. A likely description for the 251 mμ transition would be charge transfer from olefin π to the lowest alkyllithium framework, unoccupied molecular orbital. The 220 mμ transition in the hexamer probably corresponds to the predicted $A_1 \rightarrow T_2$ transition for the tetramer, although charge transfer from the highest filled butyl σ orbital to the lowest unoccupied framework orbital is an outside possibility. The charge transfer assignments might be reversed by recent results and calculations which indicate the highest σ bonding orbital of an olefin may be higher in energy than the π bonding orbital [39]. The $\pi \rightarrow \pi^*$ transitions of monoalkyl-substituted olefins normally occur below 180 mμ [33] so that the effect of coordination on this transition in but-3-enyllithium cannot be evaluated (see Addendum).

C. Spectra of Conjugated Alkyl Derivatives

Spurred by the appearance in the literature of the spectra of the corresponding carbonium ions and an interest in anionic polymerization mechanisms, several research groups have directed their attention to the spectra of phenylmethyl, allyl, fluorenyl, aryl radical ion, and similar Group I metal organometallics.

The spectra in general are found to be highly dependent on the nature of the associated metal, the solvent, and the concentration. A representative sampling of these results is presented in Table I. An interesting exception to this general solvent sensitivity is found in the spectra of benzyl and allyl organometallic compounds which are relatively insensitive to a change of solvent [48].

The ultraviolet spectra of a variety of 16 organolithium compounds,

TABLE I

Electronic Transitions of Representative Group I Organometallic Compounds

Compound	Solvent	λ_{max}, mμ (log ϵ)
(n-Butyl-Li)$_6$	Hexane	221, 225 (brd)
Phenyl-Li	THF[a]	261 (3.02), 268 (2.99), 292 (2.88)
Benzyl-Li	THF[a]	330 (3.98)
Benzyl-Na	THF[a]	355 (4.08), 485 (3.18)
9-Fluorenyl-Li	Dioxane	346, 437
9-Fluorenyl-Li	CHA[b]	452 (3.03), 477 (3.11), 510 (2.92)

[a] THF is tetrahydrofuran. [b] CHA is cyclohexylamine.

usually in THF, have been reported by Waack and Doran [47]. Streitwieser *et al.* [43, 44] have reported the spectra of some 16 organolithium and -cesium compounds, principally of the fluorenyl, benzofluorenyl, and phenylmethyl variety, in cyclohexylamine or diethyl ether. Bywater *et al.* have studied the ultraviolet spectra of polystyryl, polybutadienyl, and polyisoprenyl-lithium, sodium, and potassium in THF, cyclohexane, and benzene as solvents [10, 53]; the spectra of the living polymers are very similar to the corresponding monomeric benzyl and allyl metal derivatives [1, 47].

Polyakov *et al.* [38] have found that the UV spectrum of polystyryllithium in toluene, λ_{max} 333 (10^{-3} M) undergoes a large blue shift on decrease of concentration (λ_{max} 318 mμ at $<10^{-5}$ M). Poly(α-methyl styrene) gave similar results but the blue shift was evident at higher concentrations. This result was attributed to an equilibrium between the dimer and monomeric species, $R_2M_2 \rightleftarrows 2RM$. If this interpretation is correct, the reported spectrum at high dilution represents the first and only UV spectrum in the literature known to this author for the formal species R–Li, with the possible exception of 1,1-diphenyl-*n*-hexyllithium in *n*-hexane [46] (see also Section II-C-5). The high intensity of the transition ($\epsilon \sim 10^5$) suggest that C_p—Li σ might be a more appropriate description of the bond, rather than C_{sp^3}—Li for which overlap with the aromatic ring would be poor.

Theoretical interpretations of the UV–visible spectra have been limited to simple LCAO–MO type arguments [30, 43, 49] based on the free carbanion, with the metal ion acting as a small perturbation. The absorption maxima of ten hydrocarbon lithium salts in cyclohexylamine give fair correlation with Hückel molecular orbital transition energies (see Addendum) [43]. According to either simple Hückel or self-consistent field molecular orbital [26] theory the first transition energy for an alternant hydrocarbon anion should be the same as that for the corresponding carbonium ion. In fact the *free* carbanion transitions often seem to occur at a significantly longer wavelength than the carbonium ion transitions: compare λ_{max} $(C_6H_5)_3C^-$: 505, 432 mμ [4] in hexamethylphosphortriamide with λ_{max} $(C_6H_5)_3C^+$ 404, 431 mμ in H_2SO_4 or CH_3CN [15] or λ_{max} $(C_6H_5)_2C^-(CH_2)_4CH_3$, 484 m$\mu$ in pyridine [46] with λ_{max} $(C_6H_5)_2C^+$—CH_2CH_3, 316, 427 mμ in $FSO_3H \cdot SbF_5$ [35].

In studying the effect of methyl substitution on the ultra-violet spectra of benzyl and allyl compounds, Waack and Doran [49] found that an α-CH_3 produced a small 3 mμ *red* shift whereas a *p*-CH_3 caused a 5 mμ *blue* shift in the 330 mμ transition of benzyllithium in tetrahydrofuran. Further, an α-CH_3 substituent caused a red shift of the 315 mμ allyllithium transition. It was concluded [49] that these substituent effects could not be predicted by the use of the simple molecular orbital coefficients, C, for the rth atom in the highest filled orbital n and the *lowest* vacant orbital $n + 1$ in Eq. (9), but it was found [49] that Hückel calculations of the ω self-consistent charge type [45] would predict the correct order of substituent effects on transition energies (see Addendum).

$$\Delta E_t = (C^2_{n+1,r} - C^2_{n,r})\ \delta\alpha_r \quad (9)$$

The failure of Eq. (9) for substituted allyllithium, in view of its success in cases [11, 31] of even and nonalternant hydrocarbons, may be due to the inappropriate use of molecular orbital coefficients of the free carbanion, when in fact the species responsible for the spectra are contact ion pairs, probably with $Li \cdot (THF)_n$ strongly associated with a particular carbon. In fact, for benzyllithium, Eq. (9) does predict the correct substituent effect on the 330 mμ transition, provided the transition is correctly assigned to the $^1A \rightarrow {}^1A\ (C_{2v})$ transition from the highest filled vacant orbital to the *second* lowest vacant molecular orbital. A transition of much weaker intensity ($\epsilon \sim 1000$) at 485 mμ in the spectrum of benzylsodium in tetrahydrofuran, and 470 mμ for benzylcesium in cyclohexylamine represents the transition to the *lowest* vacant molecular orbital and should be $A_1 \rightarrow B_2$ (C_{2v} in plane polarized).

More recently Häfelinger and Streitwieser [23a] studied the ultraviolet visible spectra of the cesium salts in cyclohexyl-amine of 9-CH_3- or C_6H_5-substituted fluorenyl, methyl-substituted diphenyl and triphenyl carbanions, and diphenyl or triphenyl carbanions in which two phenyl rings are *ortho–ortho* bridged by —CH_2—, $\{C(CH_3)_2\}$, and $\{CH_2—CH_2\}$ groups. Intimate ion pairs appear to be absent in this system.

Dreiding models show the $\{CH_2—CH_2\}$ bridge to fix the $(C_6H_5)_2\ddot{C}H$ rings in a conformation *twisted* 55° out of the *sp*^{2} bond plane, a twist angle which when combined with transition intensity ratios enabled [23a] an estimate of the ring twist angle in $(C_6H_5)_2\ddot{C}H$ itself to be made of about 30° out of the C_{sp^2} bond plane. If the phenyl ring of $C_6H_5\ddot{C}H_2$ is twisted 90° out of the plane of the *sp*^{2} bonds, the $^1A_1 \rightarrow {}^1A_1$ transition becomes $^1A_1 \rightarrow {}^1A_2$ and symmetry forbidden, while the lowest energy $^1A_1 \rightarrow B_2$ remains allowed and actually may be expected to increase in intensity as the phenyl ring is twisted out of plane. Considerations such as these, along with the result that *p*-CH_3 substitution of $(C_6H_5)_2\ddot{C}H$ or $(C_6H_5)_3C$: produces a 3–8 mμ *red* shift in the absorption maximum and a 12–24 mμ red shift is obtained for α-CH_3 substitution of the diphenylcarbanions, combine to suggest that the 460 mμ transition of *ortho–ortho*-$\{CH_2\}_2C_6H_4\ddot{C}HC_6H_4$— and perhaps the 443 m$\mu$ transition of $(C_6H_5)_2\ddot{C}H$ correlate with the 470–485 mμ transition of the benzyl "anion" and not the more intense second transition near 330–350 mμ.

Häfelinger and Streitwieser [23a] in comparing the spectra of A (λ_{max} 445 mμ) and B (λ_{max} 462 mμ) with that of $(C_6H_5)_3C$:, which has two transitions at 422 and 488 mμ (cesium salt in cyclohexylamine), suggest that the unbridged phenyl ring in A and B for steric reasons is perpendicular to the plane of the central carbon sp^2 bonds, and therefore the spectra resembles that of the diphenyl derivatives C and D rather than $(C_6H_5)_3C$:. Although the conformation suggested by Häfelinger and Streitwieser is the minimum energy sterically, models in the hand of this author indicate that there is room sterically for the unbridged phenyl to assume a planar conformation providing maximum overlap with the central carbon *p* orbital. Applying an independent systems or localized transition model, we would conclude that if the 460 mμ transition of **A** is $^1A \rightarrow B_2$ and the third phenyl substituent would result in an $^1A \rightarrow {}^1A$ transition, below 340 mμ perhaps, as part of the C_6H_5—C: chromophore, the 460 mμ transition of **D** or the 447 mμ transition **C** would be only slightly perturbed as the locally excited states are of different symmetry. This is a prediction, or possibility, which could be checked by measuring the spectrum of **A** or **B** below 400 mμ, which could not be done in cyclohexylamine, to determine the presence or absence of a second intense transition.

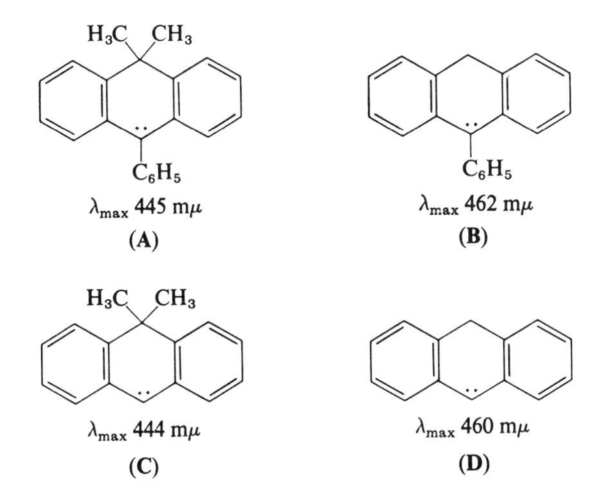

Studies of the solvent and temperature effects on the electronic transitions of 1,1-diphenyl-*n*-hexyllithium by Waack *et al.* [46], and similar studies [25] including conductance measurements [26] by Hogen-Esch and Smid on fluorenyl alkali metals, have shown that in most solvents which possess lone pair electrons capable of bonding to the metal, the major solvent effect on spectra of these compounds is due to the formation of two types of ion pairs.

The principal conclusions to be drawn from these and related studies seem to be as follows:

(1) Solvent separated ion pairs absorb at longer wavelengths than contact ion pairs, which themselves absorb at longer wavelengths than the dimer [25, 46]. Thus, successive addition to dioxane solutions of 9-fluorenyllithium

of dimethylsulfoxide converts the intimate ion pair, $F^-Li^+ \cdot S_n$ (λ_{max} 346 mμ) to the solvent-separated ion pair $F^-//Li^+ \cdot S_m$ (λ_{max} 373 mμ) ($m > n$), (see Fig. 7). The large red shift [46] in the spectrum of 1,1-diphenyl-*n*-hexyllithium in benzene (λ_{max} 415 mμ) or di-*n*-propyl ether which occurs on adding THF to the solution has a similar origin in the formation of first $R^-Li^+ \cdot (THF)_2$ (λ_{max} 450 mμ), followed by $R^-//Li(THF)_4$ (λ_{max} 496 mμ) with increasing percent composition of THF.

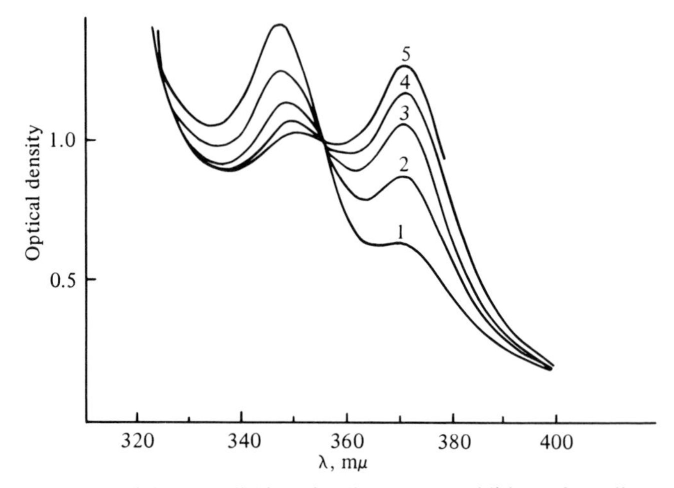

FIG. 7. Spectrum of fluorenyllithium in dioxane on addition of small quantities of dimethylsulfoxide [DMSO] × 10^3: (1) 2.36; (2) 5.10; (3) 8.49; (4) 11.4; (5) 14.7 [25, p. 312].

(2) An increase in metal ion radius, which also corresponds to an increase in ionic bond character in the organometallic compound, causes a decrease in the electronic transition energy of the contact ion pair [25]. This is the expected red spectral shift based on the ground state argument that as the metal–carbon bond becomes weaker, it is easier to promote the bonding electron to a molecular orbital delocalized over the hydrocarbon portion of the molecule (compare λ_{max}: 9-fluorenyl-Li 349 mμ; 9-fluorenyl-Cs^+ 364 mμ). Similar arguments have been used to account for the solvent effects on $n \rightarrow \pi^*$ transition in other molecules [5].

(3) Solvent separated ion pair formation [25] increases in the order $Cs^+ < K^+ < Na^+ < Li^+$; thus, although the cesium–carbon bond is the most ionic, the large cesium ion does not become sufficiently solvated to form a solvent separated ion pair. An *apparent* blue spectral shift then on changing metal ion from Li to some larger metal, Cs, may be the result of comparing the spectrum of a solvent separated ion pair with a contact ion pair [43].

(4) Solvent effects on the spectra of solvent separated ion pairs are variable in magnitude but generally an increase in dielectric constant of the solvent leads to a blue shift of the spectrum to higher energies. In THF or hexamethyl phosphoramide, 1,1-diphenyl-*n*-hexyllithium exists as a solvent separated ion pair [45]; the relative maxima are λ_{max} 485 mμ in hexamethylphosphortriamide (HMP), λ_{max} 496 mμ in THF. In contrast the transition energies of fluorenyl–metal solvent separated ion pairs are independent of solvent polarity [25] (λ_{max} 9-fluorenyllithium, 373 mμ in dimethylsulfoxide or 1,2-dimethoxyethane).

In contrast, the contact ion pairs show a general red spectral shift with increasing dielectric constant, for example [48], benzyllithium λ_{max} 328 mμ (Et_2O), 330 mμ (THF); and 9-fluorenyllithium λ_{max} 346 mμ (dioxane), 349 mμ (THF) [46]. The absence of large gross solvent effects on the spectra of benzyl, polystyryl, and allyl organometallic compounds is attributable to the considerable Lewis base strength of the carbanions, which precludes the formation of solvent separated ion pairs, even in THF [46].

(5) Temperature effects on the spectra of conjugated alkali metal compounds are frequently significant [25, 46]. A decrease in temperature of polar solvents often leads to a red spectral shift or the appearance of a new band due to a shift in equilibrium from contact ion pairs to solvent separated ion pairs. This is dramatically illustrated [25] for the absorption spectrum of fluorenylsodium in THF as a function of temperature (Fig. 8).

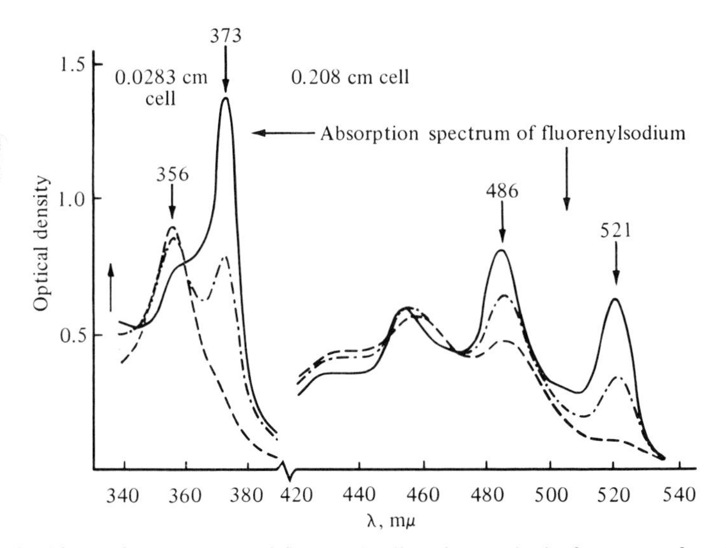

FIG. 8. Absorption spectrum of fluorenylsodium in tetrahydrofuran as a function of temperature: (––––) 25°C; (–·–·–) −30°C; (———) −50°C; path length 2.08 mm, 0.283 mm [25, p. 308].

In hexane, at room temperature the 1,1-diphenyl-*n*-hexyllithium UV spectrum has a broad maximum at 410 mμ with a shoulder near 435 mμ; at $-46°C$, the spectrum consists of two nearly equal maxima at 410 and 435 mμ. These results [46] are concentration independent, which would seem to rule out a shift of monomer (RM) dimer (R_2M_2) equilibrium to dimer as an explanation for the temperature effect, a rationale offered by Polyakov *et al.* [38]. Waack and co-workers [46] suggested dissociation of dimer (R_2M_2) to triple ion $R_2M^+M^-$ as being responsible for the increased absorption at 435 mμ; yet the results of Polyakov [38] with polystryllithium strongly imply that in hexane, 1,1-diphenyl-*n*-hexyllithium will be monomeric. The results of both workers could be explained in terms of a dissociation of the diphenyl-*n*-hexyllithium monomer into intimate ion pair (R^-M^+) at low temperature. The principal difference between the above ion pair and those previously discussed would be the absence of specific solvation of the lithium ion.

D. Spectra of Metal Adducts of Aromatic Hydrocarbons

The electronic structure and absorption spectra of mono and di negative ions from the reaction of alkali metals with aromatic hydrocarbons was reviewed [37] in 1964 by De Boer.

Since this review, Buschow *et al.* [9] have interpreted temperature and solvent effects on the spectra of aromatic hydrocarbon negative ions in terms of the dissociation of ion pairs to free ions in THF and 2-methyltetrahydrofuran. However, the ESR spectrum of naphthalenelithium in 2-methyltetrahydrofuran shows splitting by the ^7Li nucleus, indicating an intimate ion pair [3]. In contrast to the sodium derivative there was no evidence for any more highly dissociated species. Present evidence as discussed by Hogen-Esch and Smid [25] indicates that spectral shifts must be interpreted in terms of change from contact to intimate ion pairs as presented previously (Section II-C) for conjugated organometallic compounds. Boileau has suggested similar conclusions from the spectra of sodium and potassium anthracene in THF and hexamethylphosphortriamide [4].

The spectra [17–20] of the negative ions of phenanthrene, 9,10-dihydrophenanthrene, 1,2,3,4-tetrahydrophenanthrene, 1,2,3,4,5,6,7,8-octahydrophenanthrene, and 1,2,3,4,9,10-hexahydrophenanthrene have been determined by Eloranta and Vuolle.

E. Alkali Metals σ Bonded to sp^2 Carbon

The spectra of phenyl- and vinyllithium have been reported by Waack and Doran [47]. Phenyllithium in THF has absorption maxima at 261 (log ε 3.02), 268 (log ε 2.99), and 292 (log ε 2.88) mμ. The spectrum of vinyllithium has a shoulder on rising end absorption at 280 mμ. It was suggested [47] that these transitions are similar to the transitions in pyridine (λ_{max} 261, 286 mμ), vapor, and $R_2C{=}\ddot{N}{-}R$ ($\lambda_{max} \approx 230$ mμ) where the long wavelength maxima are assigned to $n \rightarrow \pi^*$ transitions.

Fraenkel *et al.* [22] more recently completed a detailed study of the UV and NMR spectra of aryllithium compounds and Grignard reagents. The NMR studies strongly support the analogy between these compounds and pyridine in that the unusually low field shift of the *ortho* hydrogens and coupling constants are similar. In each case, the chemical shift of the *ortho* hydrogens may be explained by assuming mixing between ground and a low-lying $n \rightarrow \pi^*$ state in pyridine, or $\sigma_{C-M} \rightarrow \pi^*$ in aryl–metal.

The maximum present at 292 mμ in the spectrum of phenyllithium in THF is absent in ether as a solvent. As may be seen from Fig. 9 the spectrum of

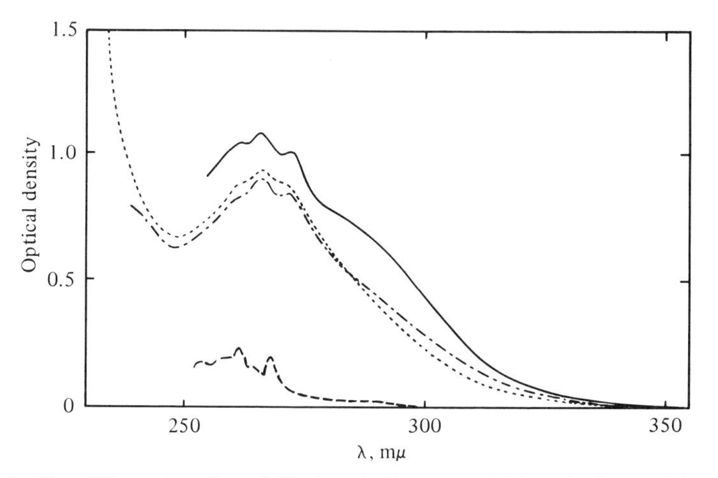

FIG. 9. The UV spectra of *p*-tolyllithium (cell path = 0.05 mm): (———) in 0.15 *M* triethylamine–ether (85:15); (·····) in 0.10 *M* pentane–ether (85:15); (–·–·–) in 0.12 *M* ether; (––––) in the hydrolysate of first solution. [G. Fraenkel *et al., J. Phys. Chem.* **72**, 953 (1968).]

p-tolyllithium in 85% pentane–15% ether exhibits maxima at 267 and 271 mμ with considerable tailing to longer wavelengths. On changing the solvent to 85% triethylamine–15% ether, the major change in the spectrum of *p*-tolyllithium is the appearance of a shoulder at 285 mμ. It seems safe to assume that the maxima near 267 and 271 mμ represent the expected benzenoid $\pi \rightarrow \pi^*$ transition.

It is possible of course that the 292 mμ maximum of phenyllithium in THF and the 285 mμ shoulder of *p*-tolyllithium in triethylamine–ether represent a change in molecular species, such as that from dimeric to monomeric aryllithium. However, the observation of a similar shoulder in the spectrum of trimesityllithium in ether where the molecular species is known to be dimeric suggests that the corresponding maxima of phenyllithium in THF and *p*-tolyllithium in triethylamine–ether result from the red shift of a transition buried beneath the $\pi \rightarrow \pi^*$ band which is responsible for the long wavelength tail in ether or pentane as a solvent. The comparable solvent effects for tetrahydrofuran and triethylamine, in spite of their vastly different basicities, then becomes more reasonable in that they both have about the same refractive index (i.e., similar polarizability). In agreement with earlier suggestions Fraenkel and co-workers have assigned the long wavelength shoulder in the spectra of the aryllithiums to a transition, analogous to the pyridine $n \rightarrow \pi^*$, transition in which the electrons associated with the metal–carbon bond are promoted to the lowest vacant π^* orbital in what is termed a $\sigma_{M-C} \rightarrow \pi^*$ transition. By coordination to the metal, strongly basic solvent ligands may be expected to increase the ionic character of the metal–carbon bond and thereby decrease the transition energy. This provides an alternate explanation to the general polarizability effect suggested earlier for the observed solvent effect. Assuming the $\sigma_{M-C} \rightarrow \pi^*$ transition assignment to be correct, the substantial red shift in the maximum of this transition in the spectrum of 2,4,6-trimethylphenyllithium (mesityllithium) as compared with *p*-tolyllithium is in keeping with the observed inductive effects of methyl groups on other $n \rightarrow \pi^*$ transitions. Compare, for example, $CH_3—C(O)CH_3$, with λ_{max} 275 mμ, and $CH_3C(O)CH(CH_3)_2$, with λ_{max} 281 mμ in chloroform.

Apparently the phenyllithium–pyridine analogy can be extended further to the diazines. The first $\pi \rightarrow \pi^*$ transition of pyrimidine occurs at slightly shorter wavelengths than that of pyridine, while the pyrimidine $n \rightarrow \pi^*$ maximum is found at 322 mμ. Similarly the $\pi \rightarrow \pi^*$ transition of *m*-dilithiobenzene in ether is found at 255 mμ and the proposed $\sigma_{M-C} \rightarrow \pi^*$ transition at 288 mμ.

Although this approach gives reasonable and satisfying answers, there is a nagging and perhaps serious defect in the assignment of long wavelength absorption of the aryllithiums as simply $\sigma_{C-M} \rightarrow \pi^*$, because we have ample experimental evidence to show that these compounds are dimeric, at least in diethyl ether. One can ask whether or not the use of two lithium orbitals and a carbon orbital to generate a three-centered bond (**1**) will not also generate vacant low-lying molecular orbitals and then push the question further to ask what the interaction of these molecular orbitals will be with the benzene orbitals, particularly the antibonding ones. This problem is very similar to that of the phenonium ion (**2**) which is also an eight-electron problem, and without more sophisticated calculations it is not possible to rule out an assignment of the long wavelength transition either to charge

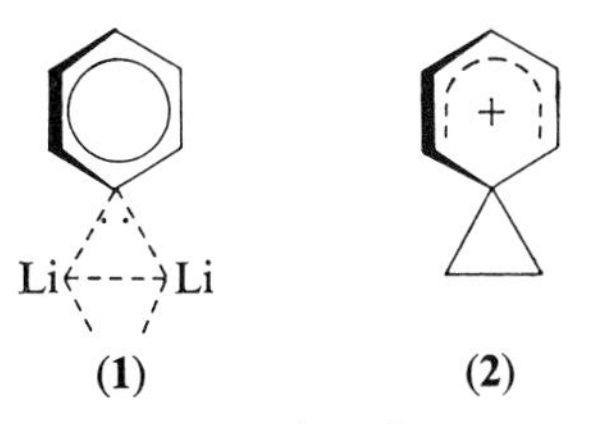

transfer from the aryl ring to a C—(Li)—C group orbital or to some transition due to the dimer as a whole.

In a consideration of the spectra of aryllithiums it is not fair to rule out the report [41] mentioned earlier (Section II-A) that freshly prepared phenyllithium exhibits a strong ESR signal, implying a diradical ground state. Considering the substantial experimental problems of ruling out trace free radical contaminants, however, some further verification of this point is highly desirable.

ORGANOMETALLIC DERIVATIVES OF GROUP II

When one considers the enormous body of literature devoted to the reactions and structure of organomagnesium compounds, the almost total absence of studies of electronic transitions represents a startling gap.

The UV spectrum of dicinnamylmagnesium in ether [λ_{max} 252 (ϵ 2×10^4), 283 (sh), 293 (sh) mμ] is reportedly [15a] very similar to the spectra of cinnamyl Grignard solution, cinnamyl alcohol, and *trans*-1-phenylpropene.

There are the interesting conflicting reports that (1) triphenylmethylmagnesium chloride (Grignard reagent) in ether forms a red color [42a] and (2) $(C_6H_5)_3CMgX$ is colorless [29a].

The spectrum of 9-fluorenylmagnesium in hexamethylphosphortriamide has been reported and compared [36a] with that of fluorenyllithium and fluorenylsodium. The spectra are quite similar.

An earlier preliminary report [22a] of a maximum near 290 mμ in the spectra of phenyl-, tolyl-, and *p*-anisylmagnesium has proved to be erroneous [22] due to the presence of phenolate ion from trace amounts of Grignard oxidation. The spectra of the Grignard reagents in tetrahydrofuran, however, do show considerable tailing absorption [22] and this in conjunction with NMR data and parallel studies on corresponding aryllithium compounds suggests the presence of a $\sigma_{M-C} \rightarrow \pi^*$ transition buried beneath the $\pi \rightarrow \pi^*$ band analogous to the pyridine $n \rightarrow \pi^*$ transition. The reported maxima are 255 mμ for C_6H_5MgBr (ϵ 1.68×10^3), 250 mμ for *p*-$CH_3C_6H_4MgBr$ (ϵ 1.41×10^3), 277 mμ for *p*-$CH_3OC_6H_4MgBr$ (ϵ 1.72×10^3) and 255 mμ for *p*-$(MgBr)_2C_6H_4$ (ϵ 1.72×10^3) in tetrahydrofuran.

REFERENCES

1. R. Asami, M. Levy, and M. Szwarc, *J. Chem. Soc.* **1962**, 361.
2. V. Astaf'ev and A. I. Shatenshtein, *Opt. Spectry. USSR* (*English Transl.*) **6**, 410 (1959).
3. N. M. Atherton, *Chem. Commun.* **1966**, 254.
4. S. Boileau and P. Sigwalt, *Compt. Rend. C* **262**, 1165 (1966).
5. G. J. Brealy and M. Kasha, *J. Am. Chem. Soc.* **77**, 4462 (1955).
6. T. L. Brown, *Advan. Organometal. Chem.* **3**, 365 (1965).
7. T. L. Brown and M. T. Rogers, *J. Am. Chem. Soc.* **86**, 2134 (1964).
8. T. L. Brown, J. A. Ladd, and G. N. Newman, *J. Organometal. Chem.* (*Amsterdam*) **3**, 1 (1965).
9. K. H. J. Buschow, J. Dielman, and G. J. Hoijtink, *J. Chem. Phys.* **42**, 1993 (1965).
10. S. Bywater, A. F. Johnson, and W. J. Worsfold, *Can. J. Chem.* **42**, 1255 (1964).
11. C. A. Coulson, *Proc. Phys. Soc.* (*London*) **A65**, 61 (1952).
12. E. De Boer, *Advan. Organometal. Chem.* **2**, 115 (1964).
13. E. De Boer and S. I. Weissman, *Rec. Trav. Chim.* **76**, 824 (1957).
14. E. De Boer and P. H. van der Meij, *Proc. Chem. Soc.* **1962**, 139.
15. N. C. Deno, *Progr. Phys. Org. Chem.* **2**, 153 (1964).

15a. R. H. DeWolfe, D. Hagmann, and W. G. Young, *J. Am. Chem. Soc.* **79**, 4795 (1957).

16. J. A.Dixon, P. A. Gwinner, and D. C. Lini, *J. Am. Chem. Soc.* **87**, 1379 (1965).
17. J. Eloranta and M. Vuolle, *Act. Chem. Scand.* **21**, 408 (1967).
18. J. Eloranta, *Suomen Kemistilehti* **B39**, 143 (1966).
19. J. Eloranta and M. Vuolle, *Suomen Kemistilehti* **B38**, 151 (1965).
20. J. Eloranta, *Acta Chem. Scand.* **18**, 2259 (1964).
21. A. G. Evans and D. B. George, *J. Chem. Soc.* **1961**, 4753.
22. G. Fraenkel, S. Dayagi, and S. Kobayashi, *J. Phys. Chem.* **72**, 953 (1968).

22a. G. Fraenkel, S. Kobayashi, and D. G. Adams, *Meeting Am. Chem. Soc., 147th, Philadelphia, April 1964.*

23. R. Grinter and S. F. Mason, *Proc. Chem. Soc.*, **1961**, 386.

23a. G. Häfelinger and A. Streitwieser, Jr., *Chem. Ber.* **101**, 657, 672 (1968).

24. K. Hafner and K. Goliasch, *Angew. Chem. Intern. Ed. Engl.* **1**, 114 (1962).
25. T. E. Hogen-Esch and J. Smid, *J. Am. Chem. Soc.* **88**, 307 (1966).
26. T. E. Hogen-Esch and J. Smid, *J. Am. Chem. Soc.* **88**, 318 (1966).
27. M. T. Jones and S. I. Weissman, *J. Am. Chem. Soc.* **84**, 4269 (1962).
28. S. N. Khanna, M. Levy, and M. Szwarc, *Trans. Faraday Soc.* **58**, 747 (1962).
29. R. Kuhn, H. Fischer, F. Neugebauer, and H. Fischer, *Ann. Chem.* **654**, 64 (1962).

29a. D. Lavie and A. Bergmann, *Bull. Soc. Chim. France* **1951**, 250.

30. H. C. Longuet-Higgins and J. Pople, *Proc. Phys. Soc.* (*London*) **68A**, 591 (1955).
31. H. C. Longuet-Higgins and R. G. Snowden, *J. Chem. Soc.* **1952**, 1404.
32. D. Margerison and J. P. Newport, *Trans. Faraday Soc.* **59**, 2058 (1963).
33. S. F. Mason, *Quart. Rev.* (*London*) **15**, 287 (1961).
34. M. Morton and L. J. Fetters, *J. Polymer Sci.* **2A**, 3311 (1964).
35. G. A. Olah and C. U. Pittman, Jr., *Advan. Phys. Org. Chem.* **4**, 344 (1966).
36. J. P. Oliver, J. B. Smart, and M. T. Emerson, *J. Am. Chem. Soc.* **88**, 4101 (1966); private communication.

36a. J. Pascault and J. Golé, *Compt. Rend. C* **264**, 326 (1967).

37. D. E. Paul, D. Lipkin, and S. I. Weissman, *J. Am. Chem. Soc.* **78**, 116 (1956).
38. D. K. Polyakov, A. R. Gantmakher, and S. S. Medvedev, *Polymer Sci. USSR* (*English Transl.*) **7**, 198 (1965).
39. M. B. Robin, R. R. Hart, and N. A. Keubler, *J. Chem. Phys.* **44**, 1803, 1644 (1966).
40. A. N. Rodionov, T. V. Talalaeva, D. N. Shigorin, G. N. Tyumofeyuk, and K. A. Kocheshkov, *Dokl. Akad. Nauk SSSR Ser. Chem.* **151**, 1131 (1963).
41. A. N. Rodionov, D. N. Shigorin, T. V. Talalaeva, V. Tsareva, and K. A. Kocheshkov, *Zh. Fiz. Khim.* **40**, 2265 (1966).
42. L. M. Seitz, unpublished observations quoted in Brown [6].
42a. D. A. Shirley, "Preparation of Organic Intermediates." Wiley, New York, 1951.
43. A. Streitwieser, Jr., and J. I. Brauman, *J. Am. Chem. Soc.* **85**, 2633 (1963).
44. A. Streitwieser, Jr., J. I. Brauman, J. H. Hammons, and A. H. Pudjaatmaka, *J. Am. Chem. Soc.* **87**, 384 (1965).
45. A. Streitwieser, Jr., "Molecular Orbital Theory for Organic Chemist." Wiley, New York, 1961.
46. R. Waack, M. A. Doran, and P. E. Stevenson, *J. Am. Chem. Soc.* **88**, 2109 (1966).
47. R. Waack and M. A. Doran, *J. Am. Chem. Soc.* **85**, 1651 (1963).
48. R. Waack and M. A. Doran, *J. Phys. Chem.* **67**, 148 (1963).
49. R. Waack and M. A. Doran, *J. Phys. Chem.* **68**, 1148 (1964).
50. M. Weiner, G. Vogel, and R. West, *Inorg. Chem.* **1**, 654 (1962).
51. E. Weiss and E. A. C. Lucken, *J. Organometal. Chem.* (*Amsterdam*) **2**, 197 (1964).
52. G. Wittig, F. J. Meyer, and G. Lange, *Ann. Chem.* **571**, 167 (1951).
53. D. J. Worsfold and S. Bywater, *Can. J. Chem.* **38**, 1891 (1960).

CHAPTER III

Organometallic Derivatives of Group III

A. General Comments

A discussion of electronic transitions in Group III organometallic compounds is limited at this time to those of organoboron. Although the considerable interest in organoaluminum [59] and organogallium [47] compounds is reflected in recent review articles, there seem to be no published studies of the UV or visible spectra of these compounds. Derivatives of indium and thallium have attracted even less attention.

The monoboranes, referred to hereafter as simply boranes, form a class of boron compounds, R_3B, in which the central atom is sp^2 hybridized with a nominally vacant boron atomic $2p$ orbital perpendicular to the plane of the three sp^2 hybrid orbitals; as a necessary consequence the substituents R and the boron are coplanar. The nature of R may be varied quite widely from alkyl or aryl to halogen, alkoxy, amino, etc., the only important exception being hydrogen. Organoboranes of the simplest formula R_2BH as a rule are found to be electron deficient dimers under normal conditions. Certainly one of the properties of boranes which make them of general interest is the fact that they are isoelectronic with the corresponding carbonium ions, R_3C^+, and analogous to carbonyls in the limited sense of the canonical form $R_2C^+—O^-$.

Another general class of organoboranes of spectroscopic interest is formed by the heterocyclic compounds in which an sp^2 hybridized boron is incorporated with its available p orbital into an aromatic system. Undoubtedly the best known example of this structural feature is borazine, often dubbed in textbooks as inorganic benzene. M. J. S. Dewar has been especially responsible for the synthesis and characterization of a variety of previously unknown boron heteroaromatic compounds [25].

Boron is frequently tetravalent, forming molecules and ions of the general formula, R_4B. These compounds, R_4B, may be regarded as the conjugate bases of the Lewis acids R_3B. Inasmuch as the vacant boron p orbital has been removed and boron may only be σ bonded to R in R_4B, the visible–ultraviolet spectra would be expected to resemble those for R_4C, modified only by the inductive effect of B^-, and would be expected to be therefore somewhat less interesting to the spectroscopist.

Organo-substituted boron hydrides and carboranes are well known, but again studies of the electronic transitions of these compounds are sparse.

B. Alkylboranes

1. *Trialkylboranes*, R_3B

Contrary to some earlier reports in the literature [90], the trialkylboranes do not have an UV or visible absorption maximum above 225 mμ. Two of three studies report that the UV spectra of the tributylboranes normal, secondary, and isobutyl in saturated hydrocarbon solvents have absorption maxima near 215 mμ ($\log \epsilon \approx 2.4$) which are nearly independent of butyl isomerism [24, 95]. A third published UV spectrum of tri-n-butylborane does not show a maximum above 170 mμ [77], but in view of the conflicting reports and the results obtained for other trialkylboranes, it is likely that this result is incorrect and that the 225 mμ absorption in tributylboranes is

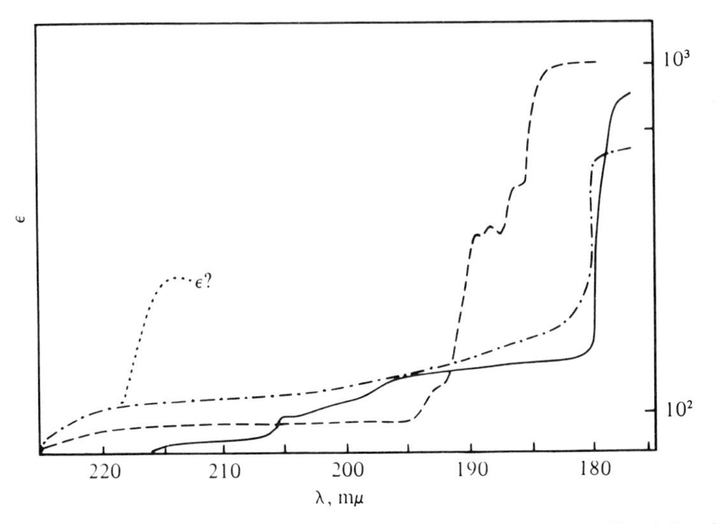

Fig. 10. The gas phase spectra of trialkylboranes: (———) $(CH_3)_3B$; (–·–·–) $(C_2H_5)_3B$; (––––) $[(CH_3)_2CH]_3B$; (·····) $(n\text{-}C_4H_9)_3B$ in hexane from the literature.

real. This borane transition of moderate intensity progresses in a regular fashion to higher energy [83], as the substituent alkyl groups are changed with increasing ionization potential of the hydrocarbon, in the order C_4H_9, C_3H_7, C_2H_5, CH_3, as may be seen from Fig. 10. It has been proposed that this UV transition series be described as an intramolecular charge transfer (CT) transition from a hydrocarbon delocalized "σ" orbital of π symmetry to the boron $2p$ atomic orbital of same symmetry. The charge transfer assignment rests on the dependence of the transition on alkyl ionization potential, and correlation with similar well characterized intramolecular CT transitions in the triarylboranes [82]. The idea that the trialkylborane excited state for this transition can be represented by the hyperconjugated structure **3** is strictly satisfactory only for trimethyl borane. The simple hyperconjugative excited state would imply that the first electronic transition energy would be about the same for all alkylboranes with a C—H bond, which is not the case. The successful accommodation of the experimental results by the charge transfer approach depends on the requirement that the positive charge in the excited state be delocalized over a large portion of the hydrocarbon framework, rather than localized in some bond α to the boron.

$$\overset{H^+}{R_2C}{=}\bar{B}R_2$$

(3)

Very rough calculations based on either the charge transfer model [82] using Eq. (10), or simple Hückel calculations [83] using the empirical heteroatom model of the methyl group as an electron pair donor, predict a transition energy for trimethylborane near 175 mμ.

$$E_{CT} = I_p - E_a - C + E_R \tag{10}$$

In Eq. (10), I_p is the ionization potential of the electron donor, E_a is the electron affinity of the acceptor C is the classical Coulombic electrostatic energy, and R represents the sum of the ground state stabilization and excited state destabilization after resonance between the two states. The values used for $(CH_3)_3B$ were $I_p = 13$ eV, $E_a = 0.0$ eV, $C = 9$ eV and $R = 3$ eV, where R was taken as twice the estimated hyperconjugation energy [72] of $(CH_3)_3B$.

As the alkylborane CT transition moves to higher energy with change in alkyl substituent, a second weaker band system ($\log \epsilon < 2$) emerges from under the CT envelope. These transitions have been assigned [83] to the promotion of an electron from a boron–carbon bond to the boron empty p orbital, a $\sigma_{B-C} \rightarrow N_B$ transition. Such an assignment is analogous to the well-recognized $n \rightarrow \pi^*$ transition of carbonyls, and the suggested $\sigma \rightarrow \pi^*$ assignment put forward by Berry [12] for the weak 200 mμ absorption of olefins. It is also interesting to find that current estimates of the Coulomb integral of boron for use in Hückel-type molecular orbital calculations, places the vacant nonbonding boron p orbital very near in energy to the antibonding orbital of an olefin [93]. This certainly simple approach would predict that the $\sigma \rightarrow \pi^*$ olefin transition should be roughly of the same energy as the $\sigma \rightarrow N_B$ borane transition. Extended Hückel calculations predict the $\sigma \rightarrow N_B$ transition energy within a few tenths of an electron volt.

The above results and assignments could have been anticipated by the following argument. The three boron–carbon localized σ orbitals may be used to construct under D_{3h} symmetry the set of delocalized molecular orbitals set forth below (Fig. 11). Two of these molecular orbitals, the E′ pair,

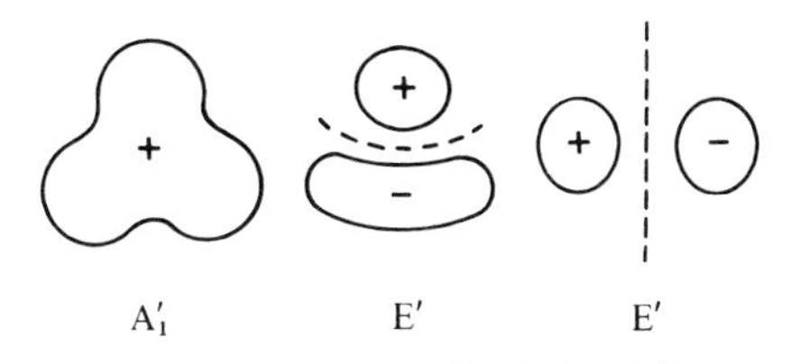

FIG. 11. Delocalized σ orbitals for BX_3 of D_{3h} symmetry.

have pseudo-π symmetry and are higher in energy than the A_1' of σ symmetry. The vacant boron p orbital is of A_2'' symmetry. The expected lowest energy transition is the doubly degenerate $A_1 \rightarrow E''$, which is allowed only if coupled to vibrations of the appropriate symmetry. A second possible transition is the total symmetry allowed $A_1 \rightarrow A_2''$ which however for reasons of local symmetry, i.e., poor spatial overlap, must also be expected to be weak. Thus we are led to expect one, perhaps two, weak transitions in the spectrum of a trialkylborane ascribable to the B—C σ bond and the vacant boron p orbital.

2. *Alkylboranes* $R_{3-n}B(\ddot{Y}R)_n$

Mono- and dialkylboranes such as alkoxy or hydroxyboranes, boron amines, and boron halides in which boron is bonded to a p–π electron donor should exhibit an intense internal CT transition from $\ddot{Y}$ to B analogous to the $\pi \rightarrow \pi^*$ transition of ketones. The ultraviolet spectra of the compounds **4–6** have been reported to 170 mμ [77]. The dipropyl-n-butoxyborane spectrum consists only of rising end absorption from 210 to 170 mμ and the

$(C_3H_7)_2BOC_4H_9$	$(C_3H_7)_2BSC_4H_9$	$(C_3H_7)_2BNH_2$
(4)	**(5)**	**(6)**

spectrum of the analogous sulfur compound possesses weak maxima at 240 (sh), 220 (sh), and 197 mμ, which seem to correspond to similar transitions in mercaptans. Alkylboronic acids, $RB(OH)_2$, are also reported to have no absorption above 220 mμ [21]. It must be assumed then that the CT or $\pi \rightarrow \pi^*$ transition of these compounds lies at a shorter wavelength than 170 mμ.

A suggestion made by this author in an earlier paper [85] is incorrect. The intense 260 mμ transition in the spectrum of air-oxidized trimethylboron was assigned to the B—O chromophore. A suggestion made by a referee of the paper [85] that the maximum in the spectrum of the oxidation product of trimethylborane might be due to a $\pi \rightarrow \pi^*$ transition of the peroxide chromophore, ⟮B—O—O⟯, now seems more attractive.

C. Aryl-, Vinyl-, and Alkynylboranes

1. *Triarylboranes*

The UV spectra of the triarylboranes exhibit broad, intense ($\epsilon > 10^4$) absorption maxima between 280 and 360 mμ, which have been described as intramolecular CT transitions promoting an electron from the aromatic ring to the boron vacant *p* orbital [82, 84, 98a, b]. The UV spectrum of trimesitylborane in Fig. 12 is representative of this class of compounds. This

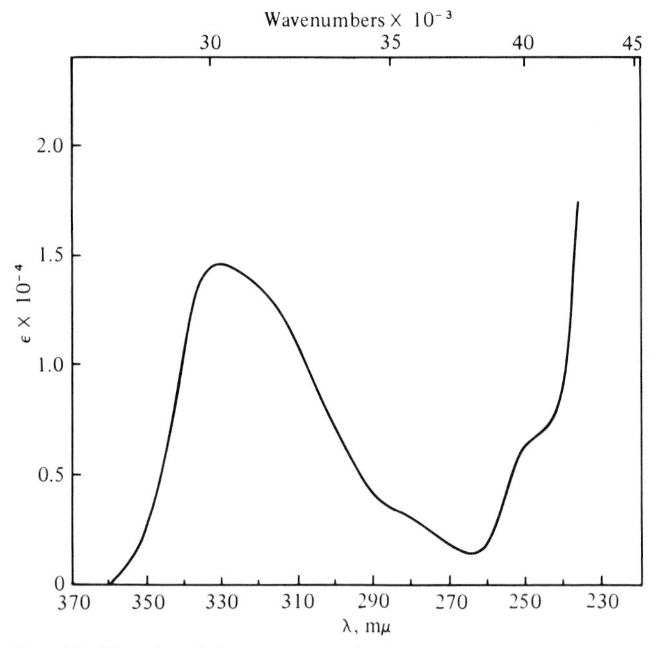

Fig. 12. The ultraviolet spectrum of trimesitylborane in isooctane [82].

280–360 mμ transition of the triarylboranes satisfies the following generally accepted criteria for a CT transition [82].

(1) The energy of the transition may be satisfactorily estimated from Eq. (8) whose terms were defined in Chapter I.

$$E_{\mathrm{CT}} = I_p - E_a - C + R \tag{8}$$

(2) As required by Eq. (8) the UV transition energy, when plotted against the ionization potential of the aromatic hydrocarbon substituent, gives a good straight line, which may also be extended to include trivinyl- and tributylborane.

(3) A plot of the UV transition energies of the triarylboranes against the corresponding intermolecular CT transition energies of hydrocarbon–iodine complexes gives an excellent linear relationship of near unity slope.

(4) The CT transition maximum may be made to disappear on complexing the borane with ammonia or acetonitrile, i.e., by removing the acceptor portion of the molecule.

(5) Steric hindrance to coplanarity of the aromatic rings and the plane of the boron sp^2 bonds causes a decrease in the intensity of the transition in a fashion similar to that found for *ortho*-substituted anilines.

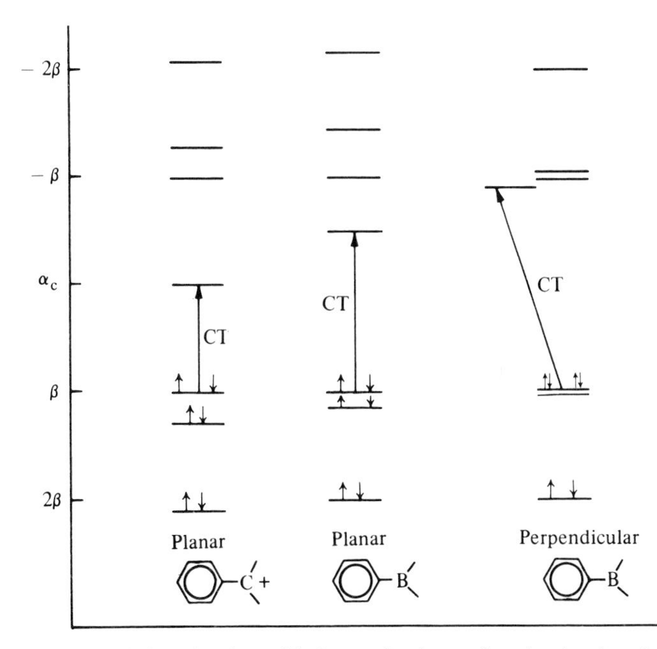

Fig. 13. Hückel molecular orbital energies for a phenylcarbonium ion and a phenylborane [82].

The simple molecular orbital description of these transitions is graphically illustrated by the molecular orbital diagram of Fig. 13 for phenylboranes and carbonium ions. Murrell has developed the argument that, given even a set of nearly degenerate donor molecular orbitals with the same acceptor, the resulting CT excited states from these orbitals may have substantial energy differences because of differing charge distribution in the derived excited states [73]. A case in point may be illustrated by the resultant charge distribution after promotion of an electron to boron from the two highest filled Hückel molecular orbitals of benzene (Fig. 13).

If the point charge distribution of structures (a) and (b) in Fig. 14 is used to calculate the Coulombic term of Eq. (8) for a CT transition, the state

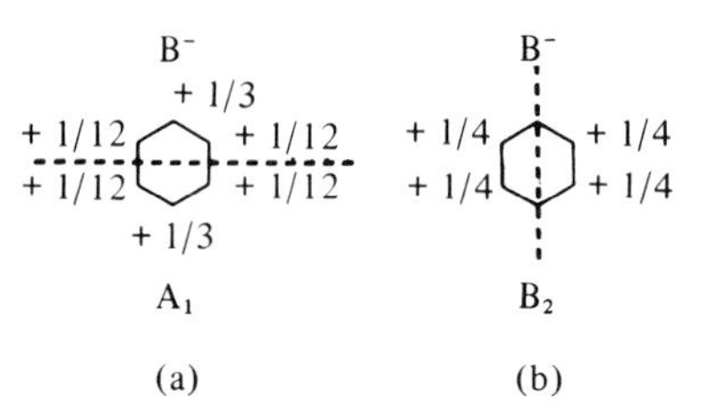

FIG. 14. Point charge distribution after promotion of an electron from two highest filled benzene orbitals to the boron $2p$ orbital (C_{2v} symmetry group).

represented by (a) is approximately 1 eV lower in energy than (b). The 282.5 mμ shoulder in the spectrum of trimesitylborane (Fig. 10) has been assigned [82] by the above argument to the expected second CT transition in the spectra of the triarylboranes. The 247 mμ shoulder also evident in the spectrum of trimesitylborane is then regarded as the aromatic L_b transition, blue shifted after configuration interaction with the second CT state at 282 mμ. Similar assignments have been made for triphenyl-, tri-*p*-tolyl-, and tri-1-naphthylborane.

A term level correlation diagram for phenylcarbonium ions and phenyl boranes is given in Fig. 15 to show the relationship between the transitions of these isoelectronic compounds. In fact, for the benefit perhaps of the unwary, it must be noted from Fig. 15 that the relative assignments, set forth above and in Ramsey [82], of the second CT and localized L_b transition in the spectra of the triarylboranes rest upon the assumption that the CT transition of the phenylcarbonium ion in going to phenylborane must remain a CT transition without violation of the symmetry noncrossing rule. The assignments could be reversed if this requirement were dropped.

The charge transfer transition of 9-phenyl-9-borafluorene is found at 405 mμ [60], a substantial red shift to that of triphenylborane at 287 mμ.

In a continuation of their calculations on the series $(\text{aryl})_n\text{BX}_{(3-n)}$, where

X = halogen or —OR, Armstrong and Perkins [1] have published Pariser-Parr-Pople self-consistent field molecular orbital calculations, including configuration interaction, on triphenyl- and tri-*p*-tolylborane. Since initial calculations using only the four highest filled and four lowest vacant orbitals were unsatisfactory, the final calculations included transitions between the six highest filled and six lowest vacant molecular orbitals in the configuration

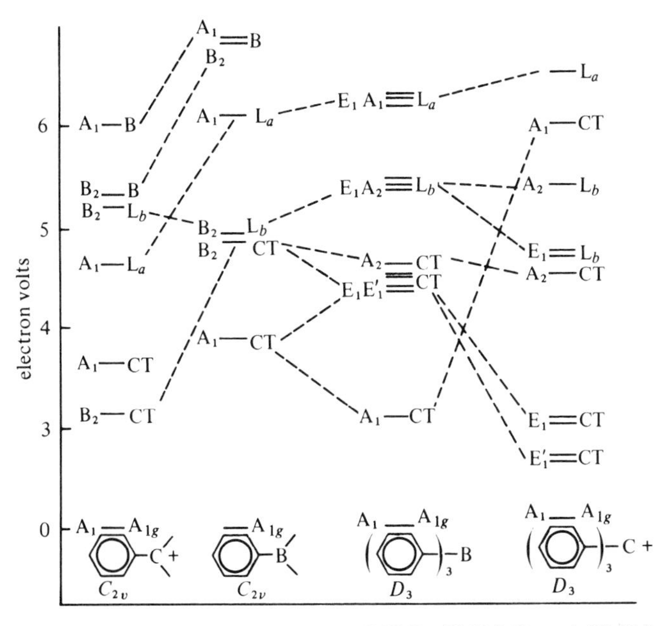

FIG. 15. Term level diagram for $C_6H_5C^+$, C_6H_5B, $(C_6H_5)_3B$, and $(C_6H_5)_3C^+$ (linear between 0 and 3) [82].

interaction. Even so, the results of the calculation appear to be in substantial conflict with the experimental data on the following points:

(1) The calculated energy of the first transition maximum is 0.3–0.4 eV too high.

(2) The lowest excited states which presumably make up the first absorption band are calculated to have little charge transfer character.

(3) Since the two lowest states of triphenylborane as calculated by Armstrong and Perkins are approximately linear combinations of the locally excited $^1B_{2u}$ benzene states, the calculated intensities are too small by a factor of almost 10^2 when compared with observed intensities.

With regard to the last point, it is suggested that the high intensity of the first transition may be due either to strong vibronic coupling or to mixing with allowed $\pi \rightarrow \sigma^*$ or $\sigma \rightarrow \pi^*$ transitions, but there is no evidence to support or eliminate either hypothesis. Agreement with higher energy transitions is satisfactory between calculated and observed maxima.

There is also a direct disagreement between Ramsey and Armstrong and Perkins on the assignments of the symmetries of the first three excited states corresponding to allowed transitions in triphenylborane. Armstrong and Perkins under D_3 symmetry make the following symmetry assignments with the indicated parenthetical oscillator strengths: A_2 (0.003) $<$ E (0.004) $<$ E (0.277). As has already been indicated, the charge transfer model with Murrell's point charge approximation used by Ramsey exactly reverses these assignments, placing the most intense $^1A_1 \rightarrow {}^1E$ transition at the lowest energy and the $^1A_1 \rightarrow {}^1A_2$ transition at the highest energy of the three. It was pointed out by Ramsey that a molecular orbital calculation would probably give the opposite result and that a conflict between the CT valence bond and the molecular orbital approach was expected here.

Studies of the solvent effect on the spectra of the arylboranes *per se* have not been reported. However, for those triarylboranes which do not, for steric reasons, form strong boron–nitrogen complexes with acetonitrile (i.e., trimesitylborane and tri-1-naphthylborane), a change in solvent from a hydrocarbon to acetonitrile results in a *blue* shift of 3.0–5.0 mμ in the maximum of the first CT transition [13]. A similar solvent effect on the spectrum of triphenylborane in ethanol has been reported [100]. This small blue shift is contrary to a common assumption that CT transitions involving increased charge separation in the excited state will exhibit large *red* shifts with increasing solvent polarity. An explanation of the blue shift may be that in the ground state the electronegativity difference between sp^2 carbon and boron results in a net dipole moment in the direction of the phenyl ring. The instantaneous transition dipole moment would be 180° in the opposite direction to this, and the polar solvent molecules previously oriented to stabilize the ground state would then actually *destabilize* the excited state, since, as stated by the Franck-Condon rule, the solvent molecules do not have sufficient time to reorient themselves within the time it takes for the electronic transition to occur. Such solvent effects are sometimes termed the result of Franck-Condon orientation strain in the excited state. An alternative explanation in ground state terms, is that any solvation of the boron which reduces the electron affinity of the boron p orbital will raise the energy of the CT transition. This may be the major effect in the case of $(C_6H_5)_3B$ but cannot be very important in trimesitylborane because of steric hindrance to the solvation of the boron by the *ortho*-methyl groups.

The phosphorescent emission spectrum of triphenylborane at 77°K in several different glasses has been reported: λ_{max} 420 mμ in isopentane; λ_{max} 384 mμ in EPA; λ_{max} 384 mμ in isopentane–ethanol, and λ_{max} 387 mμ in isopentane–diethyl ether [100].

2. *Mono- and Diarylboranes*

The author finds only one example [60] in the literature of the UV–visible spectrum of an arylalkylborane. The ultraviolet spectrum of 9-ethyl-9-borafluorene exhibits a maximum at 387 mμ, probably charge transfer, at a much lower energy than biphenyl, λ_{max} 251, or fluorene, λ_{max} 261.5 mμ. Interestingly, the substitution of chlorine for ethyl leads to a red shift λ_{max} of 9-chloroborafluorene (397 mμ) in keeping with the expected *inductive* effect of a chlorine on the CT transition.

Nikitina and co-workers [77, 78] have reported the UV absorption maxima of 21 mono- and diarylboranes in which the remaining substituents are predominantly amino, hydroxy, and alkoxy groups. For purposes of the following discussion, a brief selection of these results and others are given in Table II.

TABLE II

ABSORPTION MAXIMA OF SOME ARYLBORANES, KETONES, AND CARBONIUM IONS

Formula	λ_{max}, mμ (log ε)
$C_6H_5B(OC_4H_9)_2$	194 (4.67), 220 (4.02), 235 (3.47), 260–275 (2.79)
p-$BrC_6H_4B(OH)_2$	197 (4.6), 234 (4.2), 253 (3.8), 265–275 (2.8)
$C_6H_5B(C_4H_9)(OC_4H_9)$	199 (4.64), 225 (4.04), 235 (3.72), 263–278 (2.8)
$C_6H_5C(=O)CH_3$	245 (3.99)
$C_6H_5C(=O^+H)$—CH_3	296 (4), 335 (3)
$(C_6H_5)_2B(OC_4H_9)$	183 (4.48), 202 (4.41), 238 (4.24), 265–280 (2.87)
$(C_6H_5)_2C{=}O$	252 (4.24)
$(C_6H_5)_2C^+$—OH	344 (4), 292 (3)
$(C_6H_5)_2BCl$	230 (4.0)
$(C_6H_5)_2BNH_2$	240 (4.3), 272 (3.0)
$(\alpha$-$C_{10}H_7)_2B(OH)$	222 (5.0), 285 (3.0), 295 (4.0)

From Table II and the more extensive data compiled in Appendix B the following generalizations seem possible concerning the UV spectra of those phenyl boranes in which the boron is also bonded to a π electron pair donor, i.e., $(C_6H_5)_{3-n}B\ddot{Y}_n$, where $\ddot{Y}$ is, for example, halogen, nitrogen, or oxygen. Often a perturbed red shifted benzenoid L_b band with its multiplet structure may be discerned in the 260–280 mμ region. An intense transition appears in the 180–200 mμ region which may be regarded as a localized transition corresponding to the B bands of benzene. Finally, in addition to the above transitions, to be considered as the perhaps expected $\pi \rightarrow \pi^*$ benzenoid absorptions, a characteristic strong absorption maximum (ϵ 0.5 to 1 $\times$ 10^4) is found between 210 and 250 mμ for the phenyl boranes and at longer wavelengths for naphthylboranes. The spectrum of p-$CH_3C_6H_4B(OH)_2$ in hexane is shown in Fig. 16. It is this additional 210–250 mμ transition in the

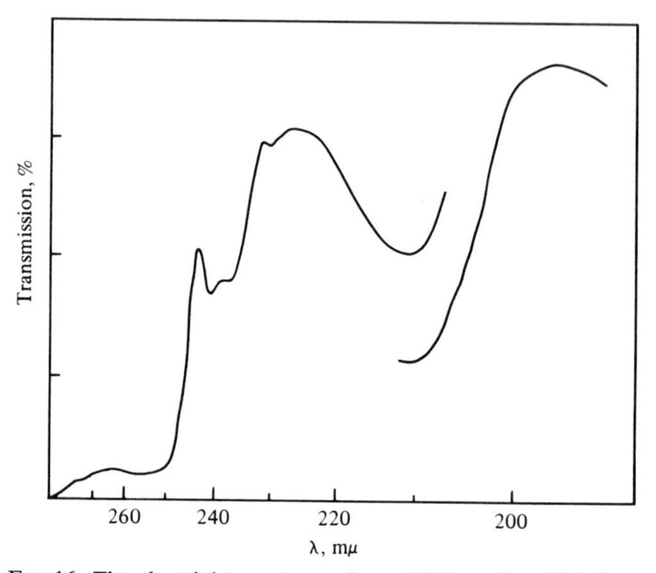

FIG. 16. The ultraviolet spectrum of paratolylboronic acid in hexane.

phenylboranes which is of particular interest as possibly corresponding to the previously discussed intramolecular CT transition observed in triarylboranes [82, 84].

Nagakura has successfully interpreted the spectra of acetophenone and benzoic acid in terms of CT transitions from the benzene ring to antibonding π^* orbitals of the carbonyl groups [75]. The extension of this approach is obvious, and a comparison of the ultraviolet spectra of related boranes, ketones and carbonium ions is given in Table II. If, however, the second transition in the spectra of the phenylboronic acids, chlorides, etc. is to be assigned to charge transfer, the transition energy should be linearly related to the ionization potential of the substituted phenyl ring according to Eq. (8), as is found for the triarylboranes. The ultraviolet spectra of the p-X-substituted phenylboronic acids have been obtained [87], where X is H, Br, F, CH_3, OCH_3, or $N(CH_3)_2$, and no linear correlation of the ionization potential of C_6H_5 with ultraviolet transition energy is obtained, even over the limited series H, Br, F, CH_3, OCH_3. A similar failure of the charge transfer approximation is apparent in the transition energies of the substituted diphenylhydroxyboranes, i.e., phenylborinic acids and esters from Table II and Appendix B.

The cause of success of the charge transfer approximation in triarylboranes and its failure when boron is bonded to oxygen or some other heteroatom π electron donor is to be found in the Hückel energy level diagram of Fig. 17, which shows the group orbitals of BO_2, C_6H_6, and the resultant

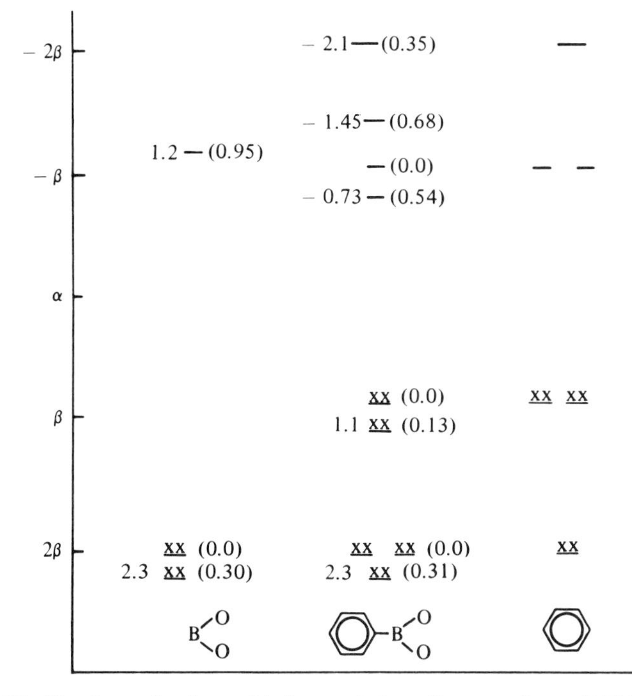

FIG. 17. Simple molecular orbital correlation diagram for —$B(O_2)$, C_6H_6, and $C_6H_5BO_2$, where (———) energy (boron atomic orbital coefficient): $\alpha_B = \alpha^\circ - 0.9\beta^\circ$.

molecular orbital energies of phenylboronic acid. In the case of the triarylboranes qualitative agreement between Hückel calculated energies and observed spectra was obtained by assuming a boron $2p$ orbital slightly *lower* in energy than the lowest vacant benzene molecular orbital.

But, the importance of π back-bonding to boron by oxygen or nitrogen, that is, the contribution to the ground state of resonance structures such as $B^-{=}O^+$ in the chemistry of boranes, seems well established, and recent NMR studies [67] of B—N rotational barriers in vinyl- and phenylborondimethylamines have been used to calculate a B—N π bond order between

0.45 and 0.63. The charge transfer transition in the boronic acids is not therefore simply to a vacant boron $2p$ orbital, but is to the lowest, vacant BO_2 group π^* orbital. The Hückel energy of this group orbital was calculated using the same Coulomb integral α as was used for the arylboranes, which is lower in energy than that of the vacant benzene orbitals, but the resultant BO_2 group orbital energy obtained lies above the vacant benzene orbitals as may be seen from Fig. 17. It should be no surprise then that when a simple Hückel molecular orbital calculation is done on $C_6H_5B(OH)_2$, the two lowest vacant molecular orbitals have boron $2p$ mixing coefficients of less than 0.5. The net result of all this leads to the conclusion that the second transition in the spectra of the hydroxy, alkoxy, and halogen phenylboranes is to be regarded as the L_a benzene transition normally found near 200 mμ, but substantially red shifted because of configuration interaction with higher-energy charge transfer states, but with some small charge transfer character. Under C_{2v} symmetry the L_a transition is ${}^1A_1 \rightarrow {}^1A_1$.

It should also be apparent from Fig. 17 that if the perturbation between the phenyl rings and the BO_2 group is reduced by rotating the rings out of the plane of the BO_2 bonds, then the lowest vacant orbital will increase in energy, moving back towards its unperturbed position and resulting in an increased L_a transition energy. This provides the rationale for the relative absorption maxima of $(p\text{-}CH_3C_6H_4)_2BOC_4H_9$ (λ_{max} 250 mμ) and $(o\text{-}CH_3C_6H_4)_2BOC_4H_9$ (λ_{max} 236 mμ), for example, where steric hindrance of the *ortho*-methyl groups force the rings out of plane.

Certainly the most complete and rigorous calculations on the mono- and diarylboranes are the LCAO–MO–SCF calculations by the Pariser-Parr-Pople method carried out by Armstrong and Perkins on diphenylboranes [6], dihalophenylboranes [5], and phenylboronic acids [7].

Calculations for diphenylboranes [6] including configuration interaction were carried out on diphenylboron chloride, diphenylhydroxyborane, diphenylaminoborane, di-*o*-tolylaminoborane, di-*p*-tolylhydroxyborane, and di-*o*-tolylhydroxyborane. The results of the calculations are in very good agreement with the observed transition energies, the calculated energy usually being within 0.1 eV of that observed. The relative magnitudes of the extinction coefficients are also predicted.

The first two transitions, $A_1 \rightarrow B_1$ and $A_1 \rightarrow A_1$ under C_{2v} symmetry, have virtually the same transition energy and comprise the first absorption band in the 260–280 mμ region in the diphenylborane spectra.

The third transition, $A_1 \rightarrow B_1$ for C_{2v} symmetry, represents the absorption between 230 and 250 mμ in the spectra of the diphenylboranes. It is satisfying that this third excited B_1 state, according to the results of the calculation, is comprised almost entirely of a single configuration in which the electron is promoted from the highest filled benzene molecular orbital to a lowest vacant molecular orbital which is stated to have a large mixing coefficient for the boron atomic orbital. The calculated energies are relatively insensitive, however, to methyl substitution of the ring or twisting of the rings up to 60° out of plane [6], but they do predict a small blue shift in the 230–250 mμ transition.

The Pariser-Parr-Pople LCAO–MO–SCF calculations by Armstrong and Perkins [5] on phenylborondihalides were carried out for fluorine, chlorine, and bromine. Agreement between the best calculated energies after configuration interaction and observed transition energies is better than within 0.2 eV for the first UV transition of phenylboron difluoride and dichloride. The agreement for the second transition is only a little less satisfactory, 0.3–0.4 eV or roughly 15–20 mμ. The lowest vacant molecular orbital from these calculations contains a boron mixing coefficient varying from 0.36 for the bromide to 0.52 for the fluoride. The second lowest starting molecular orbitals have boron mixing coefficients of zero. The first transition is calculated as the result of roughly equal mixing of two B_1 configurations from one electron excitation to the two lowest vacant molecular orbitals, and as a result has only a small amount of charge transfer character. The second transition, however, contains after configuration interaction roughly 70% of the first excited state A_1 configuration from excitation to the lowest vacant molecular orbital, and therefore possesses somewhat greater calculated charge transfer character.

An important point is, however, that neither of the first two calculated transitions is predominantly charge transfer in agreement with our earlier discussion utilizing simple Hückel calculations. In these calculations [5], it is only for configurations involving the third lowest vacant molecular orbital that better than 50% CT character is found in the excited state after configuration interaction. These states have calculated energies greater than 7 eV above the ground state and should not be observable in the quartz ultraviolet spectral region. At least qualitative agreement is obtained with Murrell's prediction that the A_1 CT state will be lower in energy than the B_1 CT state (see Section III-C-1). In most cases this is the opposite order of energies predicted from the effect of substitution on the order of the energy levels of the two highest filled molecular orbitals in the starting ground state configuration.

Very good agreement with observed transition energies was obtained by Armstrong and Perkins [7] in their Pariser-Parr-Pople SCF calculations on phenyl-, *p*-tolyl-, and *o*-, *m*-, and *p*-styrylboronic acids. The agreement is in fact much better than that reported by Armstrong and Perkins themselves because maxima at 220 mμ and 235 mμ in the spectrum of $C_6H_5B(OC_4H_9)_2$ in hexane, reported by Nikitina, *et al.* [77], were separately assigned by Armstrong and Perkins to the L_a ($^1A_1 \rightarrow {}^1A_1$) and B_a ($^1A_1 \rightarrow {}^1A_1$) transitions. However, if one examines the spectra [87] of the series p-$XC_6H_5B(OH)_2$, where X is H, CH_3, Br, or CH_3O in water and in hexane, it is found that in addition to a slight red shift of the maximum from water to hexane, a great deal of vibrational structure is observed in the absorption band of the second transition (see Fig. 16) of the phenylboronic acids. The second absorption maximum of phenylboronic acid in water is at 218 mμ, and in hexane the correct assignment for the second transition maximum in the spectra of $C_6H_5B(OH)_2$ or $C_6H_5B(OC_4H_9)_2$ is the maximum near 220 mμ; the maximum near 235 mμ in hexane is due to vibrational structure and part of the same transition envelope. The calculated and observed transition energies based on this reassignment for the L_a and B bands are: for phenylboronic acid L_a, calc. 5.5 eV, obs. 5.6 eV; B_a, B_b, calc. 6.3, 6.4 eV, obs. 6.4 eV; and for p-tolylboronic acid, for which Armstrong and Perkins did not have data, calc. L_a, 5.4 eV, obs. 5.5 eV; calc. B_a, B_b, 6.3 eV, obs. 6.3 eV. In the spectrum of p-anisylboronic acid in hexane, where the L_a band has moved to 240 mμ, the B band has also become very broad, with a pronounced shoulder at 205 mμ and a maximum at 201 mμ which may be attributed to the B_a ($A_1 \rightarrow A_1$) and B_b ($A_1 \rightarrow B_2$) transitions, respectively.

A final point of interest with regard to the Armstrong and Perkins calculation on phenylboronic acid lies in the observation that again for the L_a

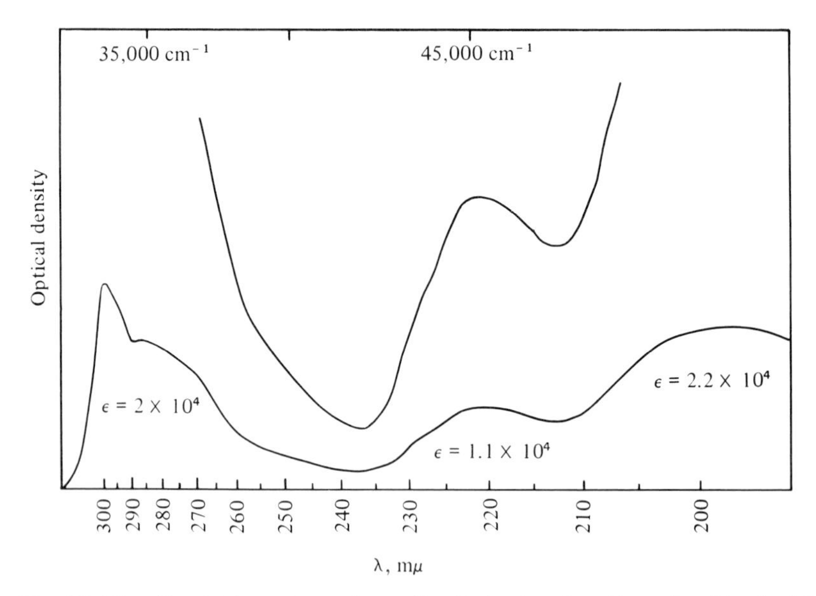

FIG. 18. The ultraviolet spectrum of paradimethylaminophenylboronic acid anhydride in hexane.

transition excited state, the configuration obtained from electron promotion to the lowest vacant molecular orbital has a mixing coefficient of 0.98, and this lowest vacant molecular orbital is stated to have a large boron mixing coefficient. Although the boron coefficient was not explicitly stated, it is presumably not greater than 0.5.

The spectrum of p-$(CH_3)_2NC_6H_9B(OH)_2$ anhydride in hexane is itself uniquely interesting [87] (see Fig. 18). The spectrum of *N,N*-dimethylaniline has an intense absorption maximum near 250 mμ which has variously been assigned to nitrogen lone pair → ring charge transfer ($A_1 \rightarrow A_1$) and perturbed L_b ($A_1 \rightarrow B_2$) transition. The corresponding transition in the p-$(CH_3)_2NC_6H_5B(OH)_2$ spectrum occurs at 290 mμ while the transition characteristic of aryl-boronic acids appears at 223 mμ, a much higher energy than would have been predicted from the progression of this transition to the red in the series p-$XC_6H_5B(OH)_2$, where X is H, CH_3, Br, or CH_3O. It appears that strong configuration interaction between the boronic acid L_a excited state, symmetry A_1 (C_{2v}), and the state corresponding to the *N,N*-dimethylaniline excited state has shifted one transition maximum to the blue and the other to the red. This requires A_1 symmetry for the aniline-like transition and enables an assignment of the 250 mμ transition of *N,N*-dimethylaniline to $A_1 \rightarrow A_1$ charge transfer.

Lockhart has studied the ultraviolet–visible spectra of compounds **7** [65]. These compounds have UV spectra with an intense absorption maxima in the region from 500 to 540 mμ. It is obvious from comparison with the spectra of other arylboron amines, chlorides, etc., that this transition is categorically unique.

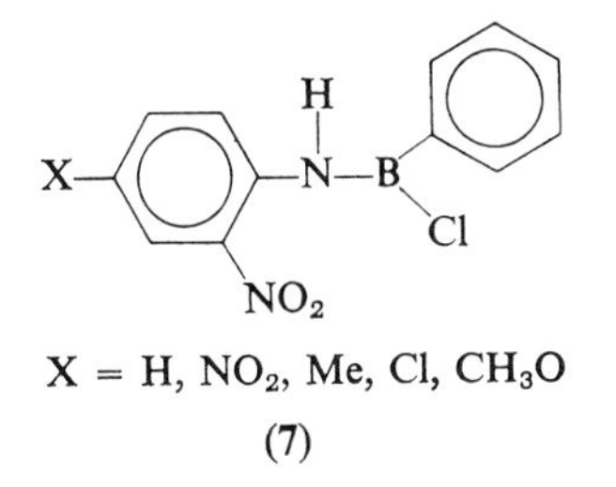

X = H, NO_2, Me, Cl, CH_3O

(**7**)

The transition shifts to the red with increasing π-electron donating ability of X, and with increasing solvent polarity. Furthermore, the transition seems to be unique to the geometry of the *ortho*-nitro group—three atoms removed from the boron. For example, the transition is not found in *o*-nitrophenylboronic acid [91]. Earlier evidence [20] has been found for ground state internal coordination of ortho aromatic nitro groups to boron in the stability of orthonitrophenyldichloroboronite (**8**) and this offers some clue to the nature of the transition.

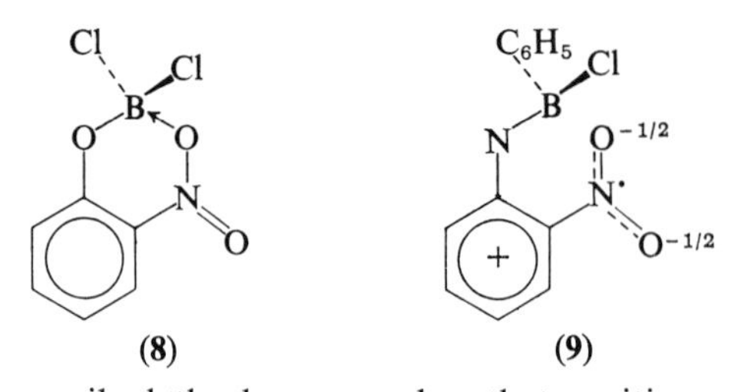

(8) (9)

Nagakura has ascribed the long wavelength transition of aromatic nitro compounds to intramolecular charge transfer from the aryl ring to the nitro group [74]. Utilizing the charge transfer description, the ground state coordination of the boron to the nitro group should increase the electron affinity of the acceptor nitro group, leading to a lower transition energy, or by a slightly different argument boron coordination will be more effective at an excited state stabilization in which the nitro group bears formal negative charge (**9**) rather than ground state stabilization. From this basis, the 500–549 mμ transition may be regarded as the normal transition of aromatic nitro compounds with charge transfer character shifted to lower energy by internal coordination of the nitro group to boron.

The spectra of compounds **10–12** below have been reported [9]. Compounds **10** and **11** were considered as being analogous to perturbed dihydroxytropylium ions and diacyclocyclopentadienide anions, respectively.

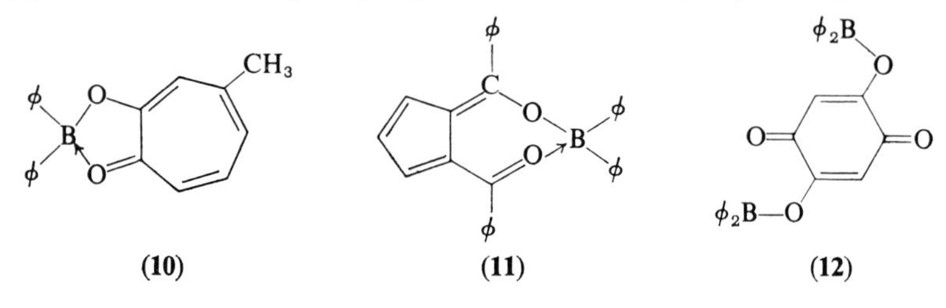

(10) (11) (12)

3. *Vinylboranes*

The vinylboranes possess one or more strong absorption maxima ($\epsilon \approx 10^4$) at much lower energies, 185–234 mμ, than the usual $\pi \rightarrow \pi^*$ transition of ethylene at 162 mμ. These electronic transitions of vinylalkylboranes in valence bond terminology represent intramolecular CT transitions from the olefin molecular orbital to boron [82].

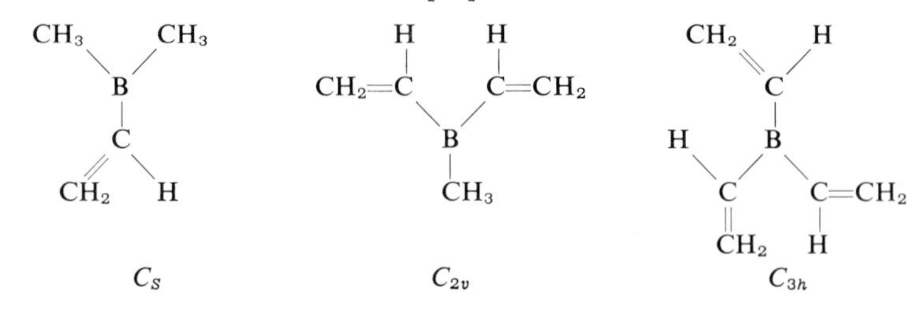

C_s C_{2v} C_{3h}

Good and Ritter [43] have obtained the UV spectra of several vinylboranes including those in Table III; their spectra were subjected to Gaussian curve analysis by Armstrong and Perkins [3] to resolve in most cases a weak shoulder on the low-energy side of the CT transition and sometimes a third transition at higher energies. In addition to the spectrum of vinylboron dichloride (λ_{max} 207 mμ), Coyle and co-workers [21] report the UV maximum of divinylboron chloride at 210 mμ, $\epsilon \sim 8000$. The results of Good and Ritter are probably correct, simply on the basis of the higher molar extinction coefficient obtained.

TABLE III

ULTRAVIOLET MAXIMA OF VINYLBORANES

Compound	λ_{max}, mμ (ϵ)[a]	λ_{max} (f)[b] Gaussian analysis
$CH_2{=}CHB(CH_3)_2$	196 (1×10^4)	228 (0.02), 195 (0.28)
$(CH_2{=}CH)_2B(CH_3)_2$	220 (1.3×10^4)	252 (0.03), 221 (0.3), 201 (0.1)
$(CH_2{=}CH)_3B$	234 (1.0×10^4)	264 (0.03), 233.5 (0.3), 215 (0.04)
$(CH_2{=}CH)_2BCl$	222 (1.7×10^4)	

[a] Molar extinction coefficient.
[b] Oscillator strength.

The CT transition of vinyldimethylborane is found at 196 mμ, and assuming C_S symmetry to be $A' \rightarrow A'$. For a divinylborane, in the first approximation we would have two degenerate CT transitions, one from each vinyl group, but configuration interaction (resonance) between the CT states will give rise to two CT states symmetrically disposed in energy above and below the degeneracy origin. If C_{2v} symmetry is assumed for divinylmethylborane, the lowest-energy CT state will be A_1, with the second CT state B_1 or B_2, depending on the choice of coordinates. The components of the transition dipoles from each vinyl of the CT state belonging to the B irreducible representation oppose each other, so that we might expect the $A_1 \rightarrow B$ transition to be weak. From the above considerations, assignment of the 221 mμ maximum to $A_1 \rightarrow A_1$ charge transfer and 201 mμ to $A_1 \rightarrow B$ charge transfer can be made.

For trivinylborane the triply degenerate CT states should split under C_{3h} symmetry into a lower-energy doubly degenerate E′ CT state and higher-energy A′ CT state, accounting for the 233.5 mμ maximum at $A' \rightarrow E'$ CT transitions and the 215 mμ maximum as the $A' \rightarrow A'$ CT transitions. Under C_{3h} symmetry only the $A' \rightarrow E'$ transition is symmetry allowed; however, coupling of A′ charge transfer with vibrations of suitable symmetry, such as E′, or symmetry reduction of the molecule could make the transition weakly allowed.

Simple Hückel calculations by Good and Ritter [43] on the vinylboranes obtain good agreement between observed and calculated transition energies when boron is given a Coulomb integral, α, equal to $\alpha^\circ - \beta^\circ$, the Hückel Coulomb and resonance integrals for carbon. The normalized boron mixing coefficients in the lowest vacant molecular orbital range between 0.5 and 0.7.

Armstrong and Perkins have treated the vinylboranes both by the free-electron [3] and the more extensive Pariser-Parr-Pople LCAO–MO–SCF methods with configuration interaction [4]. The agreement obtained between observed [43] and calculated transition energies by the Pariser-Parr-Pople LCAO–MO–SCF method for trivinylborane methyldivinylborane, and divinylchloroborane is good, i.e., 0.02–0.2 eV. The boron mixing coefficient of the lowest vacant molecular orbital varies from 0.7 for trivinyl- to 0.56 for a monovinylborane in keeping with the suggested charge transfer nature of the first transition.

The results of the calculations on the methylvinylboranes seem less encouraging for two out of three of the methylboranes considered. The calculations [4] included the methyl group both in terms of the hyperconjugative model $C{\equiv}H_3$ and as a pseudoheteroatom electron pair donor. In spite of considerable variation in methyl parameters and molecular geometry, the energies of the first $\pi \rightarrow \pi^*$ transition are not predicted within 0.3 eV, nor the second $\pi \rightarrow \pi^*$ transition within even 1.3 eV, for dimethylvinylborane and chloromethylvinylborane.

The weak bands which appear in the spectra of vinylboranes at lower energies than the $\pi \rightarrow \pi^*$ (CT) transitions (see Table III) are tentatively assigned by Armstrong and Perkins [4] to a $\sigma \rightarrow \pi^*$ transition promoting an electron in the first approximation from the vinyl C—H σ bonds to the lowest π^* molecular orbital. This assignment of these weak bands is supported by a fair linear correlation of their energies with the calculated energies of the lowest vacant π^* molecular orbital. No such correlation was found with the energy of the highest filled π molecular orbital, a result taken to rule out a $\pi \rightarrow \sigma^*$ assignment. The $\sigma \rightarrow \pi^*$ assignment is the same as that made by Berry [12] for the weak 200 mμ "mystery band" of olefins. An alternative assignment not considered by Perkins and Armstrong would be $\sigma \rightarrow \pi^*$ from the boron–carbon σ bonds (see Addendum).

Vinylboronic acids and their esters have attracted some attention [68, 101, 102]. An early report [64] that the $(CH_3)_2C{=}CHB(OH)_2$ spectrum in hexane had an absorption maximum at 220 mμ (ϵ = 10^3) is probably wrong in view of later reports that in the spectrum [68] of $CH_2{=}C(CH_3)B(OH)_2$ in water there was no maximum above 205 mμ and that $CH_2{=}CHB(OC_4H_9)_2$ absorbs [77] at 185 mμ in hexane. The ultraviolet spectrum of vinylboronate ester **13** has a maximum at 198 mμ. Woods *et al.* [101] ascribe the lower transition energy of the spectrum of compound **13** compared with other vinylboronic acid esters to an increase in the B—O overlap integral caused by planarity of the ring system. This statement was made with a reference to

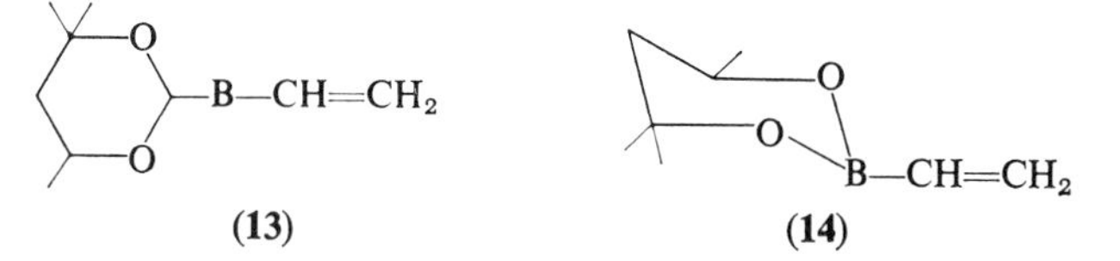

work by Matteson [68], who reports Hückel molecular orbital parameters for B and O. We are unable to tell from the literature, however, by whom or how the reported calculation was performed. Therefore, because we felt intuitively that increased B—O overlap should increase the energy of the $\pi \rightarrow \pi^*$ transition by reducing the electron affinity of boron and its interaction with the ethylene π^* orbital, we carried out simple Hückel molecular orbital calculations using the boron Coulomb integrals $\alpha_B = \alpha° - 0.9\beta°$ and $\alpha_B = \alpha° - 1.1\beta°$. In each case for α_B, the major effect of increased B—O overlap was assumed to be an increase in the B—O resonance integral, since in the simple Hückel calculation the overlap integrals themselves are set equal to zero, and in each case an increase in the B—O resonance integral led as anticipated to an *increased* transition energy. It does turn out that if a Hückel calculation is carried out including nearest neighbor overlap integrals, *but with no provision for a change in the* B—O *resonance integral with overlap*, that increased overlap produces a lower transition energy (because of overlap terms in the denominator of energy terms). This, however, does not seem to be the way to do the calculation since the same approach predicts a *decrease* in *bond* energy with overlap, contrary to experience.

From these considerations, we would rather suggest that the lower-energy transition of structure **13** relative to $CH_2{=}CHB(OC_4H_9)_2$ is due to structure **14**, which simultaneously decreases the B—O overlap and resonance integrals, and by introduction of *s* character into the vacant boron orbital which increases its electron affinity.

The spectra [76] of the vinylboronamines $[(CH_3)_2N]_2BCH{=}CH_2$ (λ_{max} 218) and $[(CH_3)_2N]BrBCH{=}CH_2$ (λ_{max} 223) are of interest because the ionization potential of ammonia (10.15 eV) is less than that of ethylene (10.5 eV), which raises the possibility that these transitions are charge transfer N $\rightarrow$ B or nitrogen $n \rightarrow \pi^*$ transitions. This is similar to Murrell's [73] assignment of the 234 mμ aniline transition to N $\rightarrow$ ring charge transfer.

4. *Alkynylboranes*

The author has been unable to find any reports of the UV spectra of simple (alkyl) (alkynyl) boranes in the literature.

The UV spectra of several monomeric conjugated dialkynylaminoboranes of the general structure **15** have been obtained by Soulie [92].

$$R_2N—B[C\equiv C—C\equiv C—R]_2$$

(15)

The spectra of these compounds have two intense, $\epsilon > 10^4$, absorption maxima (220–225 and 230–270 mμ) at energies somewhat lower than those usually found for transitions of similar intensity of conjugated polyalkynes [λ_{max} $CH_3(C\equiv C)_3CH_3$ 207 mμ (ϵ 1.3×10^5) and 268 mμ (ϵ 2×10^2)]. It seems likely by analogy with the aryl- and vinylboranes that these transitions have considerable charge transfer character; however, inasmuch as the electronic transitions of acetylene itself are not satisfactorily understood [56], little more can be said about alkynylboranes.

Absorption maxima in the spectra of $HC\equiv CB(OH)_2$ at 195 mμ and $HC\equiv CB(OCH_3)_2$ at 202 mμ (ϵ 934) in cyclohexane have been reported by Woods and Strong [102]. Unlike the result for vinylboronic acids, the acetyleneboronic acid cyclic ester (**15a**), with an absorption maximum at 202 mμ (ϵ 1590), is not red shifted with respect to the methyl ester (see Addendum).

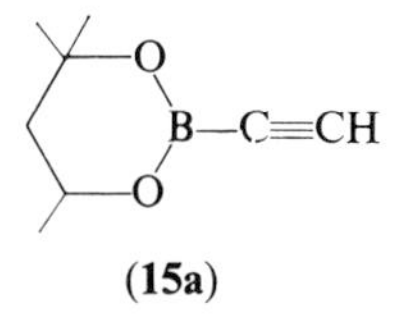

(15a)

D. Conjugated Heterocyclic Boranes

1. *Heteroaromatic Boranes*

Groups such as R_2BNR_2 and R_2BOR are π isoelectronic with the carbon–carbon double bond, and a continuing area of interest centers around the study of organoboranes derived from aromatic hydrocarbons by replacement of two carbon atoms with a boron and a nitrogen or other π electron donor atom.

TABLE IV

Electronic Transitions of Borazine

Assignment D_{3h}	$^1A_1' \rightarrow {}^1A_2'$	$^1A_1' \rightarrow {}^1A_1'$	$^1A_1 \rightarrow {}^1E'$
Borazine, λ_{max} (mμ)	199.5, 196.2, 192.8, 189.5	185[a]	171.0
Benzene, D_{6h}	$^1A_{1g} \rightarrow {}^1B_{2u}$	$A_{1g} \rightarrow {}^1B_{1u}$	$A_{1g} \rightarrow {}^1E_u$
Benzene, λ_{max} (mμ)	256	203	180

[a] Estimated, presumed hidden by $^1E'$ band.

The electronic transitions of borazine (**16**), benzene with all carbons replaced by B—N, and some substituted borazines, were reported in an early series of papers by Platt and co-workers [55, 81, 88]. The observed UV absorption maxima of borazine and their assignments along with the corresponding benzene assignments are presented in Table IV.

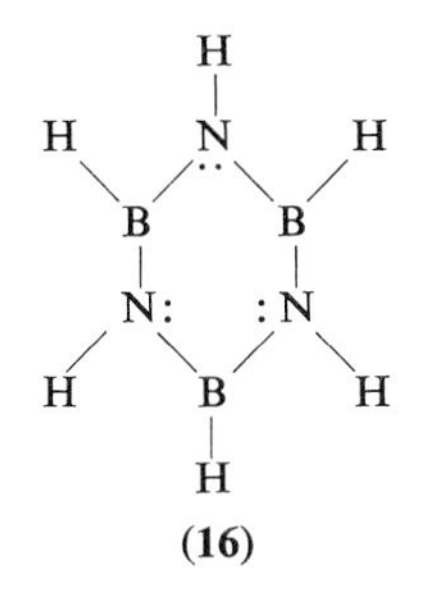

(**16**)

One interesting aspect of the UV spectra of borazines is the relative insensitivity of the transition energies to substitution at boron, compared with comparable substitution on nitrogen [23]. The *B*-trialkylborazines [50] do not absorb above 210 mμ (λ_{max} of *B*-trimethylborazine, 191, 185, and 176 mμ). By comparison [88], *N*-trimethylborazine UV absorption maxima occur at λ_{max} 227.8 (ϵ 1200), 224.2, 219.8, 217.4, 194.2 (ϵ 28,000), 190.1 and 185.9 mμ.

It seems likely from the results of UV and NMR studies [11, 54] that the phenyl groups of *N*- and B-triphenylborazine are rotated slightly out of the plane of the benzene ring by steric hindrance to coplanarity. Substitution of methyl for hydrogen in either *N*- or *B*-triphenylborazine results in a blue shift of the UV spectrum in opposition to the effect observed in borazine itself. This blue shift is attributable to increased steric hindrance by the methyl groups, forcing the substituent phenyl rings still further out of the plane of the borazine ring and thereby destroying the conjugation between the phenyl and borazine rings. The steric hindrance can be relieved by inserting an ethynyl group between the methyl and the borazine ring with the result that *B*-tri-1-propynyl-*N*-triphenylborazine UV maxima are 5–10 mμ toward the red of those of *N*-triphenylborazine [97]. An even larger spectral red shift is observed in the phenylethynylborazines in which complete coplanarity of all the rings is possible, although at least some of the red shift must be the result of extended conjugation by the alkynyl group.

Simple Hückel calculations indicate [97] that the intense (ϵ 10^5) long wavelength maximum near 280 mμ of *B*-triphenylethynylborazine represents charge transfer from phenylethynyl to borazine. A spectral blue shift of this transition on substitution at nitrogen of methyl for hydrogen is thereby accounted for in terms of reduced electron affinity of the borazole ring.

Several interesting molecular orbital calculations have appeared to supplement three earlier ones by Roothan and Mulliken [89], Patel and Basu [79], and Davies [23]. There is general agreement that in the ground state the boron is positive and the nitrogen negative and that the valence bond description —B^{-}═N^{+}— is incorrect, except in a limited sense as applied to the π electrons.

Hoffman [53] has applied his extended Hückel theory to a number of boron nitrogen compounds including borazines. According to these calculations the highest occupied orbital of borazine is a σ orbital. The calculated lowest $\pi \rightarrow \pi^*$ transitions result in a large transfer of charge from N to B and are therefore termed *internal charge transfer*.

Chalvet *et al.* [15] have carried out LCAO–MO–SCF calculations on three models of borazine and the results compared with borazine ionization potential, UV transition energies, and the diagonal element of the SCF Hamiltonian with the Hückel Coulomb integral. The best results were obtained from a borazine model in which each boron and nitrogen donates one electron to the molecular π orbitals, but the potential seen by the π electron is taken as equivalent to a free *neutral* boron or nitrogen atom. This model obtained the ionization potential of borazine within 0.5 eV and fair agreement with the UV spectra. A model in which boron was negative and nitrogen positive, —B^{-}═N^{+}—, gave unreasonable results.

Other LCAO–MO–SCF calculations on borazine, *N*-trimethylborazine, *B*-trimethylborazine, and *B*-trichloroborazine have also been reported by Perkins and Wall [80]. In contrast to nitrogen-substituted derivatives, for boron-substituted borazines the calculated one-electron energy levels were not very different from borazine itself, in agreement with the previously noted small effect of boron substitution on the spectrum of borazine. Agreement between observed and calculated transition energies is fairly good.

2. *Heteroaromatic Compounds with* C═C *Replaced by* B—$\ddot{Y}$—

According to simple molecular orbital theory, the energy levels of the even, alternate aromatic hydrocarbons are paired in energy (bonding and nonbonding) below and above the energy represented by the Coulomb integral of carbon. Furthermore, the absolute values of the coefficient of the atomic orbitals, contributing to corresponding π and π^* molecular orbitals, are the same. Considering only the two highest filled and two lowest vacant molecular orbitals (Fig. 19), we see that there are four one-electron transitions possible, the lowest energy $f \rightarrow g$, the degenerate pair $f \rightarrow h$, $e \rightarrow g$ of the same energy, and finally the $e \rightarrow h$ transition. If we now ask ourselves what will be the effect on the relative ordering of the energy levels e through h, and the resultant change in the spectrum of the hydrocarbon if two of the carbon atoms are replaced one by boron and the other by some electron pair donor

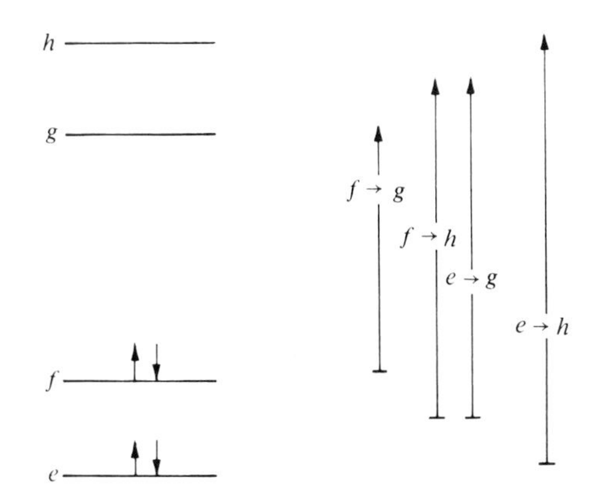

Fig. 19. Two highest filled and lowest vacant molecular orbitals of an alternant hydrocarbon.

such as N or O, the following rather qualitative but successful arguments may be made.

First, if we regard the substitution of either B or $\ddot{\text{Y}}$ for carbon as a small perturbation equivalent to a change in the Coulomb integral of the replaced carbon, the net change in the energies of the molecular orbitals for each substitution is approximated [66] by Eq. (11) where C_{ar}^2 and C_{br}^2 are the atomic orbital coefficients of the rth atom in the a and b molecular orbital, and $\delta\alpha_r$ is the change in Coulomb integral (α_C).

$$\delta\, \Delta E \cong (C_{br}^2 - C_{ar}^2)\, \delta\alpha_r \tag{11}$$

From Eq. (11) we can see that since $\delta\, \Delta E$ for corresponding molecular orbitals, i.e., f and g, e and h, will be zero because $C_{br}^2 = C_{ar}^2$, electronic transition energies between these molecular orbitals should be unaffected by the substitution of B and $\ddot{\text{Y}}$ for carbon. The value of $\delta\, \Delta E$ will not be zero for the molecular orbitals of the $e \to g$ and $f \to h$ transitions; however, we can at least say that since $\delta\alpha_r$ will be of the opposite sign for boron and for $\ddot{\text{Y}}$ the total change in relative energy may be small where the sign of $(C_{br}^2 - C_{ar}^2)$ remains the same, which will be the case if the substitution is a symmetrical one as in the 9,10-heteroatomphenanthrenes.

Second, there is the additional factor of configuration interaction between the normally degenerate states arising from the $e \to g$ and $f \to h$ transitions. The degeneracy is removed in the heteroaromatic compounds, and the large splitting of states resulting from zero-order configuration interaction will be reduced. On the other hand, the energies of the one-electron configurations themselves will be further apart so that it can be argued that the net result in terms of absorption maxima of bands may remain pretty much the same.

Third, the introduction of B and Ÿ for carbon will lower the symmetry of the molecule, generally to C_s, which means that formerly forbidden transitions such as the L_b benzene transition may now become allowed.

Taken together the above considerations, most of which are essentially those suggested by Dewar [25], lead to the expectation that the ultraviolet–visible spectra of the alternant heteroaromatic boron compounds should closely resemble those of the parent compounds, a conclusion well supported by available spectra. It is worthwhile to point out that the above is true even if the aromaticity of the hydrocarbon is considerably reduced by the substitution, since to the approximation of our arguments, the spectrum depends only on the *relative* energy differences between the molecular orbitals, i.e., one-electron configurations.

Among the heteroaromatic boron compounds for which derivatives have

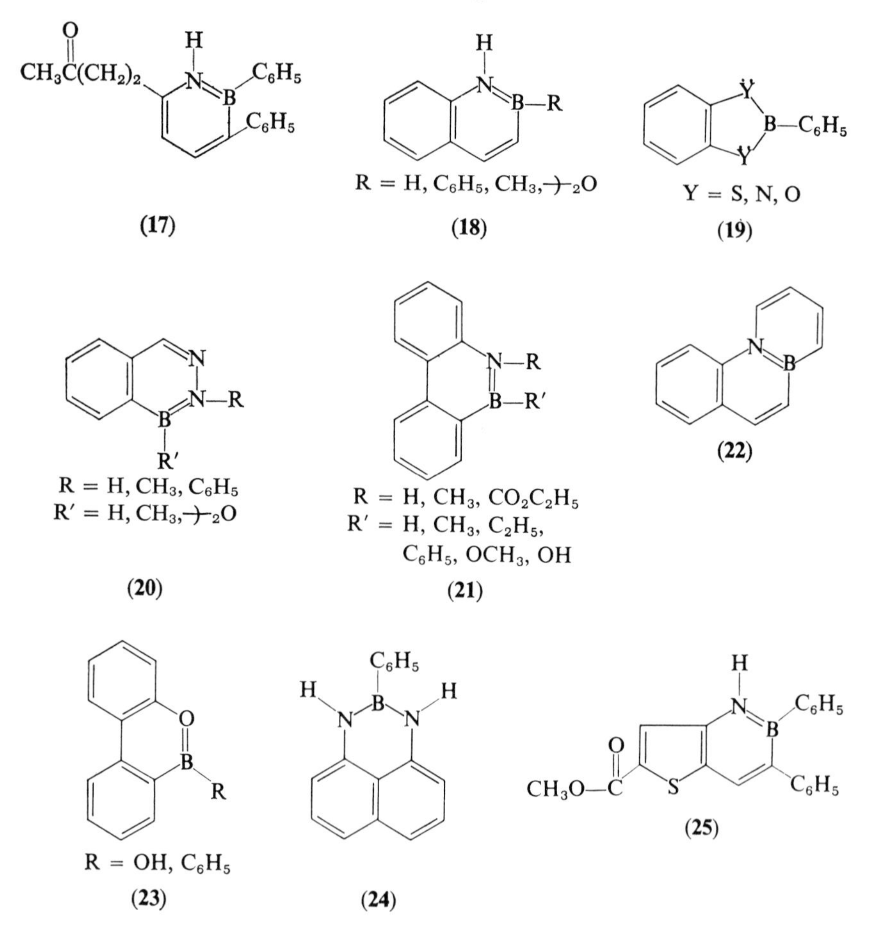

been prepared and for which ultraviolet spectra have been reported by Dewar and co-workers are structures **17–28** (see Addendum).

The correspondence of the spectra of the derived boron heteroaromatic compounds with hydrocarbon spectra is typified [25] by comparison of the

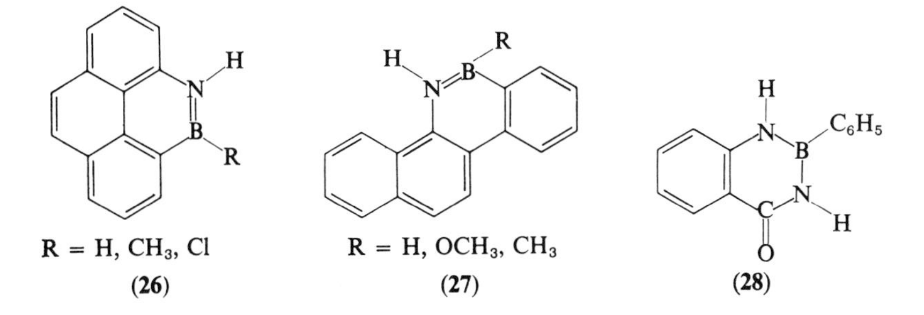

spectra of naphthalene, isoquinoline, and 2,1-borazaronaphthalene (Fig. 20). The close resemblance of the spectrum of the borane with isoquinoline is attributable in valence bond theory to the importance of the resonance

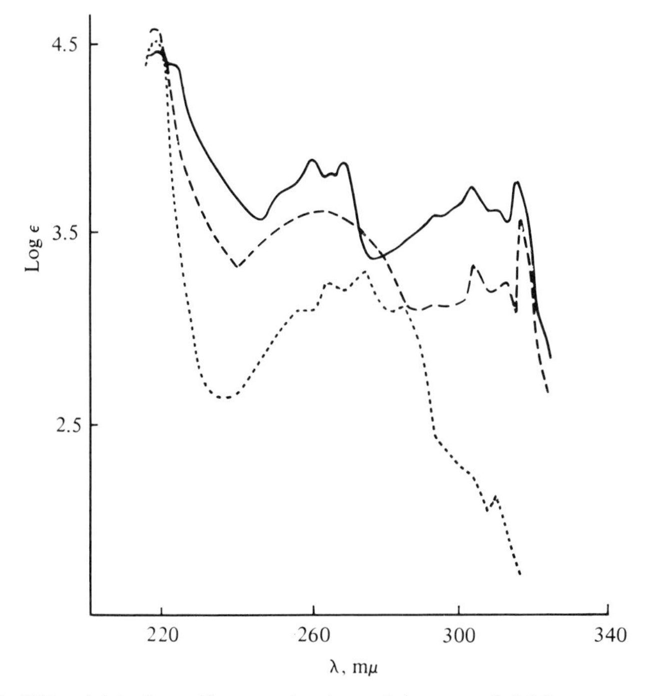

FIG. 20. Ultraviolet absorption spectra in cyclohexane of 2,1-borazaronaphthalene (———), isoquinoline (----), and naphthalene (·····).

structure **29** in the isoquinoline excited state to be compared with structure **30** for the 2,1-borazaronaphthalene.

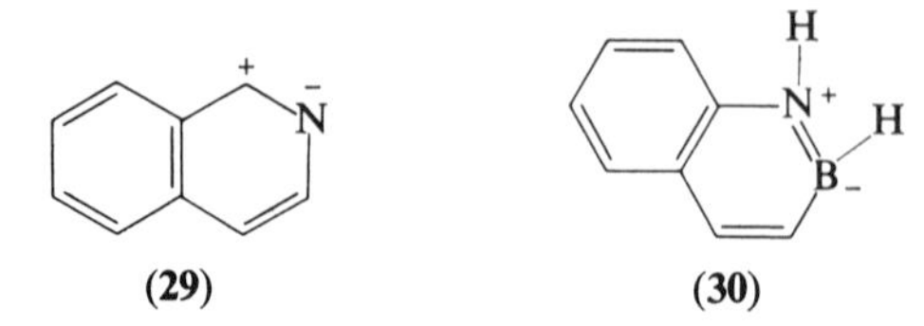

(29) **(30)**

Also interesting is the observation that for the *B*-hydroxyborazaro- [36] and boroxarophenanthrenes [27], and -naphthalenes [24], the ultraviolet spectra of these compounds in basic aqueous ethanol are most consistent with the formation of the phenolate-type anion rather than the usual tetra-coordinate borate anion (see Addendum).

This was not the case however for the *B*-hydroxy-10,9-borazarophenanthrene [67] or 10,9-boroxarophenanthrene [22] (structures **31** and **32**), in spite of the fact that in neutral solution the spectra closely resemble those of acridine and xanthone, respectively.

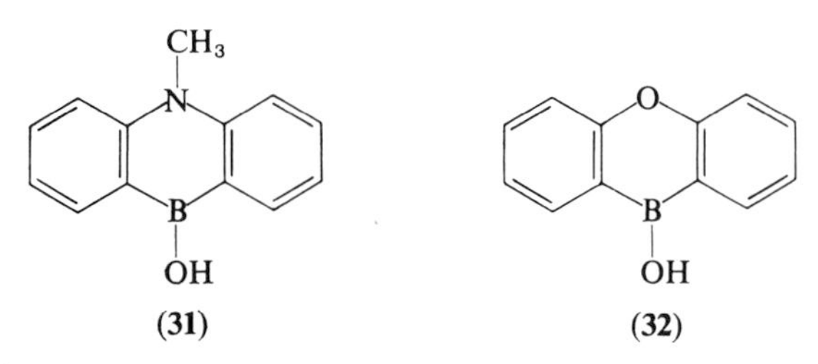

(31) **(32)**

Hemming and Johnson [51, 52] have prepared and reported the ultraviolet spectra of the diazaborolines and oxaborolens (**33–35**). Where Y and Y′ are NH, the ultraviolet spectra closely resemble the spectra of the corresponding imidazoles, but are less similar to the spectra of the isoelectronic hydrocarbon, for instance, in a comparison with structure **33** (Y and Y′ are NH, Ar is C_6H_5 with indole). Contrastingly, where Y and Y′ are O, the UV spectra are very similar to the spectra of the aryldiols, suggesting a much

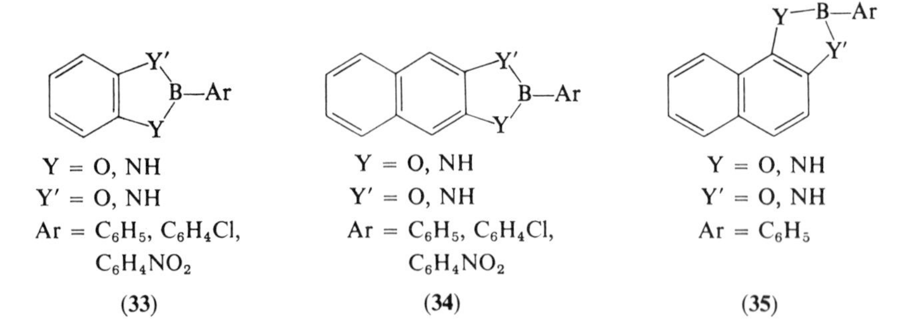

(33) **(34)** **(35)**

smaller degree of aromaticity for the boron–oxygen heterocyclic compounds than for similar boron–nitrogen compounds. Spectroscopic evidence is offered in the cases of the *o*-nitrophenylboron derivatives, for intra-molecular chelation of the boron by an oxygen of the *ortho*-nitro group.

Other potentially aromatic boron-containing heterocyclic compounds which have been made and their UV spectra reported by Gronowitz and Bugge [46] are the thienopyridines (**36–38**). Unfortunately the ultraviolet

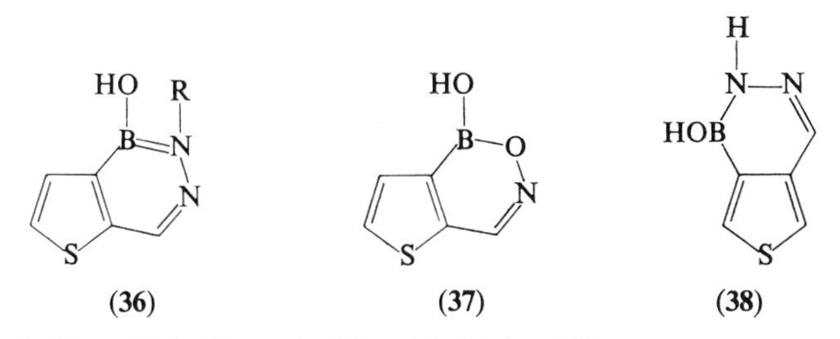

(36) **(37)** **(38)**

spectra of thieno[2,3,C]- and thieno[3,4,C]pyridine were not available for comparison.

Not much attention seems to have been paid to the ultraviolet spectra of the oxygen analogs of borazine, the boronic acid anhydrides (**39**), especially where R is an alkyl group.

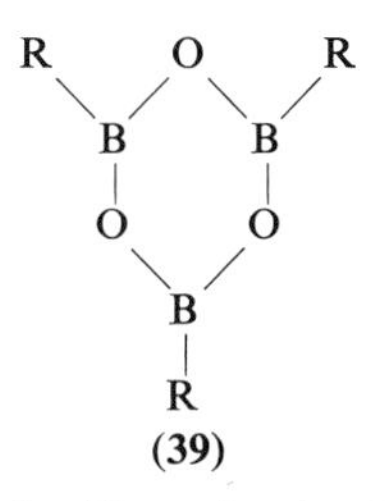

(39)

Surprisingly and perhaps significantly, in $CH_3C{\equiv}N$ and *n*-hexane, the ultraviolet spectra of the arylboronic acid anhydrides do not differ significantly from the spectra of the corresponding acids in the same solvent.

A 1,3,2-dioxaborinium ion (**40**), the boron oxygen analog of the pyrylium ion (**41**) has been obtained [8] as the perchlorate salt. The UV transition energies of ion **40** (λ_{max} 282, 374 mμ) are higher than those of the analogous pyrylium perchlorate (**41**) at 361 mμ and 408 mμ, in agreement with the prediction of Hückel LCAO–MO calculations.

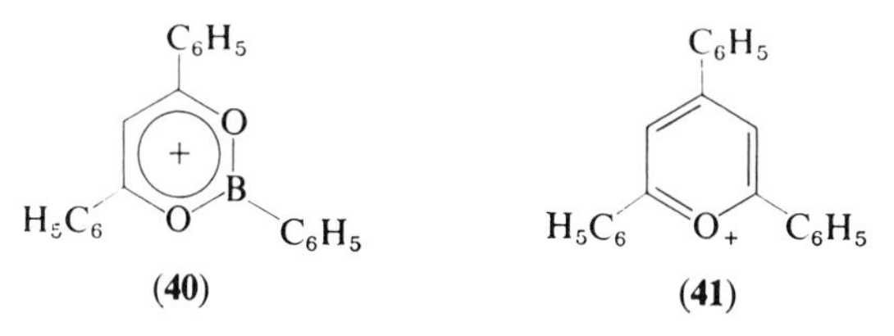

(40) **(41)**

Compounds **42** and **44** are related to the 1,3,2-dioxaborinium ion. The ultraviolet spectrum of compound **42** has an absorption maximum at 253 mμ (ϵ 9800). The spectrum of compound **43** shows no maximum above 200 mμ in water or ethanol, but in a strong base a maximum appears at 258 mμ (ϵ 9100), presumably the result of formation of the anion **44**. These transitions can be assigned to $\pi \rightarrow \pi^*$ transitions of the 6π electron aromatic ring.

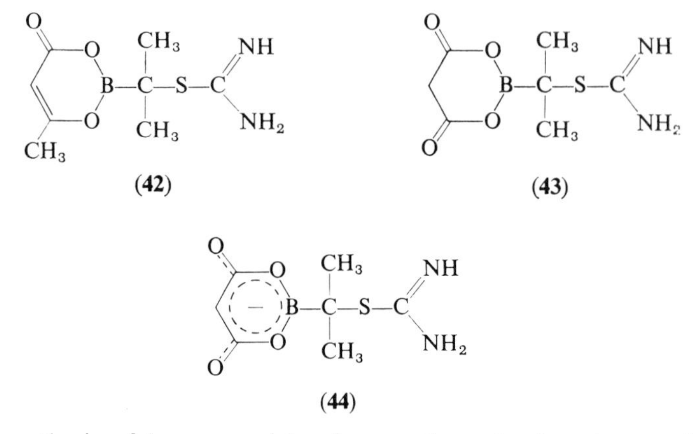

The synthesis of boron models of aromatic carbonium ions such as the cyclopropenyl and tropylium ion has been the goal of a number of research efforts. Attempts to synthesize derivatives of the borepin **45** have not been successful. Van Tamelen, Brieger and Untch, however, have reported [94] the synthesis of dibenz[*b*,*f*]borepin and the ultraviolet spectrum of the derived boronic acid ethanol amine ester (**46**), with a λ_{max} in ethanol 227 mμ (ϵ 1.6×10^4) and 298 mμ (ϵ 8.23×10^3). The spectrum of compound **46**, however, is likely to be quite different from that of an alkyl- or arylboron-substituted borepin. Of greater interest then is the recent report by van der Kerk of the ultraviolet spectrum of the benzoborepin (**47**). The spectrum of this compound (**47**) in cyclohexane has a broad absorption band with the considerable vibrational structure often found in aromatic $\pi \rightarrow \pi^*$ transitions. The

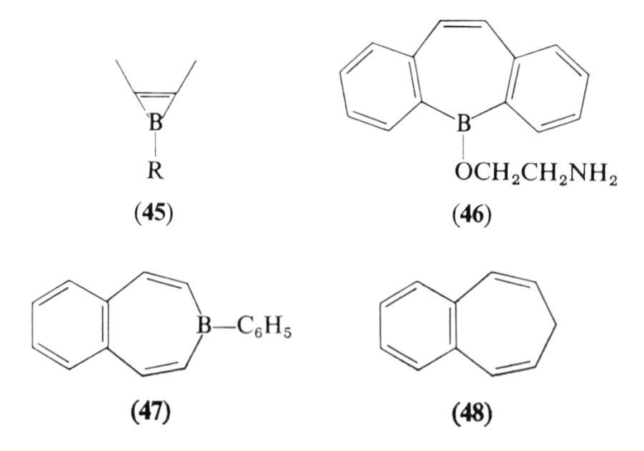

maximum absorption is at 301 mμ (log ε 4.45). The spectrum of the dimethylamine adduct of compound **47** closely resembles that of compound **48**.

3. *Nonaromatic Conjugated Heterocyclic Boranes*

An interesting organoboron analog [14] of the nonaromatic hydrocarbon cyclobutadiene has been reported: 1,4-diphenyl-2,4-bis(triethylcarbinyl)-1,3-diaza-2,4-diboretidine (**49**).

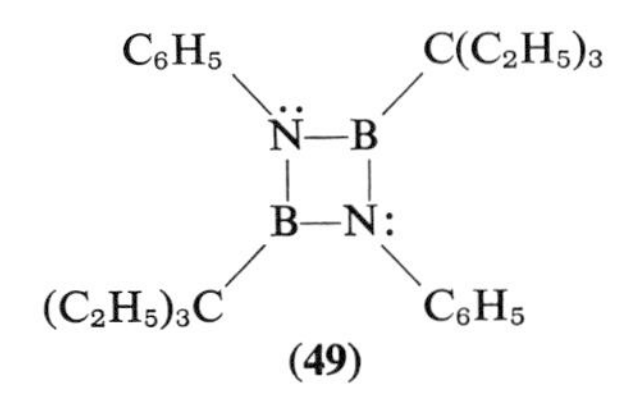

(49)

The UV absorption maxima (253, 259, 265 mμ) and molar extinction coefficients correspond closely to the maxima observed for the anilinium ion (λ_{max} 254) rather than aniline (λ_{max} 280), which suggests that (1) steric interactions force the plane of the phenyl rings out of the plane of the four-membered ring; (2) resonance structures $N^+{=}B^-$ are important; or both (1) and (2).

Another interesting heterocyclic molecule which has been reported by Watanabe and co-workers [96] is compound **50**, which has an intense absorption maximum in its spectrum at 352 mμ (log ε 4.22) indicative of considerable delocalization of electronic charge.

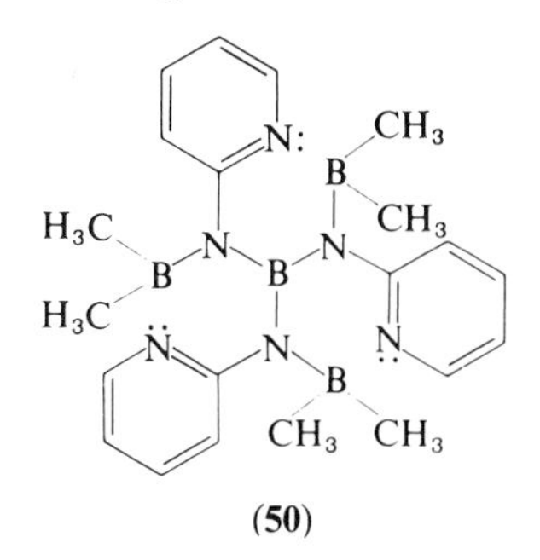

(50)

E. Triarylborane Anion Radicals and Dianions

The triarylboranes in ether, tetrahydrofuran, or benzene react with sodium in the absence of air to give highly colored mono- and disodium derivatives, represented by structures **51–53**.

$Ar_3B\cdot^{(-)}//Na^+$	$(Ar_3B^-//Na^+)_n$	$Ar_3B^{2-}://2Na^+$
(51)	**(52)**	**(53)**

Sodium reacts with triphenylboron in THF to form a diamagnetic dimer, whose visible spectrum [17, 70] shows a broad absorption maximum at 420 mμ. The fluorescence emission spectrum [17] of sodium triphenylborane dimer has maxima at 605, 645, and 675 mμ. The tri-1-naphthylborane sodium [70] is diamagnetic in ether [λ_{max} 290, 369 (sh), 420 mμ]. The visible spectrum [70] of tri-1-naphthylboranedisodium in ether has a λ_{max} of 390 and 555 mμ, and in THF 435 and 595 mμ.

Sodium trimesitylborane [80] and sodium tri-1-naphthylborane are paramagnetic monomers in tetrahydrofuran with intense broad bands in their visible spectra, such as λ_{max} (in sodium trimesitylborane) of 800 mμ. These transitions were assigned by Ramsey [83] to intramolecular charge transfer from boron to vacant aryl π^* orbitals with the further suggestion that there should be at least one additional charge transfer transition of a similar nature in the spectra of triarylborane radical ions. Weismann recently reported the spectra of a number of triarylboranes and sodium triarylboranes. An additional transition at 368 mμ (ϵ 2.2×10^4) was found in the spectrum of sodium trimesitylborane. This transition cannot correspond to the ring to boron CT transition of trimesitylborane at 331 mμ since this CT transition must be strongly blue shifted in the ion radical by the added electron density on boron. Very probably then, neglecting configuration interaction in the spectrum of sodium trimesitylborane, the 368 mμ transition may be assigned to charge transfer from boron to the *highest* vacant benzene π^* and the 800 mμ transition to charge transfer from boron to the *second* highest benzene π^* orbital (see Fig. 13).

The analogy of the above salts and their spectra with the alkali metal derivatives of conjugated hydrocarbons, as discussed in Chapter II, is readily apparent, and it seems likely that the actual species in solution are ion pairs and that the structure of the dimers and higher aggregates of the sodium triarylboranes may be similar to those of the alkyllithiums.

F. Tetravalent Boron Organometallic Compounds

In structure of the general formula R_3BR' the boron bears a formal negative charge. If R′ has a formal positive charge, of course, the molecule itself is neutral, for example, $(CH_3)_3B{:}NH_3$. If no R or R′ bears a formal charge, the structure represents an anion, for example, $(C_6H_5)_4B^-$. Finally R′ and one R may each possess a formal positive charge in which case a cation is obtained, for example, $[R_2B^-(\text{pyridine}^{2+})_2]^+$.

1. *Organoborates* $R_4B(^-)$

The absorption spectra [42] of tetraphenylborate, $[(C_6H_5)_4B]^-$ (λ_{max} 265, 274 mμ) is as expected very similar to that for tetraphenylmethane [61] (λ_{max} 210, 253 (sh), 262, 272 mμ).

2. *Neutral Complexes*, $R_3B:NR$

Inasmuch as the ultraviolet spectrum of the tetraphenylborate ion largely resembles that of tetraphenylmethane, the ultraviolet spectra of the borane amine adducts, $(aryl)_2(RO)BNR_3'$ should be very similar to diphenylmethane. Therefore, it was very surprising to find in the literature reported strong absorptions for such compounds between 230 and 240 mμ (see Table V).

TABLE V

ABSORPTION OF MAXIMA OF SOLUTIONS OF DIARYLALKOXYBORON AMINE ADDUCTS

Compound	Solvent	λ_{max}, mμ (ε)
$(C_6H_5)_2\ddot{B}(NH_3)—OC_3H_7$	Hexane	237 (8349), 270 (603)
$(C_6H_5)_2\ddot{B}(NH_3)OC_3H_7$	CH_3CN	236 (3515), 268 (203)
$(C_6H_5)_2\ddot{B}(NH_3)OC_3H_7$	$CH_3CN \cdot NH_3$	(266)
$(C_6H_5)_2BOCH_2CH_2NH_2$	CH_3CN	236 (inflection), 268 (204)

However, although the spectrum of a solution of $(C_6H_5)_2(OC_3H_7)B:NH_3$ in $CH_3C{\equiv}N$ or hexane does possess an intense transition at 237 mμ, this transition disappears if the acetonitrile solution is saturated with ammonia. The spectrum of $(C_6H_5)_2BOCH_2CH_2NH_2$ in methanol shows a pronounced broad shoulder at 230 mμ, but in acetonitrile this transition is present in the spectrum of $(C_6H_5)_2BOCH_2CH_2NH_2$ only as an inflection point. From these results (Table V) it is obvious that the appearance of strong 230–240 mμ transitions in the spectra of solutions of Ar_2BOR–ammonia or –amine complexes are the result of irreversible dissociation in solution of these complexes into esters of phenylboronic acid and amine.

The ultraviolet spectra [77] of the ammonia adducts of triphenylborane, tri-1-naphthylborane [84], and acetonitrile adduct of triphenylborane [85] exhibit only the expected well-defined L_b, L_a, and B transitions of the aromatic rings.

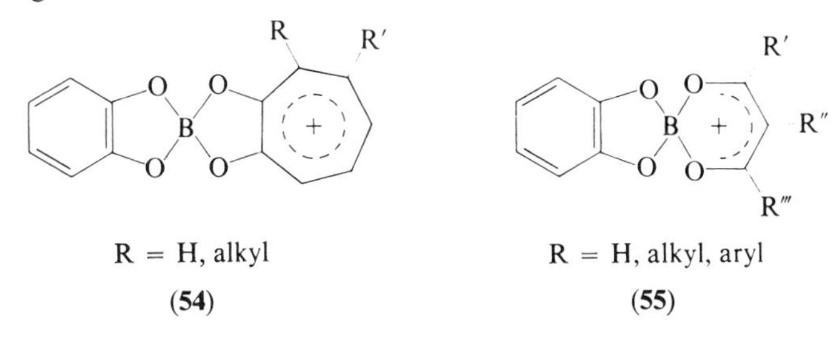

R = H, alkyl **(54)**

R = H, alkyl, aryl **(55)**

Arsene and co-workers [8] have recently reported the ultraviolet spectra of a series of spiroborates of the general structures **54**–**57** (see also Section II-C).

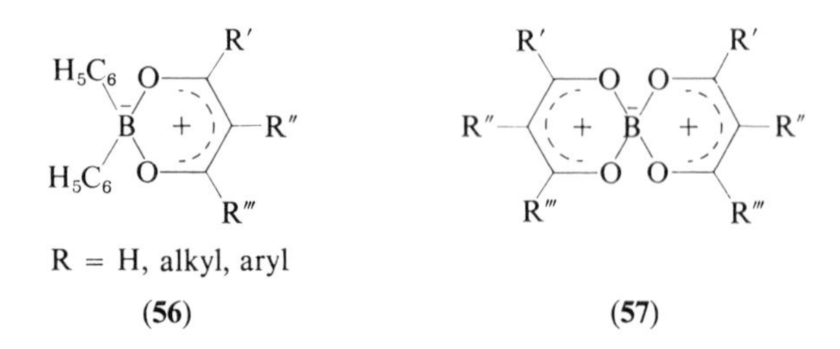

R = H, alkyl, aryl

(**56**) (**57**)

The spectra of the compounds having structure **54** are similar, with four intense ($\epsilon \approx 10^4$) absorption bands near 245, 275, 305, and 360 mμ. The 360, 305, and 245 mμ bands may be assigned to the tropolone and the 275 mμ band to the catechol moieties, as localized $\pi \rightarrow \pi$ transitions. The longest wavelength 360 mμ maximum is very solvent sensitive in that the maximum of structure **54**, where R and R′ are H, shifts from 374 mμ in cyclohexane to 355 mμ in formamide.

Each member of the series **55**, **56**, and **57** has an intense maximum in its spectrum between 300 and 400 mμ which may be assigned to a localized $\pi \rightarrow \pi^*$ transition of the chelated 1,3-dione portion of the molecule. In addition to this and a second strong absorption near 275 mμ, probably phenyl $\pi \rightarrow \pi^*$, a third transition in the spectra of several compounds with structure **55** appears as a weak ($\epsilon \sim 10^2$) shoulder near 370 mμ. In structure **55**, where R′ and R‴ are CH, R″ is H, this shoulder moves some 50 mμ toward the blue with a change in solvent from carbon tetrachloride to dimethylsulfoxide while the intense maximum of 285 mμ is essentially insensitive to solvent change. Because this shoulder is absent in other similar compounds of the series **56** and **57**, the shoulder was not assigned to an $n \rightarrow \pi^*$ transition of the nonbonding oxygen electrons, but to intramolecular charge transfer from the catechol ring to a 1,3-diketone moiety.

Where R′ and R‴ are C_6H_5 and R″ is H, the spectrum of compound **56** differs from that of the corresponding compounds **55** and **57** in that the short-wavelength absorption band (λ_{max} 300, 309 mμ) is broad and slightly more intense than the first transition at 391 mμ. Also when R′, R″, and R‴ are methyl, λ_{max} for compound **55** are 283 and 309 mμ, and λ_{max} **57** is 310 mμ; but an extra absorption appears in compound **56**, with λ_{max} 278 (sh), 310, and 338 mμ (see also Section III-C).

3. *Boronium Ions*

Even though diphenylboron chloride reacts with silver perchlorate in many solvents to give stoichiometric amounts of a boron-containing cation, past reports, interpretations, and calculations [2, 41] of the UV spectra of the diphenylboron cation or boronium ion in terms of the structure Ar_2B^+ must be regarded as erroneous and misleading, even if the structure is "hedged" by the indication of some weak specific solvation of the type Ar_2B^+:solvent. In view of the previously well-known Lewis acid strength of neutral triarylboranes and the striking difference in UV spectra between $(C_6H_5)_3B$ and $(C_6H_5)_3B:NCCH_3$ or $(C_6H_5)_3B:NH_3$, which was also pointed out some time ago [85], the description of a boron cation as R_2B^+ in any solvent capable of acting as a Lewis base must be expected to be poor. It is therefore no surprise that recent NMR and conductivity studies by Moodie and Ellul [71] of the "$(C_6H_5)_2B^+$" ion in sulfolane, ether, pyridine, or dioxane show the boron to be tetrahedrally substituted and the diphenylboronium ion to have the structure $[(C_6H_5)_2B^- (\text{solvent}^+)_2]^+$.

The reported [41] UV spectrum (λ_{max} 337 mμ) of the "$(C_6H_5)_2B^+$" ion in ethyl methyl ketone, in view of the findings of Moodie and Ellul and the now well-known stability of alkoxy-carbonium ions [86], is very probably the spectrum of the species represented by structure **58**. In which case the 337 mμ

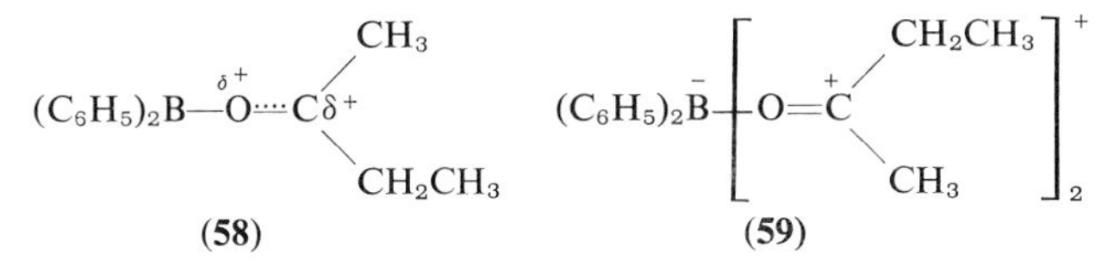

transition represents either an intramolecular CT transition from the aryl ring to the lowest vacant π molecular orbital obtained from a linear combination of the B, O, and C^+ atomic p orbitals, in which carbon is the largest contributing atomic orbital or a $\pi \rightarrow \pi^*$ transition of the B—O—C chromophore alone.

The tetracoordinate structure **59** can presumably be ruled out on the basis of the ultraviolet spectrum since neither known coordinated diarylborinates nor alkoxymethyl cations [86] have UV transitions of comparable intensity above 300 mμ.

The diphenylbipyridylboronium iodide (**60**) exhibits an intermolecular CT transition (λ_{max} $CHCl_3$, 450 mμ) in its visible spectrum [10], similar to that described by Kosower [58] for methylpyridinium iodide, in which the transition promotes an electron from the iodide ion to the pyridinium ring. The visible spectrum of bipyridylyl-*o*-phenylenedioxyboronium iodide (**61**) in methanol possess a broad absorption maximum near 400 mμ. Since the intermolecular CT band of compound **60** in methanol has shifted so far toward the blue as to be no longer discernible, the 400 mμ transition of compound **61** was assigned to an intramolecular (intraionic) CT transition by Banford and Coates [10]. The ruby red color of crystals of compound **62** was also suggested to be the result of intraionic charge transfer, as was the yellow color of the perchlorate of compound **61**.

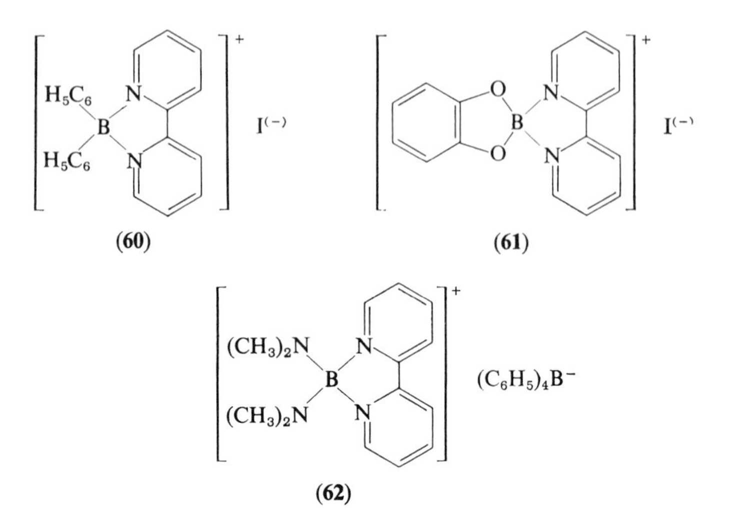

G. Substituted Boron Hydrides

The novel tropenylium undecahydroclovododecaborate (**63**) and tropenylium nonhydroclovodecaborate anions (**64**) have been synthesized by Harmon and Harmon [48, 49]. The visible ultraviolet spectra of these anions have an intense ($\epsilon > 10^4$), highly solvent-sensitive intramolecular CT band, in which the tropenylium substituent is the electron acceptor. The energy of this CT transition for compounds **63** and **64** are found to give a nonlinear correlation with Kosower's Z values [58], an empirical measure of solvent polarity based on the solvent dependence of the intermolecular CT energy of alkylpyridinium iodides. The CT transition of compound **63** is found at 513 mμ in $CH_3C{\equiv}N$ and at 373 mμ in concentrated sulfuric acid.

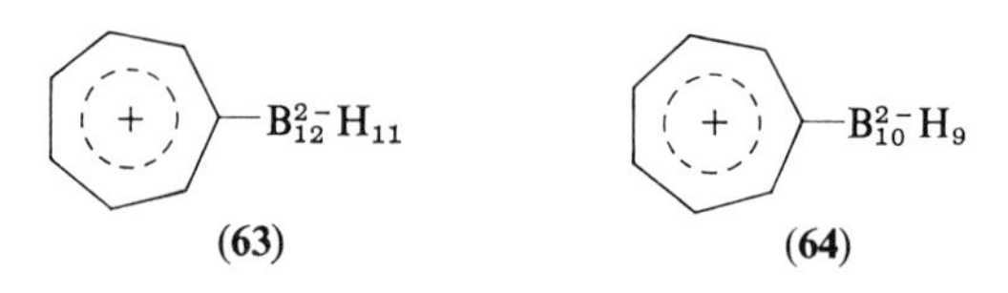

The above results of Harmon and Harmon are similar to those found in earlier studies by Graybill and Hawthorne [45] on a number of substituted 6,9-bispyridinedecaboranes, with substituents ranging in Hammett σ constant value from p-CH_3O (-0.268) to p-$C{\equiv}N$($+1.00$). All of the nine bispyridine decaboranes (**65**) exhibited a moderately intense absorption maximum between 350 and 470 mμ in their ultraviolet spectrum. The energy of this transition gives an excellent Hammett σ–ρ plot with a *negative* ρ, i.e., electron-donating substituent groups lead to an increase in the transition energy. It is concluded that the $(B_{10}H_{12})^{2-}$ unit is an electron donor in the excited state, i.e., that the transition is an intramolecular CT transition from the decaborane to pyridinium ring.

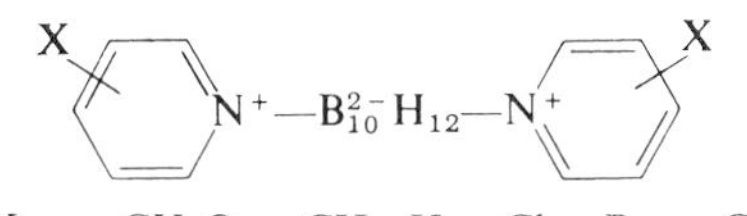

X = *p*-CH_3O, *p*-CH_3, H, *p*-Cl, *p*-Br, *m*-Cl,
p-CH_3CO, *p*-CN, 2,3-benzo, *o*-CH_3O, *o*-C_6H_5

(**65**)

Oddly enough the monopyridine derivative $PyB_{10}H_{12}NH(C_2H_5)_2$ has very nearly the same UV maxima with a slightly *smaller* intensity. If the pyridine rings behave independently as suggested, the intensity of the dipyridine transition should be roughly twice that of the monopyridine. The observed results suggest at least to this author that possibly the two pyridine rings are not in equivalent positions as required by the 6,9-structural assignment and that one ring occupies a position where the CT transition is forbidden.

H. Carboranes

The ultraviolet spectrum [92a] of *C,C'*-dimethyl-2,3-dicarbahexaborane possess a broad absorption band with shoulders at 200 and 212 mμ and a maximum at 218 mμ (ϵ 4508). In the unsubstituted 2,3-dicarbahexaborane the absorption maximum is slightly blue shifted to 215 mμ. Single-crystal x-ray diffraction studies and molecular orbital calculations by Lipscomb *et al.* [93a] led to structure A below in which donation from the π bond is allowed along the arrow toward an available sp^2 orbital of the apex B atom. The spectrum of *C*-methyl-2-carbahexaborane shows only a shoulder at

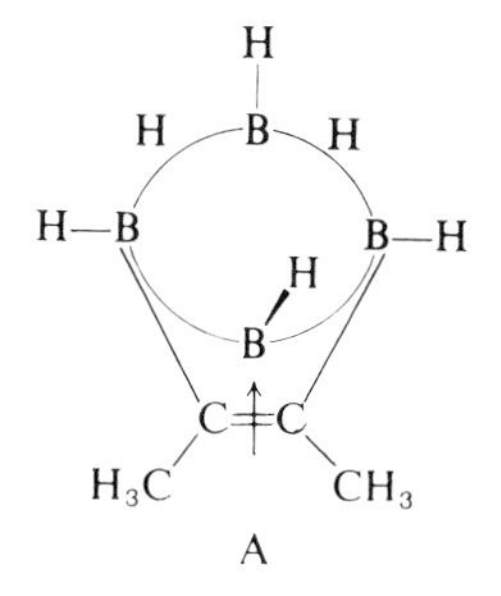

A

205 mμ on rising end absorption to 190 mμ [92a]; therefore, the 218 mμ transition of *C,C'*-dimethyl-2,3-dicarbahexaborane seems associated with the —C═C— π bond and may arise from a charge transfer-type transition from the —C═C— π bond to the boron hydride framework.

REFERENCES

1. D. R. Armstrong and P. G. Perkins, *Theoret. Chim. Acta* **8**, 138 (1967).
2. D. R. Armstrong and P. G. Perkins, *J. Chem. Soc. A* **1966**, 1026.
3. D. R. Armstrong and P. G. Perkins, *Theoret. Chim. Acta* **4**, 69 (1966).
4. D. R. Armstrong and P. G. Perkins, *Theoret. Chim. Acta* **5**, 11 (1966).
5. D. R. Armstrong and P. G. Perkins, *Theoret. Chim. Acta* **5**, 215 (1966).
6. D. R. Armstrong and P. G. Perkins, *Theoret. Chim. Acta* **5**, 222 (1966).
7. D. R. Armstrong and P. G. Perkins, *J. Chem. Soc.* **1967**, 123.
8. A. Arsene, A. T. Balaban, I. Bally, A. Barabas, M. Paraschiv, and C. N. Rentea, *Spectrochim. Acta* **23A**, 1373 (1967).
9. I. Bally, A. Arsene, M. Paraschiv, E. Romas, and A. T. Balaban, *Rev. Roumaine Chim.* **11**, 1409 (1966).
10. L. Banford and G. E. Coates, *J. Chem. Soc.* **1964**, 3564.
11. H. J. Becher and S. Frick, *Z. Physik. Chem. (Frankfurt)* **12**, 241 (1957).
12. R. S. Berry, *J. Chem. Phys.* **38**, 1934 (1963).
13. G. Bianchi, A. Cogoli, and P. Grunanger, *J. Organometal. Chem. (Amsterdam)* **6**, 598 (1966).
14. J. Casanova, Jr., H. R. Kiefer, Daniel Kuwada, and A. Boulton, *Tetrahedron Letters*, **1965**, 703.
15. O. Chalvet, R. Daudel, and J. Kaufman, *J. Am. Chem. Soc.* **87**, 399 (1965).
16. S. S. Chisaick, M. J. S. Dewar, and P. M. Maitlis, *J. Am. Chem. Soc.* **83**, 2708 (1961).
17. T. L. Chu, *J. Am. Chem. Soc.* **75**, 1730 (1953).
18. T. L. Chu and T. J. Weismann, *J. Am. Chem. Soc.* **78**, 23 (1956).
19. G. E. Coates and J. G. Livingston, *J. Chem. Soc.* **1961**, 1000.
20. T. Colclough, W. Gerrard, and M. F. Lappert, *J. Chem. Soc.* **1956**, 3006.
21. T. D. Coyle, S. L. Stafford, and F. G. A. Stone, *J. Chem. Soc.* **1961**, 3103.
22. J. M. Davidson and C. M. French, *J. Chem. Soc.* **1960**, 191.
23. D. W. Davies, *Trans. Faraday Soc.* **56**, 1713 (1960).
24. A. G. Davies, D. G. Hare, and L. F. Larkworthy, *Chem. Ind. (London)* **1959**, 1519.
25. M. J. S. Dewar, *in* "Advances in Chemistry" (R. F. Gould, ed.), Vol. 42, p. 227. Am. Chem. Soc., Washington, D.C., 1964.
26. M. J. S. Dewar and R. Dietz, *J. Chem. Soc.* **1959**, 2728.
27. M. J. S. Dewar and R. Dietz, *J. Chem. Soc.* **1960**, 1344.
28. M. J. S. Dewar and R. Dietz, *J. Org. Chem.* **26**, 3253 (1961).
29. M. J. S. Dewar and R. Dietz, *Tetrahedron* **15**, 26 (1964).
30. M. J. S. Dewar and R. C. Dougherty, *J. Am. Chem. Soc.* **84**, 2648 (1962).
31. M. J. S. Dewar and R. C. Dougherty, *J. Am. Chem. Soc.* **86**, 433 (1964).
32. M. J. S. Dewar, R. C. Dougherty, and E. B. Fleischer, *J. Am. Chem. Soc.* **84**, 4882 (1962).

33. M. J. S. Dewar, C. Kaneko, and M. Bhattacharjee, *J. Am. Chem. Soc.* **84**, 4884 (1962).
34. M. J. S. Dewar and V. P. Kubba, *J. Am. Chem. Soc.* **83**, 1757 (1961).
35. M. J. S. Dewar and V. P. Kubba, *Tetrahedron* **7**, 213 (1959).
36. M. J. S. Dewar, V. P. Kubba, and R. Pettit, *J. Chem. Soc.* **1958**, 3073, 3076.
37. M. J. S. Dewar and P. M. Maitlis, *J. Am. Chem. Soc.* **83**, 187 (1961).
38. M. J. S. Dewar and P. M. Maitlis, *Tetrahedron* **15**, 26, 35 (1961).
39. M. J. S. Dewar and P. Marr, *J. Am. Chem. Soc.* **84**, 3782 (1962).
40. M. J. S. Dewar and W. Poesche, *J. Org. Chem.* **29**, 1757 (1964).
41. C. M. French and J. M. Davidson, *J. Chem. Soc.* **1958**, 114.
42. D. H. Geske, *J. Phys. Chem.* **63**, 1062 (1959).
43. C. D. Good and D. M. Ritter, *J. Am. Chem. Soc.* **84**, 1162 (1962).
44. L. Goodman and P. Love, private communication reported in Ramsey [83], 1966.
45. B. M. Graybill and M. F. Hawthorne, *J. Am. Chem. Soc.* **83**, 2673 (1961).
46. S. Gronowitz and A. Bugge, *Acta Chem. Scand.* **19**, 1271 (1965).
47. N. N. Greenwood, *Advan. Inorg. Radiochem.* **5**, 91 (1962).
48. K. M. Harmon and A. B. Harmon, *J. Am. Chem. Soc.* **86**, 5036 (1964).
49. K. M. Harmon and A. B. Harmon, *J. Am. Chem. Soc.* **88**, 4093 (1966).
50. M. F. Hawthorne, *J. Am. Chem. Soc.* **83**, 833 (1961).
51. R. Hemming and D. G. Johnson, *J. Chem. Soc.* **1964**, 466.
52. R. Hemming and D. G. Johnson, *J. Chem. Soc. B*, **1966**, 314.
53. R. Hoffmann, *J. Chem. Phys.* **40**, 2474 (1964).
54. K. Ito, H. Watanabe, and M. Kubo, *J. Chem. Phys.* **34**, 1043 (1961).
55. L. E. Jacobs, J. R. Platt, and G. W. Schaeffer, *J. Chem. Phys.* **16**, 116 (1948).
56. H. H. Jaffe and M. Orchin, "Theory and Applications of Ultraviolet Spectroscopy." Wiley, New York, 1962.
57. G. J. M. Van der Kerk, *Intern. Symp. Organometal. Chem., 3rd, Munich, 1967.*
58. E. Kosower, *J. Am. Chem. Soc.* **80**, 3253 (1958).
59. R. Köster and P. Binger, *Advan. Inorg. Chem. Radiochem.* **7**, 263 (1965).
60. R. Köster and G. Benedikt, *Angew. Chem. Intern. Ed. Engl.* **2**, 323 (1963).
61. S. R. LaPaglia, *J. Mol. Spectry.* **7**, 461 (1961).
62. R. Letsinger, *in* "Advances in Chemistry" (R. F. Gould, ed.), Vol. 42, p. 1. Am. Chem. Soc., Washington, D.C., 1964.
63. R. L. Letsinger and S. B. Hamilton, *J. Org. Chem.* **25**, 592 (1960).
64. R. L. Letsinger and J. Skoog, *J. Org. Chem.* **18**, 895 (1953).
65. J. C. Lockhart, *J. Chem. Soc.* **1962**, 3737.
66. H. C. Longuet-Higgins and R. Sowden, *J. Chem. Soc.* **1952**, 1404.
67. P. M. Maitlis, *J. Chem. Soc.* **1961**, 425.
68. D. S. Matteson, *J. Am. Chem. Soc.* **82**, 4231 (1960).
69. D. S. Matteson and G. D. Schaumberg, *J. Organometal. Chem. (Amsterdam)* **8**, 359 (1960).
70. C. W. Moeller and W. K. Wilmarth, *J. Am. Chem. Soc.* **81**, 2638 (1959).
71. R. B. Moodie and B. Ellul, *Chem. Ind. (London)* **1966**, 767.
72. R. S. Mulliken, *Chem. Rev.* **41**, 215 (1947).
73. J. N. Murrell, *Proc. Phys. Soc. (London)* **A68**, 969 (1955).
74. S. Nagakura, *J. Chem. Phys.* **23**, 1411 (1955).
75. S. Nagakura and J. Tanaka, *J. Chem. Phys.* **22**, 236 (1954).
76. K. Niedenzu, J. W. Dawson, G. A. Neece, W. Sawodny, D. Squire, and W. Weber, *Inorg. Chem.* **5**, 2161 (1966).
77. A. N. Nikitina, V. A. Petukhov, A. F. Galkin, N. S. Fedotov, Yu. N. Bubnov, and P. M. Aronovich, *Opt. i Spektroskopiya* **16**, 976 (1964).

78. A. N. Nikitina, V. A. Vaver, N. S. Fedotov, and B. M. Mikhailov, *Opt. i Spektroskopiva* **7**, 644 (1959).
79. J. C. Patel and S. Basu, *Naturwissenschaften* **47**, 302 (1960).
80. P. G. Perkins and D. A. Wall, *J. Chem. Soc. A* **1966**, 235.
81. J. R. Platt, H. B. Klevens and G. W. Schaeffer, *J. Chem. Phys.* **15**, 598 (1947).
82. B. G. Ramsey, *J. Phys. Chem.* **70**, 611 (1966).
83. B. G. Ramsey, *J. Phys. Chem.* **70**, 4097 (1966).
84. B. G. Ramsey, M. Ashraf El-Bayoumi, and M. Kasha, *J. Chem. Phys.* **35**, 1502 (1961).
85. B. G. Ramsey and J. E. Leffler, *J. Phys. Chem.* **67**, 2242 (1963).
86. B. G. Ramsey and R. W. Taft, *J. Am. Chem. Soc.* **88**, 3058 (1966).
87. B. G. Ramsey, unpublished results, 1968.
88. C. W. Rector, G. W. Schaeffer, and J. R. Platt, *J. Chem. Phys.* **17**, 460 (1949).
89. C. C. J. Roothan and R. S. Mulliken, *J. Chem. Phys.* **16**, 118 (1948).
90. J. Rosenbaum and M. C. R. Symons, *Proc. Chem. Soc.* **1959**, 92.
91. W. Seaman and J. R. Johnson, *J. Am. Chem. Soc.* **53**, 711 (1931).
92. J. Soulie, *Compt. Rend.* **262C**, 376 (1966).
92a. J. R. Spielman, private communication, California State College, Los Angeles, 1968.
93. A. Streitwieser, Jr., "Molecular Orbital Theory for Organic Chemist." Wiley, New York, 1961.
93a. W. E. Strieb, F. P. Boer, and W. N. Lipscomb, *J. Am. Chem. Soc.* **85**, 2331 (1963).
94. E. E. Van Tamelen, G. Brieger, and K. Untch, *Tetrahedron Letters* **8**, 14 (1960).
95. M. Wahab, A. Trutia, R. Titeria, and V. Vinter, *Rev. Roumaine Phys.* **9**, 599 (1964).
96. H. Watanabe, K. Nagasawa, T. Totani, T. Yoshizaki, and T. Nakagawa, *in* "Advances in Chemistry" (R. F. Gould, ed.), Vol. 42, p. 116. Am. Chem. Soc., Washington, D.C., 1964.
97. H. Watanabe, T. Totani, and T. Yoshizaki, *Inorg. Chem.* **4**, 657 (1965).
98a. T. J. Weismann, Meeting Am. Chem. Soc., 150th, Atlantic City, N.J., 1965.
98b. T. J. Weismann, Meeting Am. Chem. Soc., 155th, San Francisco (1968).
99. S. I. Weissman, J. Townsend, D. Paul, and G. Pake, *J. Chem. Phys.* **21**, 227 (1953).
100. J. L. R. Williams, P. J. Grisdale, and J. C. Doty, *J. Am. Chem. Soc.* **89**, 4538 (1967).
101. W. G. Woods, I. S. Bengelsdorf, and D. L. Hunter, *J. Org. Chem.* **31**, 2767 (1966).
102. W. G. Woods and P. L. Strong, *J. Organometal. Chem.* (*Amsterdam*) **7**, 371 (1967).

CHAPTER IV

Organometallic Derivatives of Group IV

A. The Role of *d* Orbitals

For the first time in this discussion of electronic transitions it becomes necessary to consider the role of vacant *d* or *f* orbitals in ground and excited states of nontransition metal organometallic compounds. The "shapes" of the *d* orbitals are shown in Fig. 21. Since in this and succeeding chapters we will be primarily concerned with the π overlap of *p* and *d* orbitals and overlap between *d* orbitals, *p–d* π, *d–d* π, and *d–d* δ type bonds are illustrated in Fig. 22. Furthermore, since for *d* orbitals of principal quantum number *n* the

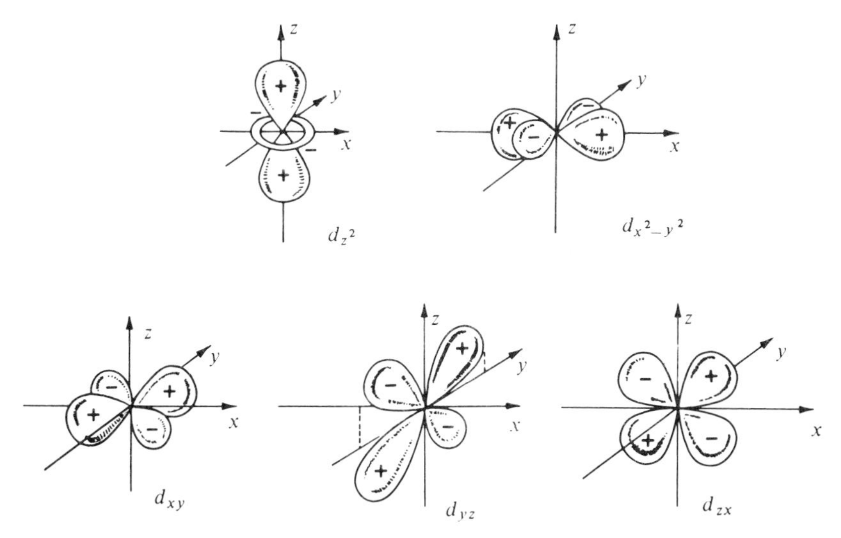

Fig. 21. The shapes and orientation of *d* atomic orbitals.

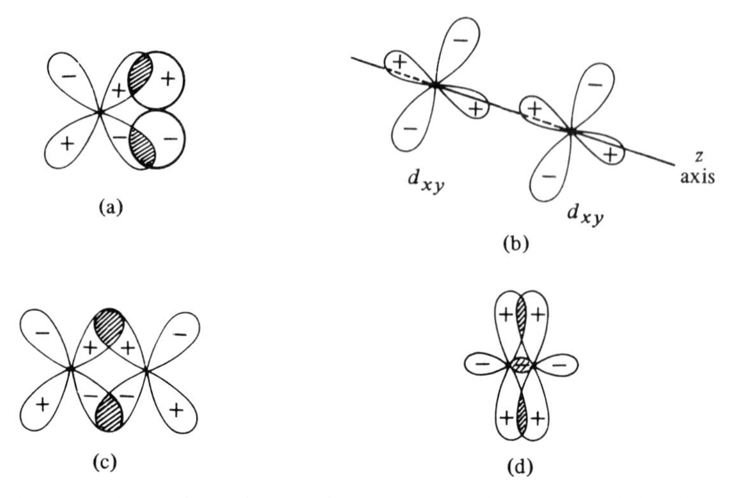

FIG. 22. Some types of bonding possible between p and d atomic orbitals: (a) $3p\,\pi$–$3d\,\pi$; (b) δ bonding; (c) $3d\,\pi$–$3d\,\pi$; (d) $3d_{z^2}$–$3d_{z^2}$.

$(n + 1)s$ and p orbitals are certainly not much greater and quite possibly lower in energy than the d orbitals [38], hybridization energies will be small and we must consider the possibility of hybridized orbitals in bonding. In an early paper on chemical bonds involving d orbitals, Craig *et al.* [34] calculated a number of overlap integrals involving $3s$, $3p$, $3d$, $4s$, and of $4p$ orbitals and, among other things, concluded that in the case of p–$d\ \pi$ bonding, overlap is appreciably increased by the addition of p character to d orbitals. The p–d and s–d hybrid orbitals of interest are depicted in Fig. 23.

Although in the past there has been some disagreement as to the magnitude or importance of d–d or d–p interaction in organometallic derivatives of Groups IV and Va, the present evidence and consensus seem to agree that both ground and excited state properties are best explained by the inclusion of d orbitals (or hybrids where appropriate) in descriptions of bonding and electronic transitions.

Examples of earlier experimental results pointing to the importance of

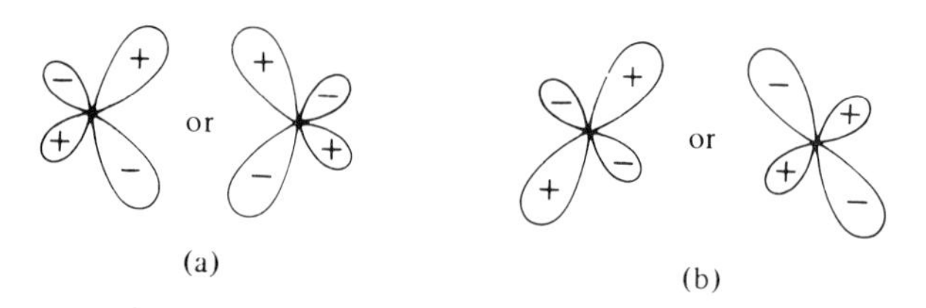

FIG. 23. The shapes and orientation of (a) p–d and (b) s–d hybrid orbitals.

d–p π bonding are found in the measurements of acid strengths of *p*-$(CH_3)_3MC_6H_4CO_2H$ (where M is Si, Ge, or Sn) by Chatt and Williams [31] which indicated *d–p* π bonding in the ground state. Similarly, first-order rate constants for the acid-catalyzed hydrolysis of fluoroalkyl-, ω-cyanoalkyl-, and *n*-alkylsilicon hydrides have been interpreted [131] in terms of *d–p* π bonding in $(CH_3)_2SiHCH_2Cl$ between Cl lone pair electrons and a Si *d* orbital, even though the Cl is one nonconjugating substituent removed from the Si.

More recent evidence is found in the ESR studies by Bedford and co-workers [10] on the radical anions of phenyltrimethylsilane and -trimethylgermane, and by Husk and West [72] on the radical ion of dodecamethylcyclohexasilane $[(CH_3)_2Si]_6 \cdot Na)$, in which the unpaired electron was found to be delocalized over all six Si atoms. The work of Husk and West makes the important point of the delocalized nature of the lowest vacant orbitals arising in part at least from *d–d* overlap in otherwise saturated molecules of Group IV organometallics.

The unequivocal observation of pentacoordinate silicon [118] species would most logically require the involvement of *d* orbitals, although in fairness it should also be recognized that CH_5^+ also is a stable species thermodynamically and in this case the hybridization is most probably $2sp^22p3s$. One might therefore argue that by analogy pentavalent silicon may be the result of 4*s* hybridization.

A spectroscopic method of elegant simplicity and general applicability for determining the presence of excited state *d*-orbital π bonding in phenyl derivatives has been developed by Goodman and co-workers [60]. The criteria to be applied experimentally is that if X and Y are potential π electron donors or acceptors, the intensity of the $A \rightarrow L_b$ benzene transition, normally occurring near 260 mμ, of the *para*-X,Y-disubstituted benzene will be *less* than that of the most intense X- or Y-monosubstituted benzene L_b transition if X is a net π donor, and Y a net π acceptor. As a corollary, if both X and Y are π donor or π acceptor, the L_b transition intensity of *p*-XC_6H_4—Y will be greater than in either X—C_6H_5 or Y—C_6H_5. Therefore, presumably if we know the character of X the character of Y may be found by what has been termed a *substituent interference experiment* which determines the $A \rightarrow L_b$ transition intensities of the appropriately substituted benzenes. Thus, the $A \rightarrow L_b$ transition intensity of *p*-methoxyphenylsilane (ϵ 1075) is less than that of anisole (ϵ 1560) and *p*-methylphenylsilane (ϵ 216) is less than phenylsilane (ϵ 242). Inasmuch as we know the methoxyl group and presumably the methyl group to be π electron donors, silicon is established in the excited state as a π electron acceptor through *d* orbital overlap with the benzene ring.

According to Goodman, the dipole strength, *D*, for the L_b transition of a *para*-disubstituted benzene is given by Eq. (12). The E_{ij} refers to the energy of the configuration from the transition $i \rightarrow j$, in which $\bar{1}$ and $\bar{2}$ are highest filled and lowest unfilled molecular orbitals, nodal through the substituent

$$D = D_R + D_S + \sin\theta_R \sin\theta_S D_{E_{1u}}, \qquad (12)$$

where

$$\sin\theta = \frac{\langle \Psi_{B_{2a}}|H|\Psi_{E_{1u}}\rangle}{E_{B_{2u}} - E_{1u}} = \frac{E_{\bar{1}2} - E_{1\bar{2}}}{2(E_{B_{2u}} - E_{1u})},$$

axis, and 1 and 2 are perturbed highest filled and lowest vacant benzene molecular orbitals. It can then be demonstrated that *for electron donors whose effective Coulomb integral is less than that of the lowest occupied benzene degenerate orbitals,* $\sin\theta > 0$, *and for electron acceptors whose effective Coulomb integral is greater than (i.e., less negative) than that of the lowest vacant benzene degenerate molecular orbitals,* $\sin\theta < 0$. Insofar as the above requirements are met by the usual p–π electron donors and π electron acceptors, such as d atomic orbitals, the substituent interference rule above is of general applicability. Although not specifically pointed out by Goodman and co-workers, the rule given above would be expected to fail in a case in which the substituent vacant acceptor orbital lies in energy somewhere between the lowest vacant degenerate benzene molecular orbitals and α, the sp^2 carbon $2p$ Coulomb integral. One possible example of this kind of an exception might be found in the phenylboranes, consider L_b of $(C_6H_5)_3B$ (ϵ 1.9 × 10^4) and L_b of $(p\text{-}CH_3C_6H_4)_3B$ ($\epsilon = 2 \times 10^4$) from Section III-C.

The Eq. (12) obtained by Goodman is similar to expressions derived both by Matsen [88] and by Platt [112] in which the transition moment of a polysubstituted benzene is obtained as the vector sum of the monosubstituted benzene transition moments, with the important restrictions that the substituents do not strongly perturb the benzene ring. In this framework, the transition intensity for a *para*-disubstituted benzene is proportional to the expression $(m_X{}^2 + m_Y{}^2 + m_X m_Y)$, if m represents a positive transition moment for a π electron releasing substituent and is negative for a π electron withdrawing substituent. It is claimed however that the derivation of Eq. (12) is not dependent on assumptions inherent in the spectroscopic moment theory and hence Eq. (12) remains valid, even when spectroscopic moment additivity rules fail.

The substituent interference effect may be appreciated in a qualitative way from simple valence bond structures. For mono- or *para*-disubstituted benzenes the benzene L_b transition is in plane polarized and perpendicular to the substituent axis of structure **66**. For either electron donating substituents X or withdrawing substituents Y we see that contributions to the excited state by structure **67** or **68** will increase the intensity of the transition. Incorporating both X and Y *para* into the same ring leads to the valence bond structure **69** and cancellation of the component substituent transition dipole moments which contribute to the $A \rightarrow L_b$ transition intensity. Of course,

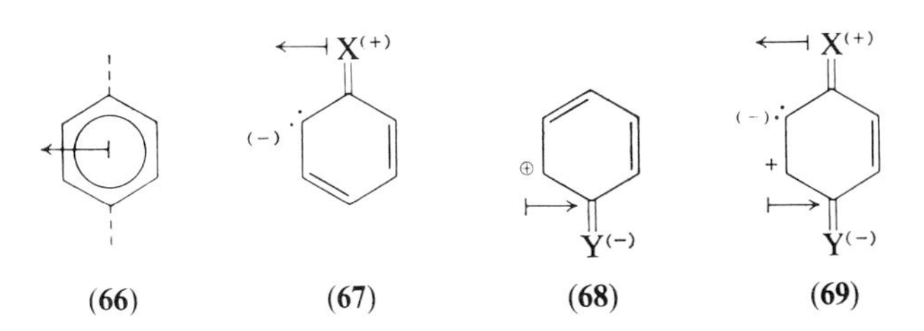

valence bond structures such as **70** or **71** make no contribution to the intensity of the transition.

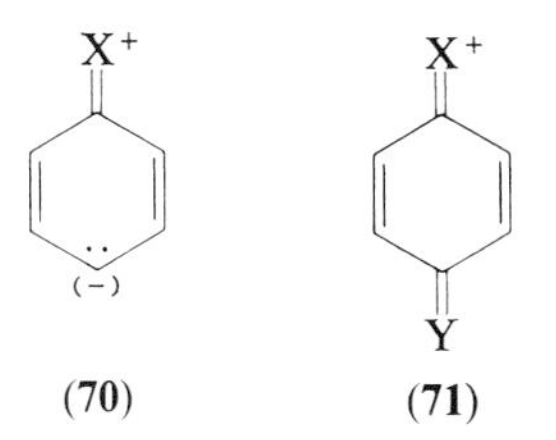

The substituent interference method of Goodman has been applied to the series p-$CH_3OC_6H_4X$, where X is H, $(CH_3)_3C$, $(CH_3)_3Si$, $(CH_3)_3Ge$, or $(CH_3)_3Sn$, by Musker and Savitsky [94]. The ratios of the p-$(CH_3)_3M$-substituted anisole L_b oscillator strengths to that of anisole were found to be Si (0.80), Ge (0.86), and Sn (1.0) in keeping with the expected decrease in $d\pi$–ring π orbital overlap in the order Si > Ge > Sn. Participation by Sn 5*d* orbitals appears negligible, although in a similar series of halogen-substituted anisole 5*d*–ring π overlap was found to be significant for iodine.

Reikshsfel'd and co-workers [116, 125] have obtained NMR evidence for complex formation between $(C_6H_5)_3SiH$ acetone, dioxane, and diethyl ether, and $C_6H_5SiH_3$ with *N,N*-dimethylformamide, indicating that some more detailed studies of solvent effects on the UV and visible transitions of Group IV organometallics presumed to involve vacant *d* orbitals might be worthwhile.

At the risk of stating the obvious, before leaving the subject of d–p π overlap, perhaps we should remind the reader that unlike p–p π overlap, the overlap of *d* orbitals is largely insensitive to rotation about the σ bond because of the fourfold symmetry of the *d* orbitals. Ebsworth [43] has also pointed out that other things being equal, $d\pi$–sp^3 overlap may be as great as $\sqrt{\frac{2}{3}}(d\text{–}p\ \pi)$ overlap. This means that d–p π overlap does not confer conformational stability on a molecule and "steric inhibition of resonance" is not observed in d–p π as it is for p–p π interaction.

In view of the importance of inductive effects in the spectra of Group IV metal organometallic compounds, it is worth recalling the various orders of electronegativity of the Group IV metals. Cotton and Wilkinson, for example [33], prefer the following increasing order of electronegativity Si (1.90) $<$ Sn (1.96) $<$ Ge (2.01) $<$ Pb (2.33). The Allred-Rochow order, on the other hand, is Pb (1.55) $<$ Sn (1.72) $<$ Si (1.74) $<$ Ge (2.02). In fact, since polarizability may be in some transitions the controlling factor (as it appears to be in the electronic transitions of *p*-halonitrobenzenes and *p*-haloanisoles [73]) an electron releasing "inductive," i.e., polarizability, substituent effect order Pb $>$ Sn $>$ Ge $>$ Si on transition energies may also be observed.

The anticipated inductive and *d* orbital effects of metal substitution on the electronic transitions of simple chromophores such as C=C, C=O, CO_2, and N=N, have been outlined in terms of qualitative molecular orbital arguments by R. West [138]. These arguments are presented further on in the course of our discussion of specific chromophores.

B. Alkyl and Aryl Noncatenated Organometallic Compounds

The far ultraviolet spectra of $(CH_3)_{4-n}SiH_4$ ($n = 0, 1, 2, 4$) have been obtained by Harada and co-workers [68a]. Successive methylation of SiH_4 produces a 2 mμ/CH_3 red shift in the first absorption band, for example, λ_{max}, mμ ($\epsilon \times 10^{-4}$): SiH_4, 115 (2.1), 125 (sh), 138 (sh), 155 (sh); $Si(CH_3)_4$, 115, (2.6), 138 (1.4), 167 (0.7). No obvious bands could be assigned to transitions of the $\sigma \rightarrow 3d$ type. Tetraalkylgermanium compounds reportedly have a shoulder in their spectrum near 182 mμ which becomes a maximum in the spectra of $RGeCl_3$ [107]. The $RGeBr_3$ show an intense broad absorption close to 2150 Å. There are scattered reports [36, 100, 107] in the literature of absorption maxima near 200 mμ in the ultraviolet spectra of tetraalkyltins, for example [36], $(n\text{-}C_4H_9)_4Sn$, λ_{max} 202 mμ (log ϵ 3.99). No assignment of these transitions to either $\sigma \rightarrow \sigma^*$ or Rydberg transitions has been made.

The spectra of $(n\text{-}C_4H_9)_nSnX_{(4-n)}$, where X is Cl, OCH_3, CH_3CO_2, or SC_4H_9, also contain a strong band in the 202–212 mμ region. Where X is SC_4H_9, an additional absorption is found near 240–250 mμ [59].

Bell and Walsh [11] have obtained the vacuum UV spectra of disiloxane, hexamethyldisiloxane, chlorosilane, trisilylamine, and disilylmethylamine. The spectrum of each silyl compound studied exhibits a large blue shift relative to the spectrum of the corresponding methyl compound. This result is attributed to "loss of nonbonding, lone pair character" because of *d* orbital overlap of the Si atom with line pair electrons resulting in bonding character and an increase of the lone pair electron ionization potential.

More recently Pitt and Fowler [110] have also commented on the result that the increasing substitution of trialkylsilyl groups for hydrogen leads to a blue shift of the first absorption maximum, whereas tertiary amines absorb at longer wavelengths than primary amines. Compare, for example, $(C_2H_5)_3$-$SiNH_2$, λ_{max} 209 mμ (ϵ 1780) and $(CH_3)_3Si_3N$, λ_{max} 200 mμ (ϵ 4850) with $CH_3CH_2NH_2$, λ_{max} 212 mμ (ϵ 800) and $(CH_3)_3N$, λ_{max} 228 mμ (ϵ 900). The spectra of amines show, in general, two transitions above 170 mμ. The lowest-energy transition remains relatively constant near 215–230 mμ and is assigned to $n \rightarrow 3s$, since inductive effects on the n and $3s$ atomic orbitals should be comparable. The second transition found at 190 mμ in the spectrum of CH_3NH_2 moves to 213 mμ in $(CH_3CH_2)_3N$ and is evidently $n \rightarrow \sigma^*$. Pitt and Fowler suggested that the transition in the aminosilanes would be $n \rightarrow \sigma^*$ although considering the relative insensitivity of the transition to the number of silyl groups, again $n \rightarrow 3s$ might be more appropriate. The intriguing suggestion was made that the long wavelength transition could be attributed as an alternative to the $\pi \rightarrow \pi^*$ transition of the p–d π Si–N bond, or $n \rightarrow 3d$ silicon charge transfer.

Interestingly enough silyl substitution on nitrogen resulted in a *red* shift of the typical aniline spectrum. Apparently the nitrogen orbital is so lowered in energy by its interaction with the phenyl ring that only the inductive effect of the silicon remains operative.

The invariance of the transition energy (200–202 mμ) of the linear and cyclic polydimethylsilazanes, $CH_3[(CH_3)_2SiNH]_nSiCH_3$ or $[(CH_3)_2SiNH]_n$, is evidence that there is little aromaticity or delocalization associated with the total Si—N framework [110]. This could be the result of either different Si d orbitals for each Si—N $(d\text{–}p)\pi$ bond or alternatively just the result of the large difference in the coulomb integrals of Si and N and a small Si–N resonance integral.

The first ultraviolet absorption maximum of the *N*-halohexamethyldisilazanes, $[(CH_3)_3SiNX$, where X is I, Br, Cl] have been assigned by Bailey and West [6] to $n_X \rightarrow \sigma^*_{N-X}$. The transition energies decrease in the order Cl > Br > I from 295 to 408 mμ.

The gem disulfides, $R_2C(SR')_2$, also have an absorption band near 235 mμ which Jaffe and Orchin [73] ascribe to a transition promoting a lone pair electron to the lowest molecular orbital resulting from d–d overlap of the vacant S d orbitals. In the series $[(CH_3)_3X]_2S$ the 215 mμ band of (X is C) is shifted to 202 mμ, where X is Si, the absorption maximum of $[(CH_3)_3Ge]_2S$ at (215 mμ) is, on the other hand, quite close to that of the di-*n*-butyl sulfide. Cumper *et al.* offer the explanation [36] that, "if a lone pair electron of sulphur is being promoted to a $3d$ orbital by this absorption then an interaction between the $3d$ orbitals of the sulphur and silicon atoms could explain the shift towards lower wavelengths both in $(Me_3Si)_2S$ and in the polymethylthiosilanes." The above rationale seems unlikely, since the lowest molecular orbital formed from the sulfur and silicon $3d$ atomic orbitals

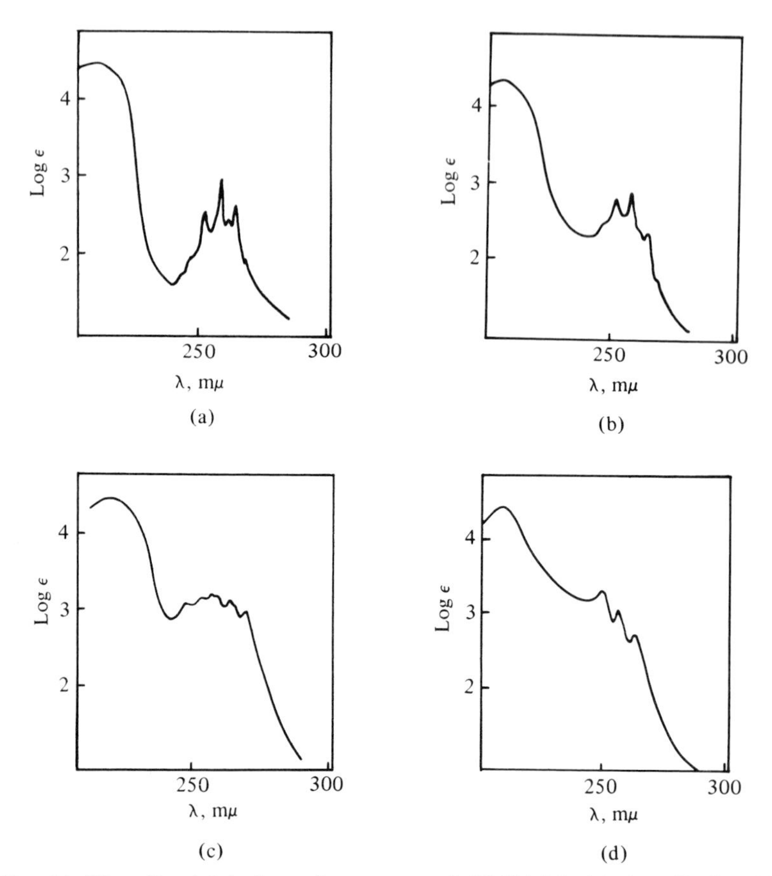

FIG. 24. The ultraviolet absorption spectra of $(C_6H_5)_4M$: (a) Ge; (b) Sn; (c) Si; (d) Pb [82].

should be lower in energy than the sulfur *d* orbital, thereby resulting in a spectral shift to longer wavelengths, as in the case of polysilanes, *not a shift to shorter wavelengths*. A blue shift may be obtained by this mechanism only if the transition to the lowest group molecular orbital were forbidden. The explanation offered previously by Bell and Walsh [11] with regard to the spectra of ethers and amines seems more likely, i.e., metal *d–p* π bonding with the sulfur lone pair electrons in the order Si > Ge.

In passing it is interesting to note that the ultraviolet spectrum [8] of $(CH_3)_3SiCH_2Li$ shows no absorption maximum above 200 mμ, in isooctane where the compound is highly associated. Unfortunately no attempt was made to obtain the spectrum in a more basic dissociating solvent.

Because of the continuing interest in the possible role of vacant *d* orbitals and their conjugation properties, there has been a small but consistent interest in the ultraviolet spectra of aryl derivatives of the Group IV metals. Since the first studies by Milazzo [89], the ultraviolet spectra of the tetraphenyl derivatives of Si, Ge, Sn, and Pb have been reinvestigated by C. N. R. Rao and co-workers [114], and by LaPaglia [82] who has also obtained the corresponding emission spectra.

The absorption spectra of the compounds $(C_6H_5)_4M$, where M is Si, Ge, Sn, and Pb, are given in Fig. 24 and may be compared with the spectrum of $(C_6H_5)_4C$ in Fig. 25. The major perturbation of the usual benzene absorption in the spectra of $(C_6H_5)_4M$, which is reflected in loss of vibrational structure in the ^1L_b band and the enhanced ϵ per phenyl ring, is primarily due to steric crowding of the phenyl rings, and configuration interaction between the benzene locally excited states. Therefore, as the covalent radius of the central metal atom becomes larger in the order $C < Si < Ge < Sn$, the ^1L_b band becomes more typical in appearance of the usual ^1L_b transition. There is a few millimicron red shift of the ^1L_b absorption maximum in the spectra of $(C_6H_5)_4M$, with M as Si having the largest shift, and this has been suggested as possible evidence of *d* orbital participation in silicon. *The red shifts are so small, however, that it is important to recall that our expected spectral red shift on extending the conjugation of the aromatic ring by d orbitals is strictly valid only for the zero–zero transition, that is, for the transition from the zero-level vibrational ground state to the zero vibrational level of the excited state.*

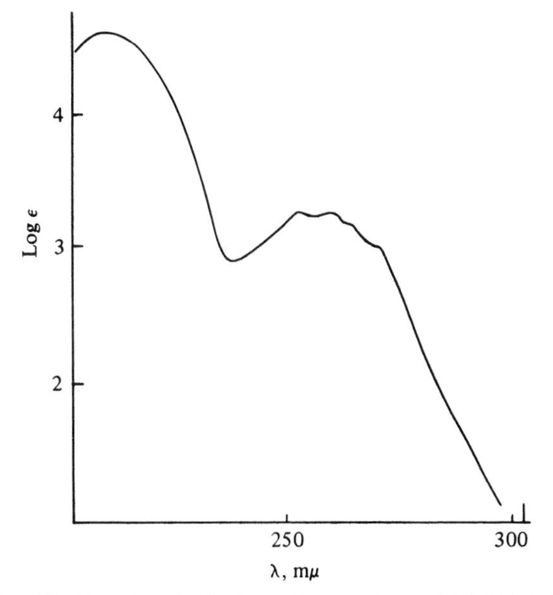

FIG. 25. The ultraviolet absorption spectrum of $(C_6H_5)_4C$ [82].

The structural maxima of the first two bands of the ^1L_b transition in $(C_6H_5)_4M$ are given [82] in Table VI. Milazzo [89] originally assigned λ_2 (Table VI) to the O–O transition in which case the tetraphenylsilane O–O transition is at lower energy than other $(C_6H_5)_4M$ ^1L_b transitions, but more recently LaPaglia has reassigned the O–O transition to λ_1 (Table VI), which places the $(C_6H_5)_4M$ transition energy in the order Sn > Ge > Si > C.

TABLE VI

VIBRATIONAL STRUCTURE OF $(C_6H_5)_4M$ ^1L_b TRANSITION, FIRST TWO BANDS (IN CM^{-1})

λ_{max} (cm^{-1})	$C(C_6H_5)_4$	$Si(C_6H_5)_4$	$Ge(C_6H_5)_4$	$Sn(C_6H_5)_4$	$Pb(C_6H_5)_4$
1	36,700	36,850	37,160	37,240	—
2	(37,520)	37,400	37,690	37,740	37,840

One very real consequence however of increasing the atomic weight of the central metal atom in the series $(C_6H_5)_4M$ is the increasing importance of spin–orbit coupling of angular momentum, which allows the mixing of singlet and triplet states. As a result, in the case of the heavier metal tetraphenyls the singlet–triplet ^1L_b absorption may be observed in concentrated chloroform solution with the ϵ's approaching 1. In terms of the emission spectra of the heavier metal tetraphenyls there is a decrease in the phosphoresence life time to the order of 1×10^{-4} sec and an increase in the singlet → triplet crossing rate [82, 83].

If strongly π electron donating groups were substituted in the *ortho–para* positions of $C_6H_5MR_2$ derivatives, one might expect a greater chance of observing what should be an increased conjugation between the ring and Group IV metal *d* orbitals. This expectation seems to be borne out in triphenylsilanol by the findings [55] that the 270 mμ absorption maxima of *p*-dimethylamino-substituted triphenylsilanols are significantly red shifted with respect to the corresponding maxima near 260 mμ of the corresponding carbinols. This red shift is attributed to Si *d* orbital participation enhanced by the *para*-nitrogen lone pair electrons. In a similar fashion [114] a maximum appearing near 237 mμ in the spectrum of the *p*-tri-phenylsilylphenol, but absent in tritylphenols or *m*-triphenylsilylphenol, has been suggested to be the ^1L_a band red shifted by silicon *d* orbital conjugation with the ring. If the ^1L_a assignment of the 237 mμ band is used to estimate the $(C_6H_5)_3Si$ Hammet σ constant by means of the Rao correlation [113], a reasonable value of 0.31 ± 0.1 is obtained in agreement with an earlier value of 0.31 suggested by Benkeser *et al.* [12], all of which perhaps leads some support to the interpretation of the 237 mμ absorption as a red shifted ^1L_a transition.

The ultraviolet absorption maxima of the series $(C_6F_5)_nSi(C_6H_5)_{4-n}$ has been published by Fearon and Gilman [46] with the finding that when $n = 4$, the ^1L_b transition is a smooth band at 271 mμ (ε 4720).

Nearly simultaneous detailed studies of the ultraviolet spectra of the series $C_6H_5MC(CH_3)_3$ and $C_6H_5CH_2MC(CH_3)_3$, where M is C, Si, Ge, or Sn, have been carried out by Nagy *et al.* [98] and Egorov and Loktionova [45]. The absorption maxima of these and other compounds are listed in Table VII, from which it is obvious that the most surprising result is that the greatest

TABLE VII

TRANSITION ENERGIES OF RC_6H_5 (ETHANOL)

R	λ_{max}, mμ (ε)	λ_{max}, mμ (ε)
$C(CH_3)_3$	208 (8500)	258 (198)
$Si(CH_3)_3$	211 (10,600)	260 (300)
$Ge(CH_3)_3$	208 (12,000)	258 (198)
$Sn(CH_3)_3$	209 (12,400)	252 (600)
$CH_2C(CH_3)_3$	211 (4630)	259 (174)
$CH_2Si(CH_3)_3$	227 (8690)	267 (430)
$CH_2Ge(CH_3)_3$	225 (7200)	269 (339)
$CH_2Sn(CH_3)_3$	236 (5520)	272 (494)

red shift in both the benzene ^1L_b and ^1L_b transitions is obtained *not* when the metal is directly conjugated with the ring, but when there is an intervening —CH_2— group. As a matter of fact, in the series $C_6H_5MC(CH_3)_3$, only where M is Si is there any significant red shift of the L_a and L_b absorption maxima at all. A larger effect is observed in the spectrum of *p*-$(CH_3)_3SiC_6H_4$-$Si(CH_3)_3$, where the ^1L_a transition is shifted to 226 mμ and the ^1L_b to 270 mμ [3], compared with the di-*p*-*tert*-butylbenzene maxima at 214 and 263 mμ. These results are similar to the earlier work of Eaborn and Parker [42a, f] who found in studying the spectra of 11 substituted *o*-, *m*-, and *p*-benzyltrimethylsilanes that the red shift of the ^1L_a transition of the silane relative to the substituted toluene varied from 24 mμ for *p*-NO_2 to 0.0 mμ for *m*-$CONH(C_6H_5)$. Among *para*-substituted $RC_6H_4CH_2Si(CH_3)_3$ the smallest ^1L_a red shift, 6 mμ, was found where R is CH_3O. According to traditional arguments such a result would have been interpreted to mean that the —$CH_2Si(CH_3)_3$ group was a net electron donor. It has been shown more recently [63], however, that when spectral shifts are calculated in linear energy units, there is little difference between shifts calculated on the basis of monosubstituted benzene as to whether the *para* groups are both electron donors or acceptors, or whether one substituent is a donor and the other an electron acceptor and they thereby oppose each other.

Further, the red shift of the $C_6H_5CH_2M(CH_3)_3$ ^1L_a and ^1L_b absorption maxima, when M is carbon, *increases* in the order Sn > Ge > Si, which is the opposite order of the usual importance of *d* orbital participation when the metal is directly bonded to the aromatic ring (see Section IV-A). Similar effects have been found in spectra of Group IV organometallic compounds in which the metal is conjugated or homoconjugated with an olefinic chromophore (see Section IV-C-3).

Nagy suggests that the red shift in the spectra of the benzyl derivatives is the result of an extended hyperconjugation provided by *d* orbital overlap of the metal with the $C{=}(H_2)$ pseudo-π orbital as in structure **72**. It seems that such overlap of the *d* orbital would lower the energy of the $C{=}(H_2)$ π, or $C{\equiv}[(H_2)(Si)]$, orbitals thus decreasing their interaction with the filled π orbitals of the aromatic ring, not the reverse, as suggested. Whether *d* bonding combined with the antibonding hyperconjugating orbitals will lower the benzene π^* orbitals significantly is a different question, but it seems unlikely.

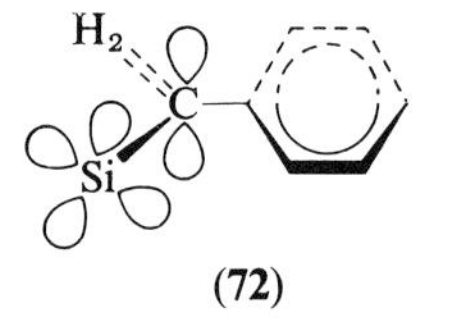

(72)

Two other alternative explanations appear more intuitively attractive. The first, originally suggested by Eaborn, is that in valence bond terms structures such as **73** are important in the excited state [42a]. The second possibility, illustrated by structures **74a** and **74b**, is based on the argument that the large, diffuse 3*d* orbitals of the metal may extend far enough through space to effectively overlap the π orbitals of the benzene ring. When the metal is directly substituted on the aromatic ring, one of the ring π^* antibonding orbitals possesses a nodal plane passing through the Si and is therefore not affected in energy by the substitution. In the benzyl case, the Si *d* orbitals could effect the lowering of *both* benzene lowest vacant π^* orbitals, resulting in a larger net red shift than that observed for the directly conjugated metal. Similar explanations have been offered by Jaffe to explain the spectra of allyl sulfides and are discussed further in connection with the spectra of vinyl and allyl Group IV metals (Section IV-C-3).

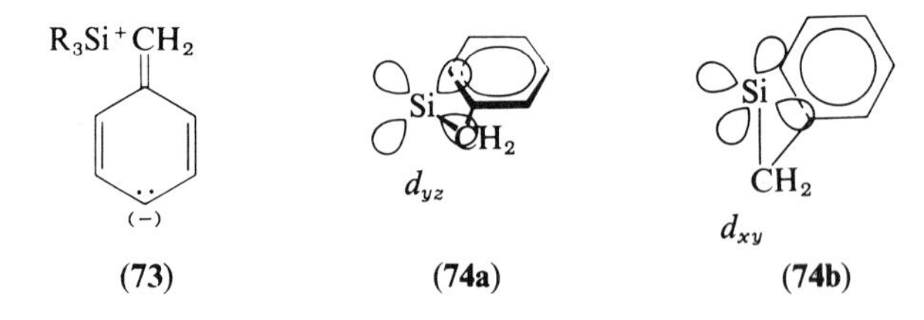

(73) **(74a)** **(74b)**

Considerable support for the hyperconjugative interaction of structure **73** can be found in the work of Bock and Alt [14] on the intermolecular charge transfer transitions of *para*-substituted benzene–RC_6H_5–tetracyanoethylene (TCE) complexes. Where R is $Si(CH_3)_3$ and $C(CH_3)_3$, the silyl complex absorbs at a higher energy than *tert*-butylbenzene, in spite of the fact that the greater electropositive character of Si should have resulted in the opposite order. This result is taken as evidence of lowering of the energy of the benzene filled donor molecular orbitals by *d–p* π bonding. Of particular interest however, is the red shift of the charge transfer transition energy by 3500 cm^{-1} (0.43 eV) in the spectrum of $(CH_3)_3SiCH_2C_6H_5 \cdot TCE$ compared with its position in $(CH_3)_3CCH_2C_6H_5 \cdot TCE$, a result completely compatible with a lower ionization potential as a result of contributing structure **73** but incompatible with Si *d* orbital participation, which would lower the energy of the filled phenyl molecular orbitals. The red shift in the spectra of the tetracyanoethylene complexes (3500 cm^{-1}) is comparable to a red shift of the corresponding ^1L_a transitions which amounts to some 3200 cm^{-1}.

The observation by Nametkin that the ^1L_a absorption maximum of 2-sila-2,3-dihydroindene is only 3 mμ lower in energy than that of 2,3-dihydroindene is in agreement with the hyperconjugation model since the silicon–carbon bond in 2-sila-2,3-dihydroindene is not in the correct conformation for hyperconjugation. The result, however, is not incompatible with the *d* orbital overlap model **74**, particularly structure **74a**, since where the Si lies in the plane of the benzene ring, as in the siladihydroindene, the d_{yz} orbital is orthogonal (nonbonding) to one of the benzene π^* orbitals and the d_{xy} orbital is orthogonal to all of the π^* orbitals. The structure **74b** represents a rotation of the silicon slightly out of the plane of the phenyl ring.

The ionization potential of $C_6H_5CH_2Si(CH_3)_3$, 0.7 eV lower than that of $C_6H_5C(CH_3)_3$, can only be explained in terms of a hyperconjugation resonance model which would raise the energy of the benzene π orbital by 0.7 eV and the π^* orbital to a smaller extent.

The conclusion which seems dictated by the presently available data is that the major decrease in the ^1L_a transition energy between $C_6H_5MR_3$ and $C_6H_5CH_2MR_3$ is due to the importance of metal–carbon bond hyperconjugation with the phenyl ring. It is possible that metal *d* orbital interactions with the ring may be of minor importance.

In the spectra of the series $C_6H_5M(CH_3)_3$ the order of ^1L_a transition energies Ge > Sn > Si may be accounted for by decreasing order of *d* orbital participation which alone would lead to the order of transition energies Sn > Ge > Si, opposed by an inductive electron release in the order Sn > Si > Ge which by itself would lead to transition energies in the order Ge > Si > Sn. The observed order of ^1L_a transition energies could be regarded as the net result of these two opposing trends, whereas inductive electron release in the order Si $\geq$ Sn > Ge > C would be enough in itself. The apparent blue shift in Table VII of the ^1L_b transition maximum of $C_6H_5Sn(CH_3)_3$ relative to that of $C_6H_5C(CH_3)_3$ is probably false. The appearance [97] of ^1L_b bands strongly suggest that if anything the O–O transition of the $C_6H_5Sn(CH_3)_3$ is shifted slightly toward the red, in keeping with the usual parallel substituent effects on ^1L_a and ^1L_b transitions. Therefore, at attempt by Nagy and Hencsei [97] to rationalize the blue shifted 252 mμ transition of $C_6H_5Sn(CH_3)_3$ on the basis of opposing resonance and inductive effects is not only unnecessary, but conceptually incorrect unless we assume $(CH_3)_3Sn$ to be electron withdrawing with respect to $—C(CH_3)_3$.

Simple molecular orbital calculations of the ω self-consistent charge variety, using the parameters $\alpha_{Si} = \alpha^\circ - 2\beta^\circ$ and $\beta_{C—Si} = 0.7\beta^\circ$, carried out by Nagy and co-workers on C_6H_5Si, predicted a red shift of the benzene spectrum by about $0.1\beta^\circ$ or roughly 0.23 eV

The UV spectrum of $(CH_3)_3Si(CH_2)_2C_6H_5$ does not differ appreciably from that of $(CH_3)_3CC_6H_5$, emphasizing the uniqueness of the $R_3SiCH_2C_6H_5$ structure [45].

Another approach to studying the role of *d* orbitals and aromatic rings is to increase the electronegativity of the substituent R_3M of $C_6H_5MR_3$ in an orderly fashion and see what happens to the ultraviolet spectra. For example, we would predict that if the overlap between ring and metal *d* orbitals were important, an increase in the electronegativity of M by suitable substituents R would lower the energy of the *d* orbitals and at the same time make them less diffuse. Marrot and co-workers [85] studied the ^1L_b transitions of the series $(C_6H_5)_nSnCl_{4-n}$ and $(C_6H_5)_nGeCl_{(n-4)}$ and tried to interpret their results in terms of combined inductive and resonance effects. It is difficult to say much with confidence, however, because the spectral shifts are so small. The differences between the vibrational structure maxima for corresponding Ge and Sn compounds are less than 1 mμ, the Ge compound usually having the longer wavelength absorption. For corresponding maxima in the ^1L_b transition of the series $(C_6H_5)_nMCl_{4-n}$ the differences between $n = 0$ and $n = 4$ are about 2 mμ, and depending on whether one accepts the Milazzo [89] or the La Paglia [82] assignment of the O–O transition, the spectral shift from $n = 0$ to $n = 4$ may either be to the red or blue. Trying to rationalize the order of oscillator strength per phenyl (f/n) in $(C_6H_5)_n$-$MCl_{(4-n)}$ is an interesting exercise (Table VIII). Since Ge and Sn have about the same electronegativities and covalent radii, one may suggest that the factor 2.4 in the difference between $(C_6H_5)_4Ge$ and $(C_6H_5)_4Sn$ f/n may be accounted for by *d* orbital interaction of the Ge 4*d* orbital with the phenyl ring, which is ineffective for the 5*d* orbitals of Sn. The gradual increase in f/n from $(C_6H_5)_4Sn$ to $(C_6H_5)_2SnCl_2$ could be attributed to an inductive effect. Finally, it would then be reasonable to suggest that the large increase

TABLE VIII

OSCILLATOR STRENGTH OF ^1L_b BAND OF $(C_6H_5)_nMCl_{4-n}$

Compound	$(C_6H_5)_4Sn$	$(C_6H_5)_3SnCl$	$(C_6H_5)_2SnCl_2$	$C_6H_5SnCl_3$
$f \times 10^5$	1742	1526	1051	1152
$(f/n) \times 10^5$	435	509	525	1152
Compound	$(C_6H_5)_4Ge$	$(C_6H_5)_3GeCl$	$(C_6H_5)_2GeCl_2$	$C_6H_5GeCl_3$
$f \times 10^5$	4084	1221	1388	605
$(f/n) \times 10^5$	1021	407	694	605

in ^1L_b oscillator strength in going from $(C_6H_5)_2SnCl_2$ to $(C_6H_5)SnCl_3$ was now the result of d–ring π interaction brought about by the inductive effect of Cl in lowering the energy of the Sn $5d$ orbitals. To make all of the above arguments consistent now with the *decrease* in f/n by a factor of 2.5 in the series $(C_6H_5)_4Ge$ to $(C_6H_5)GeCl_3$ requires considerable gymnastics, but we can try. In addition to being inductively electron withdrawing, Cl is also π electron donating, and interaction between the Ge d orbitals and chlorine lone pair p may be postulated to raise the energy of the resultant "d" orbital and reduce its interaction with the aromatic ring. In valence bond terminology structure **75** becomes more important than structure **76**. This argument has already been involved to explain the blue shift of the charge

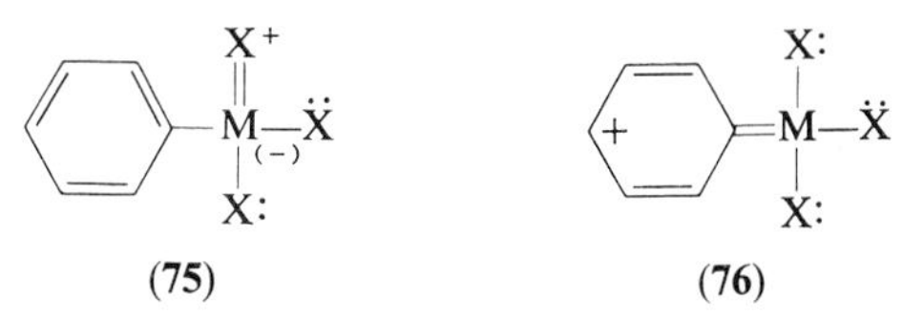

(75) **(76)**

transfer transition of $(C_6H_5)_3B$ relative to that of $(C_6H_5)_2BCl$ or $(C_6H_5)BCl_2$ (Section III-C). Why then does the resonance effect of Cl predominate in the case of $C_6H_5GeCl_3$, while the inductive effect prevails for $C_6H_5SnCl_3$, in the substituent effect on the intensity of the ^1L_b transition? Possibly the d orbitals of $—SnCl_3$ or $—Ge(C_6H_5)_4$ represent a sort of optimum energy for interaction with the phenyl ring as opposed to bonding with the halogen lone pair electrons. As an alternative approach one might emphasize the importance of the Cl electron donation ability decreasing the importance of d orbital overlap with the phenyl ring but utilizing Sn $4f$–$5d$ hybrids in Sn compounds such that some of these hybrid orbitals interact with the halogen lone pairs, while the remaining orbitals are left free to delocalize the π electrons of the ring. All of the above are rather speculative, however, in the absence of molecular orbital calculations which include not only the valence electron orbitals of carbon chlorine and metal atom, but also the d orbitals of chlorine, germanium, and tin along with its vacant $4f$ orbital. What is the effect, for example, of overlap between the inner filled metal d orbitals and the vacant chlorine $3d$ orbital?

Brown and Prescott [28] have made a very interesting analysis along similar lines of the intensity of the ^1L_b band for the phenylsilanes $C_6H_5Si(CH_3)_3$, $C_6H_5SiCH_3(OCH_3)_2$, $C_6H_5Si(OCH_3)_3$; siloxanes, $(C_6H_5SiO_{3/2})_n$ where $n = 8, 10, 12$; and diphenylsiloxanes, $(CH_3)_3SiO(C_6H_5SiO)_nOSi(CH_3)_3$ where $n = 1, 2, 3$; $(C_6H_5)_2SiO_3$; and related compounds. Again, there is almost no variation in the wavelength of absorption for the O–O ^1L_b transition. The phenylsilane and phenylsiloxane O–O transition maxima lie between 269 and 270 mμ, whereas the diphenylsiloxane O–O maxima fall in the range 270–271 mμ. As previously found, however, for the phenylgermanium and -tin halides, there is a significant and orderly variation in the ^1L_b transition oscillator strengths, f. Interestingly, it is found to be possible to separate the total oscillator strength into the sum of a vibrationally induced f_q, and substituent component f_v, such that $f = f_v + f_q$. As might be expected within a series of related compounds, the vibrational component f_v remains roughly constant over large variations in f_q. The substituent component $f_q \times 10^4$ increases in the expected fashion based on increasing Si d orbital interaction with the ring as the substituents become more electronegative in the order $C_6H_5Si(CH_3)_3$ (9.6), $C_6H_5SiCH_3(OCH_3)_3$ (17.2), $C_6H_5Si(OCH_3)_2$ (19.3), $(C_6H_5SiO_{3/2})_8$ (31.0). If substituent transition moment parameters, q which measure charge displacement from substituent to ring, are calculated from the formula, $q = (3hf_q/8\pi^2mc\nu^0)^{1/2}$; q's are obtained in the range of -5×10^{-10} cm. for example, for —$SiO_{3/2}$. By way of comparison the electron withdrawing ability of CF_3 in the ^1L_b transition is expressed by a q of -9×10^{-10} cm. Brown and Prescott also looked in some detail at the intramolecular dipole–dipole and the dipole–induced dipole interactions between phenyls in the polyphenylsilsesquioxanes $(C_6H_5SiO_{3/2})_n$ and diphenylsiloxanes. Variations in f_q and q of $(C_6H_5SiO_{3/2})_n$ ^1L_b transitions with n would be accounted for by theoretical calculation of the polarizability (dipole–dipole) and field effect (dipole–induced dipole) interactions between nonbonded phenyl rings in a fixed geometry.

The ultraviolet spectra of $(CH_3)_nSi(OC_6H_5)_{4n}$, where $n = 0$–3, have been compared with simple molecular orbital calculations on C_6H_5O and C_6H_5OSi by Nagy and Hencsei [97]. The absorption maxima of the L_a transitions of the alkylphenoxysilanes are *blue* shifted with respect to the L_a transition of *tert*-butyl phenyl ether or anisole: compare, for example, λ_{max}: $CH_3Si(OC_6H_5)_3$, 211, 273 mμ; $(CH_3)_3SiOC_6H_5$, 218, 272 mμ; $CH_3OC_6H_5$, 219, 271 mμ; and $(CH_3)_3COC_6H_5$, 221, 270 mμ. A comparison of the wavelengths of the *center* absorption maxima indicates a small red shift of the ^1L_b transition in the silyl ethers relative to anisole, but this seems suspect in view of the more pronounced blue shift of the ^1L_a transition, and close examination of the published spectra suggest that the O–O transition may also be blue shifted. The molecular orbital calculations are apparently carried out in a fashion which forces the prediction of a 2 mμ red shift of $C_6H_5OSi\equiv$ spectra relative to the C_6H_5OR spectrum: to obtain this calculated red shift a C—O resonance integral of 0.80 is assumed for C_6H_5COR and it is required that the inductive effect of Si increase the C—O resonance integral to 1.0β, which is the *same* value of k_{C-O} utilized on the same page by the authors for the $C-O^-$ resonance integral in the phenolate anion $C_7H_5O^-$. This does not seem realistic. It would rather seem that the correct C—O resonance integral to be used should have been obtained from the linear relationship developed on the previous page of the article between β_{C-O} and the transition energy of C_6H_5OR, which gave an approximate value of $k=0.82$ for β_{C-O} in $C_6H_5OSi\equiv$. The experimental results seem most consistent with the predominant effect of back bonding of the oxygen lone pair electrons to the vacant Si *d* orbitals. This has the effect of decreasing relative to anisole the perturbation of the benzene orbitals by the oxygen lone pair electrons. In other words, the Si *d* orbitals decrease the importance of resonance forms of the type $R-O^+=C_6H_5^-$ in the excited state.

The same problem of *d–n* π interaction in silyl aryl ethers has been examined in a more imaginative fashion by Bock and Alt [15] in studying the charge transfer maxima of the tetracyanoethylene complexes of C_6H_5X, di-1,4-XC_6H_4, 1-$XC_{10}H_7$, and 2-$XC_{10}H_7$, where X is H, OCH_3, $OCH(CH_3)_2$, and $OSi(CH_3)_3$. A qualitative energy level diagram is given in Fig. 26 for the tetracyanoethylene charge transfer complexes with the alkyl and silyl phenyl ethers. The transitions of interest are those which promote an electron from an occupied molecular orbital of benzene to the lowest vacant molecular orbital of tetracyanoethylene. Making the legitimate assumption that small changes in structure of the donor molecule will not affect the energy of the acceptor orbital simplifies things because we can interpret changes in charge transfer transition energy *only* in terms of the substituent effect on the filled orbitals of the aromatic ring. The nodal properties of the two highest filled molecular orbitals of benzene degenerate and naphthalene nondegenerate are given in structures **77–80**. From these it is easily seen, as indicated in Fig. 26, that substitution in the 1 and 4 positions of naphthalene or benzene by OR will raise by resonance effects the energy of one molecular orbital

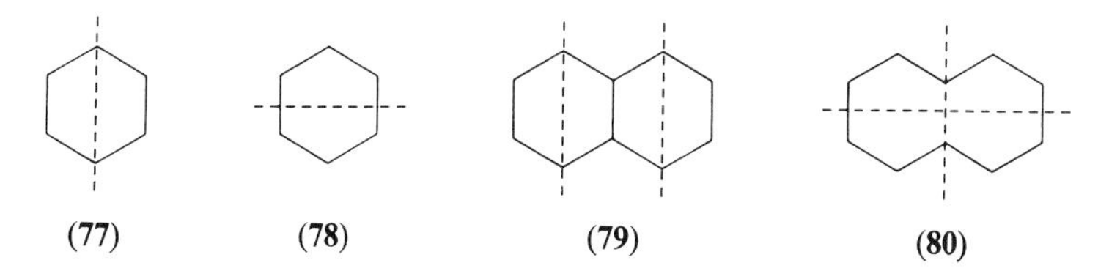

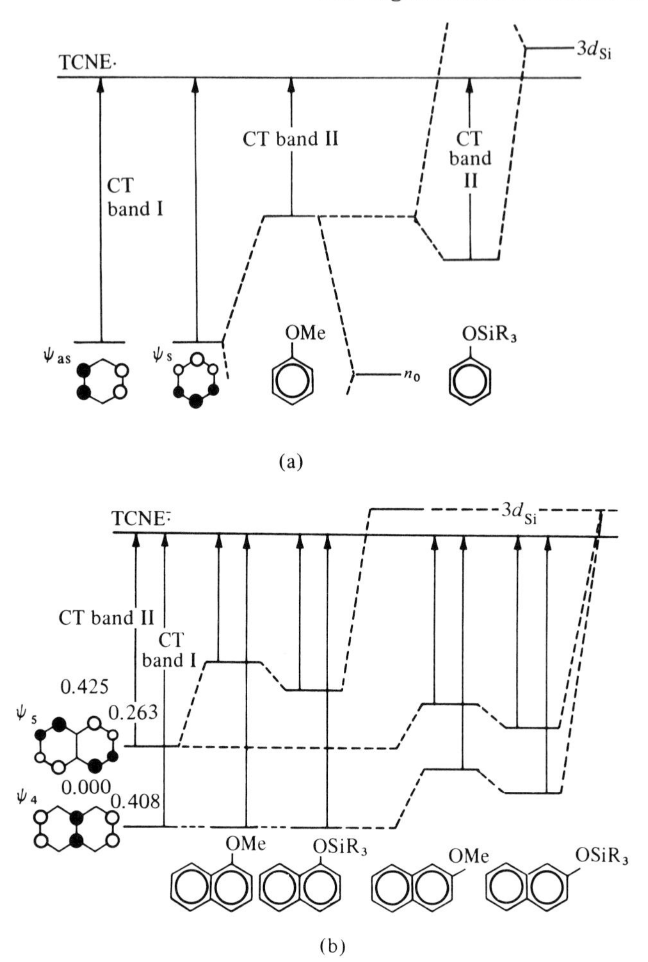

Fig. 26. Qualitative molecular orbital energies for (a) alkylphenyl and silylphenyl and (b) naphthyl ether tetracyanoethylene charge transfer complex (TCNE$\cdot^-$) [15].

while leaving the second unchanged. This leads to the appearance of two charge transfer bands in the spectra of those C_6H_4X, where X is other than hydrogen, rather than the one observed for benzene–tetracyanoethylene.

We would predict the substituent effect of changing R from CH_3 to $Si(CH_3)_3$ on the charge transfer transitions of $C_6H_5OR \cdot (CN)_2C{=}C(CN)_2$ in the following fashions, as indicated in Fig. 26. First, the charge transfer transition originating from structure **77** will remain unchanged since it possesses a nodal plane through the substituent. The electron releasing inductive effect of R_3Si will serve to *raise* the energy of the molecular orbital in structure **78**.

Since the tetracyanoethylene acceptor orbital remains unchanged in energy, this would lead to a red shift, when R is methyl. Superimposed on the inductive effect is the interaction between the silicon $3d$ orbital (or appropriate $3d$–$4p$ hybrid) with this highest filled orbital which will lower its energy. Therefore, depending on whether the *d* orbital electron withdrawing resonance effect or electron donating inductive predominates, we would predict a red or blue shift of the appropriate charge transfer maximum with the structural change aryl—O—CH_3 to aryl—O—$Si(CH_3)_3$. The arguments expressed in Fig. 26 may, of course, be applied to naphthalene. As may be seen from the selected representative data in Table IX, a blue shift of the

TABLE IX

COMPOUND–TETRACYANOETHYLENE CHARGE TRANSFER COMPLEX TRANSITIONS IN CH_2Cl_2

Compound	λ_{max} (mμ)	Compound	λ_{max} (mμ)
$C_6H_5OCH(CH_3)_2$	390, 516	$C_6H_5OSi(CH_3)_3$	393, 467
1-$C_7H_7OCH_3$	427, 660	1-$C_7H_7OSi(CH_3)_3$	427, 631
2-$C_7H_7OCH_3$	484, 610	2-$C_7H_7OSi(CH_3)_3$	475, 589

charge transfer absorption in the spectra of the aryl silyl ether complexes relative to the methyl ethers is actually observed. The conclusion then is that at least in describing electronic transitions, *d*–*n* π bonding is important between silicon and oxygen.

The ultraviolet spectrum of silazarophenanthrene (**81**) has been found by Gaidis and West [49] to be similar to that of phenanthrene and the borazarophenenthrene (Section III-D-2) which led to the suggestion that at least one *d* orbital is used to overlap both the nitrogen lone pair and ring π orbitals. However, NMR measurements of ring currents in compound **81** and the corresponding P—N compound suggest that, at least in the ground state, cyclic delocalization does not take place (J. M. Gaidis, Ph.D. thesis, Univ. of Wisconsin, 1967).

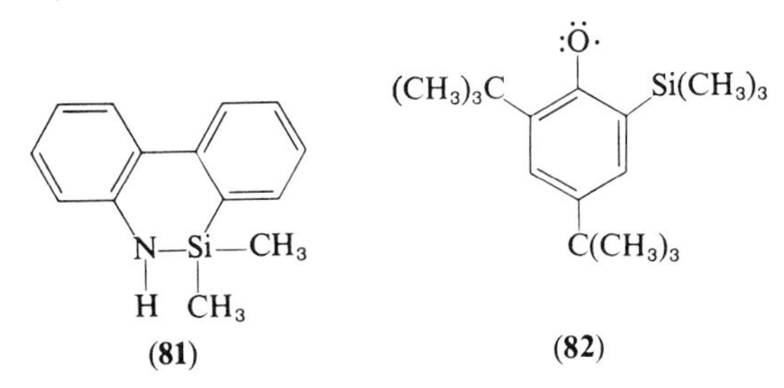

Among anions and free radicals there are the interesting reports of the ultraviolet spectra of triphenylsilyllithium [137] in THF, λ_{max} 335 mμ (log ϵ 4.0) and the silicon-substituted [95] radical (**82**) (λ_{max} 580 mμ). The absorption of $(C_6H_5)_3SiLi$ may be compared with the absorption maxima of $(C_6H_5)_3CLi$ at 425 and 500 mμ, also in THF. Certainly, if we were dealing with the free anions, the $(C_6H_5)_3Si:^{(-)}$ should absorb at longer wavelength than $(C_6H_5)_3C:^{(-)}$ since the valence state ionization potential of sp^2-hybridized silicon is some 2 eV smaller than that of sp^2 carbon. Either the transitions of $(C_6H_5)_3Si:^{(-)}$ which correlate with the 425 and 450 mμ absorptions of $(C_6H_5)_3C$: are to be found in the near infrared, or the Li—Si bond is so much more covalent than the C—Li bond that there is a blue shift of the transition of $(C_6H_5)_3SiLi$ relative to $(C_6H_5)_3CLi$. The NMR chemical shifts [137] of $(C_6H_5)_3SiLi$, however, indicate considerable delocalization of the Si electrons into the phenyl rings.

The novel analogs (**83a**) of *p*-cyclophanes (**83b**) have been synthesized and their ultraviolet spectra discussed by Haiduc *et al.* [68a]. In the spectra of *p*-cyclophanes, where $m + n \leq 7$, the ^1L_a and ^1L_b bands undergo pronounced increases in intensity and shift to a lower energy due to deformation of the rings and transannular π interactions. A similar although less dramatic effect is observed in the spectra of structure **83**, where X is Cl, when compared with $Cl_5C_6Si(CH_3)_2OSi(CH_3)_3$, for example. The smaller effect is probably due to the fact that the chlorines have already introduced considerable steric deformation of the ring; it is known [7] that "crowded" pentachlorobenzenes exhibit spectral properties similar to the *p*-cyclophanes.

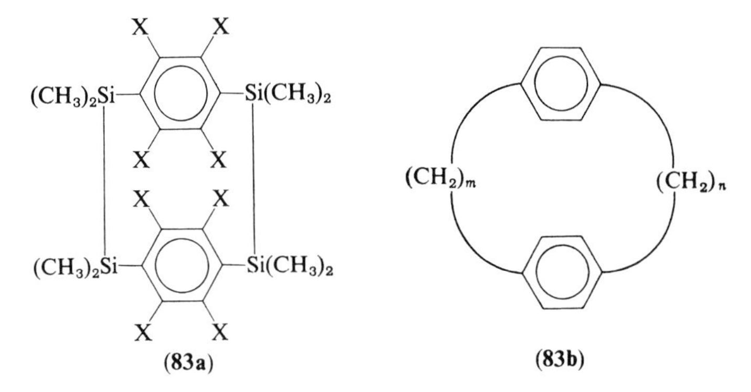

(**83a**) (**83b**)

C. Alkenyl and Alkynyl Organometallic Compounds

1. *Acyclic Unsaturated Organometallic Compounds*

Chromophores of the structure M—A—B, such as M—C=C or M—C≡C—, whether neither atom of the sequence A—B bears an electron lone pair, have been designated as Class I chromophores by West [138].

The molecular orbital correlation diagram by West, based on simple qualitative arguments and showing the inductive and resonance effects of the metal substituent on a Class I chromophore, is given in Fig. 27. The inductive effects raise the π and π^* orbitals equally while lowering the metal orbital in energy. The interaction between the metal d orbital and the original π and π^* orbitals displaces both the original antibonding and bonding orbitals to

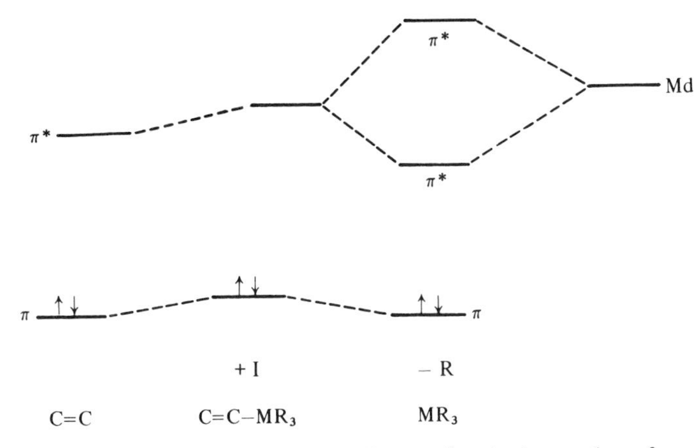

Fig. 27. A qualitative molecular orbital diagram for the interaction of a metal with a *d* orbital with an olefin by inductive (I) and resonance (R) effects.

lower energies, while creating a third antibonding orbital at higher energy than the original metal atomic orbital. Since the starting *d* orbital is closer in energy to the π^* than the π orbital, the predicted net effect is to lower the π^* orbital more than the molecular orbital, thus decreasing the $\pi \rightarrow \pi^*$ excitation energy. From the ionization potential of ethylene (10.45 eV) minus the $\pi \rightarrow \pi^*$ transition energy (7.68 eV), the ionization potential of silicon (8.15 eV), and its average $3p \rightarrow d$ promotional energy (6.20 eV), the energy of a typical $3d$ level is estimated by West† to be about 0.8 eV above the ethylene π^* orbital. The relative energies of the $3d$ and π^* orbitals therefore are not expected to be too large for significant interaction between them.

The ultraviolet spectra of a variety of vinyl and allyl derivatives of Group IV metals have been studied by Petukhov and co-workers [106], however, and the attractively simple arguments of *d–p* π overlap seem quite inadequate at this point. The ultraviolet absorption maxima of representative vinyl compounds can be found in Table X. It is true that the ultraviolet maxima of $Me_3SiCH{=}CH_2$ (178 mμ) is substantially red shifted compared with

† A more sophisticated estimate of the energy of Si $3d$ orbital by Armstrong and Perkins [4] places the energy at 0.95 eV or 1.8 eV above the π^* orbital of ethylene.

TABLE X

ABSORPTION MAXIMA OF SOME VINYL ORGANOMETALLIC COMPOUNDS

Compound	λ_{max}, mμ (ϵ)	Compound	λ_{max}, mμ (ϵ)
$(CH_3)_3SiCH{=}CH_2$	178 (1.8×10^4)	$F_3SiCH{=}CH_2$	<170
$(CH_3)_3GeCH{=}CH_2$	182 (2×10^4)	$Cl_3SiCH{=}CH_2$	177 (1.4×10^4)
$(CH_3)_3SnCH{=}CH_2$	186 (2.8×10^4)	$F_3CCH{=}CH_2Si(CH_3)_3$	181 (1.2×10^4)
$Cl_3GeCH{=}CH_2$	180 (2×10^4)	$(CH_3)_3SiCH{=}CH_2SiMe_3$	195 (2.5×10^4)

ethylene (λ_{max} 165 mμ), but it is virtually the same as $(CH_3)_3CCH{=}CH_2$ (λ_{max} 178 mμ). Furthermore, with increasing energy of the metal *d* orbital if *d–p* π overlap were the controlling, factor, a blue shift in the absorption maximum of $MCH{=}CH_2$ should be obtained in the order Si < Ge < Sn. On the contrary, the spectral shift is found to be in the opposite direction, and we must conclude that *d–p* π conjugation is of minor importance in determining the first-transition energy of vinyl derivatives of *single* Group IV metals. It can be anticipated, though, that the lower-energy *d* π molecular orbitals of catenated metal derivatives would approach more closely the hydrocarbon π and π^* orbitals and thereby have a larger effect than a single *d* orbital.

Petukhov and co-workers have argued that conjugation of the metal–vinyl σ bond with the bonding orbital is of primary importance in determining R—M bond substituent effects on olefin $\pi \rightarrow \pi^*$ transitions. The argument rests primarily on the observation that the UV maximum of $F_3CCH{=}CHSi(CH_3)_3$ is virtually the same as that of $CH_2{=}CHSi(CH_3)_3$ and therefore pure inductive effects must be small. It was also felt that the observation of an absorption maximum in the spectra of Br_3Ge derivatives to the red of R_3Ge compounds was not in agreement with an inductive order effect on the spectrum of ethylene.

The arguments against the inductive effect do not, however, seem compelling, and by and large the available spectral data may be accommodated on the basis of a simple inductive effect. For example, there is a progressive blue shift in the absorption maxima of the series $R_3SiCH{=}CHSiR_3$, with R varying as CH_3, Cl, or F in the order of increasing electronegativity of R_3Si. Similarly, the UV maximum of $Cl_3GeCH{=}CH_2$ is at a slightly lower wavelength than $(CH_3)_3GeCH{=}CH_2$. The blue shift in the spectrum of $F_3CCH{=}CH_2Si(CH_3)_3$ relative to $(CH_3)_3SiCH{=}CHSi(CH_3)_3$ seems in keeping with the expected inductive effect of CF_3. Replacement of H by either an electron withdrawing or electron donating group usually results in a small red shift of a $\pi \rightarrow \pi^*$ transition, so that the small red shift of $CF_3CH{=}CHSi(CH_3)_3$ relative to $CH_2{=}CHSi(CH_3)_3$ seems acceptable. True, the ultraviolet spectra of Br_3M derivatives do exhibit absorption maxima at much higher wavelengths than R_3M derivatives; these maxima are most probably associated, however, with the transition found in $CHBr_3$ [76] (λ_{max} 225 mμ) or CH_3GeBr_3 (λ_{max} 215 mμ), and do not necessarily constitute an exception to the inductive effect model. Deviations in the expected magnitude of the blue shift or even a small red shift in the spectra of Cl_3M derivatives can be rationalized by the argument that up to a point, electron withdrawing groups on the metal M will decrease the energy of the *d* orbitals, making *d–p* π overlap more effective and imposing a bathochromic effect on the inductive blue shift. The reader may well wonder why the inductive effect of $(CH_3)_3Si$ should be the same as that of $(CH_3)_3C$ on the spectrum of ethylene. With respect to this, we can only point to the relative Hammett σ_{meta}-substituent constants of the two groups, $C(CH_3)_3$ (σ_{meta} -0.10 ± 0.03) and $Si(CH_3)_3$ (σ_{meta} -0.04 ± 0.1), and suggest that their inductive effect may be comparable after all, in spite of differences in electronegativity.

Although we do not see how the available data rule out an inductive effect explanation for the substituent effect of X_3M on the spectra of olefins, this does not mean that the suggestion of σ–π (or μ–π) conjugation by Petukhov has no validity. In terms of simple molecular orbital theory inductive effects on the transition maxima of simple olefins should be negligible. A number [13, 117] of careful molecular orbital calculations now indicate that the lowest $\sigma \rightarrow \pi^*$ transition of an olefin should not be much greater in energy than the $\pi \rightarrow \pi^*$ transition. Furthermore, it is well known that the carbon–metal σ-bond energy decreases in the order Si > Ge > Sn. It is certainly not unreasonable to suggest then that substituent effect of R_3M on the spectrum of ethylene is the result of weak configuration interaction between the $\sigma \rightarrow \pi^*$ and $\pi \rightarrow \pi^*$ excited states. A decrease in the $\sigma \rightarrow \pi^*$ transition energy would thus lower the $\pi \rightarrow \pi^*$ transition. This description would appear to be the same as saying that structure **84** contributes to the first excited state of $R_3MCH{=}CH_2$, which is what we assume the suggestion of Petukhov *et al.* on σ–π conjugation to mean (see Addendum).

Substituent effects on the spectra of acetylenic compounds [16, 42g] for the few cases studied seem to be similar to those discussed for alkenes. The spectrum of $(C_6H_5)_3SiC{\equiv}C{-}C(CH_3){=}CHCH_3$ shows a 15 mμ red shift relative to that of the unsubstituted compound [17]. The reported [44]

$$R_3\overset{(+)\cdot\cdot}{M}\ddot{C}H{-}\overset{\cdot\cdot(-)}{C}H_2$$

(84)

absorption maxima of $(CH_3)_3CC{\equiv}CH$ at 190 mμ (ϵ 330), 214 mμ (ϵ 184), and 220 mμ (ϵ 198) may be compared with those of $(CH_3)_3SiC{\equiv}CH$ at 194 mμ (ϵ 420), 196 mμ (ϵ 260), and 262 mμ (weak) (see Addendum).

2. *Heterocyclic Organometallic Compounds*

The ultraviolet spectra of several phenyl-substituted silacylopentadienes have been reported [5, 19]. The ultraviolet spectrum of compound **85**, λ_{max} 230 mμ (ϵ 1.3 × 10^4) and 370 mμ (ϵ 2 × 10^4) may be related to the intense transitions [128] of tetraphenylcyclopentadiene at 245, 269, and 346 mμ. Compound **86** absorbs at 246 and 362 mμ, and cyclopentadiene at 238 mμ (ϵ 4 × 10^3).

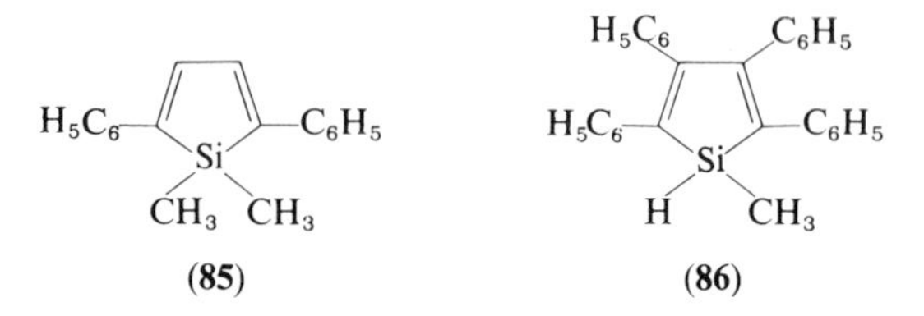

(85) **(86)**

A number of compounds originally thought to have structure **87** have now been found [74] to be of structure **88** instead. When M is Ge, R is C_6H_5, and R′ is CH_3, λ_{max} of 227 mμ (ϵ 4.6 × 10^4) and 270 mμ (ϵ 9.5 × 10^3) are reported [74]. The spectrum of compound **88**, where M is Ge, R is H, and R′ is CH_3, shows [135] a single strong absorption at 206 mμ which may be compared with a similar absorption of $(CH_3)_3GeCH{=}CHGe(CH_3)_3$ at 195 mμ; however, there is not a significant red shift relative to the spectrum [108] of 1,4-cyclohexadiene.

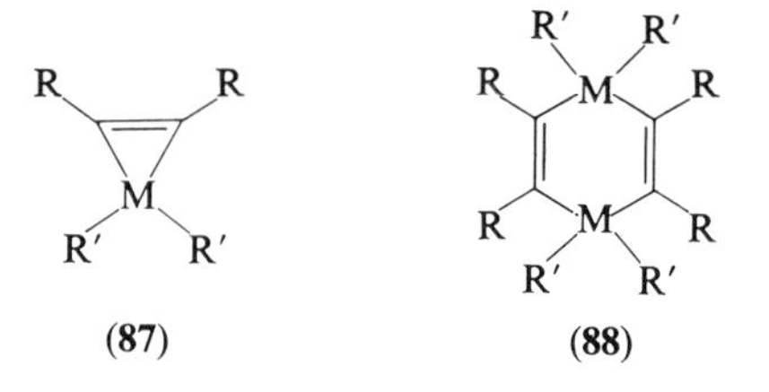

(87) **(88)**

3. *Homoconjugated Organometallic Compounds*

Most surprising at first thought is the result [107] that the organometallic compounds $R_3MCH_2CH{=}CH_2$ (M is Si, Ge; Sn and R are alkyl or halogen) have two intense absorption maxima in their UV spectrum, one in the region 170–180 mμ and a second between 190 and 210 mμ. This second transition, at a significantly longer wavelength than that of the conjugated vinyl derivative, progresses to lower energy in the order Si > Ge > Sn (see Table XI). This red shift is very similar to that discussed previously for benzyl derivatives of Group IV metals, and we will begin by assuming the long wavelength transitions to be $\pi \rightarrow \pi^*$.

In the introduction to this chapter we have already indicated some of the evidence for Si *d–p* π interaction across an intervening —CH_2— group. The assignment of the 235 mμ maximum of gem disulfides $R_2C(SR_2)_2$ to a transition to a molecular orbital resulting from overlap of the Si 3*d* orbitals has also been mentioned previously. The 210 mμ transition of dialkyl sulfides is red shifted to about 221 mμ in allyl sulfides, which is further evidence of interaction between sulfur and π bond across an intervening CH_2 group [78], probably by means of the vacant sulfur *d* orbitals [73].

With the above precedents, it is very tempting to assign the long wavelength transitions of the $R_3MCH_2CH{=}CH_2$ class of compounds to a transition to a molecular orbital derived from interaction of the metal vacant *d* orbitals with the olefin double bond. For purposes of further discussion let us consider the case where the Si is in the nodal plane of the carbon–carbon π bond on the *x* axis perpendicular to and passing through the midpoint of the C—C double bond. For this molecular symmetry the bonding π

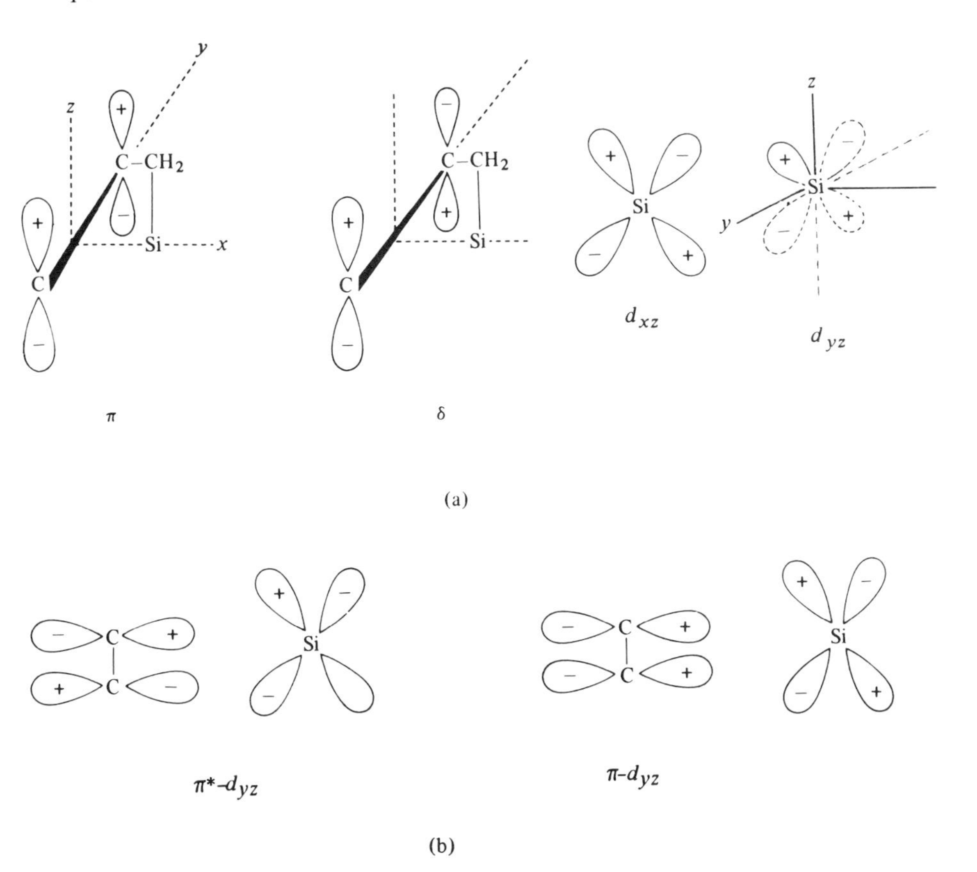

FIG. 28. Orientations of π and *d* orbitals of $C{=}C{-}CH_2Si$: (a) with Si in the *xy* plane; (b) with Si in the *yz* plane.

orbital (a) has π symmetry and the π^* (b) antibonding orbital δ symmetry in Fig. 28. From this idealized geometry we can readily see that molecular orbitals of lower energy may be constructed from combinations of the d_{zx} and π and from the d_{yz} and π^* atomic and molecular orbitals. The d_{zx}–π interaction must be small, however, because of the large energy difference between the starting orbitals. The energy difference for the d_{yz}–$\delta(\pi^*)$ combination is more favorable but the δ or lateral overlap integral cannot be large and the resultant molecular orbital would not be greatly lower in energy than the original π^* orbital. Certainly this overlap should be less effective than the d–p π overlap for M—CH=CH_2 and cannot be used to explain the larger spectral red shift of $MCH_2CH{=}CH_2$ relative to MCH=CH_2. On the other hand, if the silicon atom is rotated above the x–y plane more nearly in the y–z nonnodal symmetry plane of the carbon–carbon double bond, d–p π interaction of the d_{xy} and/or d_{yz} atomic orbitals may substantially lower the energy π^* orbital *without affecting the bonding* π *orbital* (see Fig. 28). In keeping with these geometric requirements is the reported absence of the transition near 180 mμ in cyclopent-3-enylgermanes of structure **89**, whereas in compounds of structure **90** a broad intense absorption maximum near 230 mμ is red shifted in comparison with the acyclic

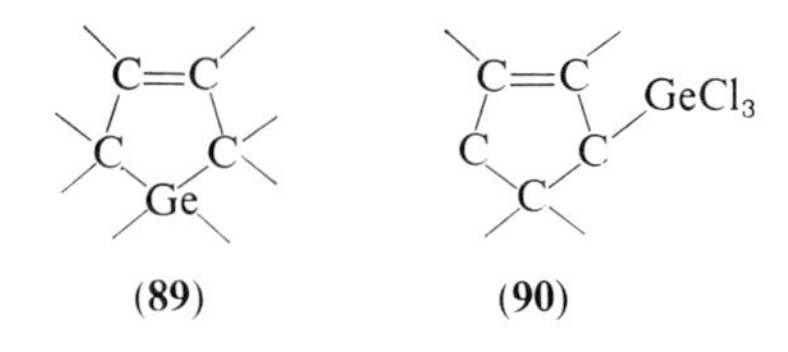

compounds. Inasmuch as d–p π bonding for the vinylmetallic compounds lowers *both* the bonding and the antibonding molecular orbital, the greater red shift in the spectrum of allyl organometallic compounds can also possibly be rationalized by in this fashion.

The arguments of the previous paragraphs fail, however, to account in a simple way for a shift of the long wavelength transition to lower energy in the order Si > Ge > Sn, which is again the reverse order of expected d orbital overlap. The blue shift in the silane spectra on the substitution of halogen for hydrogen in the SiR_3 group requires that we postulate either an unreasonably large inductive effect for the halogens, or that competitive d–n π bonding of the silicon with the lone pair electrons of the halogens greatly reduces the metal d orbital interaction with the carbon–carbon π orbitals. The latter argument was also used earlier to account for the smaller perturbation of the benzene spectrum by $GeCl_3$ than $Ge(C_6H_5)_3$. Contradicting this, however, is the report that the transition maximum of Cl_3Ge-$CH_2CH{=}CH_2$ at 206 mμ is red shifted compared with the absorption of $(CH_3)_3GeCH_2CH{=}CH_2$ at 195 mμ, although the opposite order is found for the vinylgermanes. A complex argument balancing the overall inductive effect of X_3M on the $\pi \rightarrow \pi^*$ transition energy against the p π electron donating and inductive effect of halogen on the metal d orbitals can be constructed which rationalizes the observed substituent effects but which is highly speculative.

As an alternative, Petukhov and co-workers have made the suggestion that the long wavelength transition of $R_3MCH_2CH{=}CH_2$ compounds may be explained as the result of σ_{M-CH_2}–π interaction. This interaction, it is argued by Petukhov, is supported by the order of decreasing transition energies of the saturated σ-bonded Group IV derivatives $Me_4Si > Me_4Ge > Me_4Sn$ and $R_4Si > RSiCl_3$, but $RGeCl_3 > R_4Ge$ which is the same order observed for the allyl derivatives. The exact role of this σ–π interaction was not defined, although a similar suggestion, in terms of hyperconjugation, was made some time ago by Eaborn [42a] to account for an observed red shift of the benzene $^1A \rightarrow {}^1L_b$ transition of benzylsilanes. As outlined below, the exact role of this σ–π interaction, if any, needs to be more clearly defined.

In one case the $\sigma \rightarrow \pi$ interaction may be thought of as a small contribution in valence bond terms of the form $R_3M^+CH_2{=}CH{-}C^-H_2$ to the stabilization of the excited state of the $\pi \rightarrow \pi^*$ transition. This translates, in molecular orbital terms, to an interaction between the π molecular and pseudo-π hyperconjugative molecular orbital, which is of *lower* energy and principally made up of metal–carbon σ bond. This type of interaction, by raising the energy of the π orbital with little effect on the π^* orbital, would decrease the energy of the $\pi \rightarrow \pi^*$ transition.

Such an argument accounts in a very satisfactory manner for the $\pi \rightarrow \pi^*$ transition energy of $R_3MCH_2CH{=}CH_2$ in the order $(CH_3)_3Sn < (CH_3)_3Ge < (CH_3)_3Si < Cl_3Si < F_3Si$ (if a 180 mμ maximum is $\pi \rightarrow \pi^*$) $\leq (CH_3)_3C$. It is important to this approach that the 180 mμ transition of $F_3SiCH_2CH{=}CH_2$ correspond to the 190–210 mμ absorption series, $\pi \rightarrow \pi^*$, of the other derivatives, rather than the 170–180 mμ absorption maxima. This is necessary because the hyperconjugative mechanism under consideration does not in itself provide for a *blue* shift of the absorption maximum relative to that of $(CH_3)_3CCH_2CH{=}CH_2$ (λ_{max} 178 mμ), since C—C hyperconjugation itself is normally considered minimal with respect to C—H hyperconjugation.

It is now well established that in organic chemistry the steric and geometric requirements for C—H hyperconjugation or resonance are similar to those of other types of resonance. The maximum interaction is obtained when the σ bond is in the same plane as the π bond. In this fashion, the $\sigma \rightarrow \pi$ interaction postulate would account for the dependence of the transition energy on the relative geometry of the metal–CH_2 π bond and the σ bond, as illustrated by the previously mentioned cyclic structures **89** and **90**. Note that in structure **89** the metal–CH_2 bond lies in the *nodal* plane of the π bond.

Hyperconjugation is not enough, however. When we come to the spectrum of trimethylsilylcyclopentadiene [47], we discover that its absorption maximum at 241 mμ is only slightly red shifted relative to cyclopentadiene (λ_{max} 238 mμ) [105]. If molecular models are examined, it is found that the rigidity of the cyclopentadiene ring requires the planar structure **91** and the accompanying Newman projection.

Compounds of type **90** may however assume the slightly twisted structure giving the conformation of structure **92** and its Newman projection.

Comparison of the conformations of structures **91** and **92** indicates that hyperconjugation of the metal–silicon bond of trimethylsilylcyclopentadiene with the C=C double bond should be only a little less effective than that in other allylsilanes or structures such as **92**; however, the possibility of silicon *d* orbital π overlap, as indicated in Fig. 28, which remains possible for structure **92**, is completely removed in the rigid structure of cyclopentadiene ring.

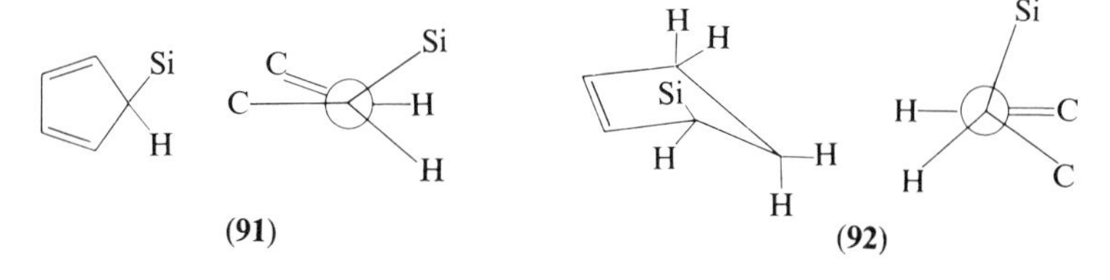

In conclusion, *if* the long wavelength transition of the allylorganometallic compounds in Table XI and similar compounds is to be regarded as essen-

TABLE XI

ALLYL GROUP IV ORGANOMETALLIC COMPOUNDS

Compound	λ_{max}, mμ (ε)	λ_{max}, mμ (ε)
$(CH_3)_3CCH_2CH{=}CH_2$		179 (11,000)
$(CH_3)_3SiCH_2CH{=}CH_2$	178 (sh) (6000)	190 (10,000)
$(CH_3)_3GeCH_2CH{=}CH_2$	174 (11,000)	195 (11,000)
$(CH_3)_3SnCH_2CH{=}CH_2$	181 (22,000)	210 (14,000)
$Cl_3SiCH_2CH{=}CH_2$	180 (sh) (10,000)	186 (11,000)
$F_3SiCH_2CH{=}CH_2$		180 (5000)
$Cl_3GeCH_2CH{=}CH_2$	182 (15,000)	206 (5000)

tially type $\pi \rightarrow \pi^*$, the best interpretation of substituent effects is one which invokes *both* homoconjugation of the metal *d* orbitals and metal–carbon bond hyperconjugation with the olefin π electrons. Thus, the $R_3M^+CH_2{=}CH{-}CH_2^{(-)}$ hyperconjugation serves to raise the energy of the filled π orbital while at the same time interaction between the metal *d* orbitals and antibonding π* orbital results in lowering of the π^* orbital, and a spectral red shift larger than could be accounted for by either effect alone (see Addendum).

If the long wavelength transition is assigned to $\pi \rightarrow \pi^*$, the short wavelength absorption maxima (Table XI) of the allylorganometallic compounds can be assigned to the same transition which occurs in this region in the spectra of alkyltin and alkylgermanium compounds, i.e., $\sigma \rightarrow \sigma^*$ or $\sigma \rightarrow 3d$.

Based on absorption maxima or extinction coefficients at 170 mμ for the compounds R_4M (R is alkyl or halogen) the transition energy of R_4M for promotion of a σ electron is in the order $RSiCl_3 \gg R_4Si > R_4Ge > RGeCl_3 > R_4Sn$. From a glance at Table XI we quickly see that the energies of the short wavelength transitions do *not* fall in this order, but that the long wavelength maxima fit this order exactly. *This raises the possibility that the short wavelength transitions of* $R_3MCH_2CH{=}CH_2$ *should properly be assigned to the* $\pi \rightarrow \pi^*$ *transitions and the long wavelength absorption to a* σ *electron transition.* Several descriptions of the σ transition are possible which are roughly equivalent. The transition might be designated as $\sigma \rightarrow \pi^*$ in which the highest filled σ orbital is primarily the localized metal–carbon bond, or alternatively the transition could be designated charge transfer from R_3M to the vacant antibonding π orbital. In valence bond terms these descriptions amount to labeling the excited state as primarily $(CH_3)_3M \cdot CH_2{=}CH{=}\dot{C}H_2 \leftrightarrow (CH_3)_3M^+CH_2{=}CH{-}CH_2^{(-)}$. In specific molecular orbital terms, assignment of the long wavelength transition to $\sigma \rightarrow \pi^*$ implies the assumption that the highest filled hyperconjugating oribtal of $[(CH_3)_3M(H_2)]{\equiv}C$ is *higher* in energy than the bonding π orbital. The $\sigma \rightarrow \pi^*$ or charge transfer assignment here is similar to that made for the short wavelength transition of the trialkylboranes. The greater intensity is easily rationalized on the basis of the involvement of the metal *d* orbitals in the lowest molecular π^* orbitals. It is also evident that, as formulated here, the $\sigma \rightarrow \pi^*$ (charge transfer) assignment will have similar geometric requirements as those set forth in our discussion of compound **92** and trimethylsilylcyclopentadiene. Perhaps the principal objection to assignment of $\sigma \rightarrow \pi^*$ for the long wavelength maxima of $R_3MCH_2C{=}C$ is that it is unnecessary, since the effect of R_3MCH_2 substitution on the spectrum of $C{=}C$ is similar to that for the spectra of $R_3MCH_2C_6H_5$, where there seems to be no question of the $\pi \rightarrow \pi^*$ character of the ^1L_b band near 260 mμ.

To summarize our discussion of the $R_3MCH_2C{=}C$ chromophore we can say the following. Allylorganometallic compounds of Group IV exhibit two absorption maxima in their spectra between 170 and 215 mμ. The long wavelength absorption is most probably $\pi \rightarrow \pi^*$, correlating with the $\pi \rightarrow \pi^*$ transition of neopentylethylene at 179 mμ. The short wavelength transition may be assigned to a transition involving promotion of metal–carbon σ electrons; $\sigma \rightarrow \pi^*$ is suggested. A reasonable case with some supporting

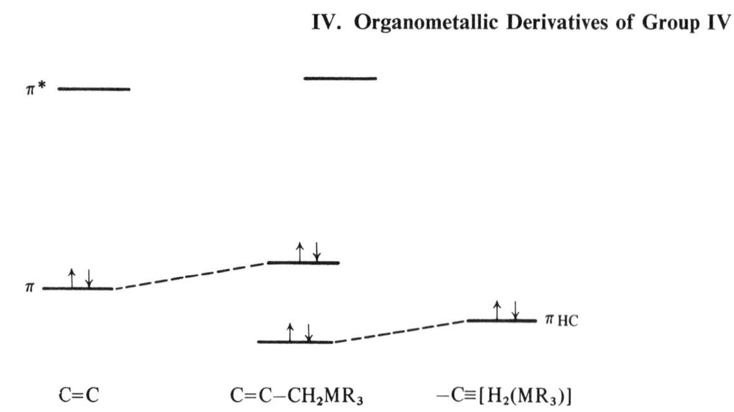

FIG. 29. A qualitative molecular orbital description for $R_3MCH_2CH{=}CH_2$ assigning the longest wavelength transition to $\pi \rightarrow \pi^*$.

evidence can be made for reversing these assignments, however. Figures 29 and 30 summarize the molecular orbital descriptions corresponding to the alternative assignments of the $\pi \rightarrow \pi^*$ and $\sigma \rightarrow \pi^*$ transitions.

The ultraviolet spectra of dialkyldi-2-propynylsilanes, -germanes, and -stannanes show [62] a pronounced spectral red shift to lower energy and increase in extinction coefficient in the order Sn > Ge > Si. The results therefore parallel those already discussed for allyl organometallic compounds.

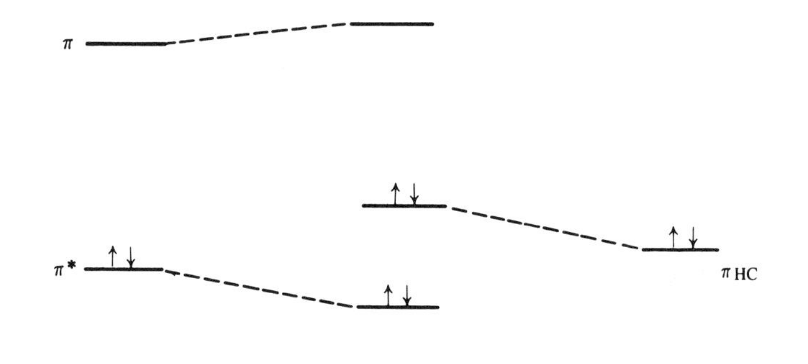

FIG. 30. A qualitative molecular orbital description for $R_3MCH_2CH{=}CH_2$ assigning the longest wavelength transition to $\sigma_\pi \rightarrow \pi^*$.

D. THE R_3M—C(X)R AND $R_3M(CH_2)_nC(X)R$ CHROMOPHORES

Since Brook [20] first noticed that benzoyltriphenylsilane was yellow in color and then published somewhat later with co-workers [26] the first study of the spectral properties of α-silyl ketones, a variety of explanations have been offered for the large red shift of the $n \rightarrow \pi^*$ transition in these and other Group IV metal ketones relative to their carbon analogs. Before examining any theoretical arguments relating to the spectra of Group IV metal ketones, however, it would be well to collect experimental facts.

Absorption maxima and extinction coefficients of compounds representative of the general structure $R_3'M—C(O)R$ are gathered in Table XII.

TABLE XII

ABSORPTION MAXIMA OF R_3MCOR'

Compound	λ_{max}, mμ (ϵ)
$(CH_3)_3CC(O)CH_3$	277 (23); 186 (1100)
$(CH_3)_3SiC(O)CH_3$	388 (93), 372 (126), 358 (100), 346 (62), 333 (34), 323 (18); 194 (4200)
$[(C_6H_5)_3Si]_2C{=}O$	554, 524, 478; 262–273
$(C_6H_5)_3SiC(O)CH_3$	363, 376 (406), 392; 266 (706); 195 (110,000)
$(C_6H_5)_3GeC(O)CH_3$	352, 365 (385), 380; 260 (1450); 192 (91,000)
$(C_6H_5)_3SnC(O)CH_3$	363 (sh), 375, 391
$(CH_3)_3SiC(O)C_6H_5$	424 (101); 252 (11,700); 198 (30,200)
$(C_6H_5)_3SiC(O)C_6H_5$	425 (295); 257 (16,200); 194 (162,000)
$(C_6H_5)_3SiC(O)C_6H_4OCH_3$	414 (335); 290 (20,800)

The most striking point is that the $n \rightarrow \pi^*$ transition of aliphatic ketones (λ_{max} $CH_3C(O)CH_3$, 279 mμ; ϵ 15) and aromatic ketones (λ_{max} $C_6H_5C(O)$-CH_3, 325 mμ; ϵ 50) which is usually weak, broad, and structureless, has become more intense, with considerable structure even in polar solvents, and, most important, the $n \rightarrow \pi^*$ transition energy is substantially lowered (λ_{max} $(C_6H_5)_3SiC(O)CH_3$, 376 mμ; λ_{max} $(C_6H_5)SiC(O)C_6H_5$, 425 mμ). The benzoyl compounds $XC_6H_4C(O)M(alkyl)_3$ also exhibit intense absorption maxima in their spectra, usually in the neighborhood of 250 mμ, which are easily identified as the slightly red shifted characteristic $\pi \rightarrow \pi^*$ transition of aromatic ketones, and which has also been assigned [96] to intramolecular charge transfer from ring to carbonyl. We can compare these absorption maxima with that in the spectra of $(CH_3)_3C(O)CC_6H_5$ at 240 mμ.

It is important that we are correct in our identification of the long wavelength transition of $R_3'MC(O)R$ as $n \rightarrow \pi^*$. It is well known that $n \rightarrow \pi^*$ transitions are shifted to a *higher* energy by a change to more polar solvent whereas the $\pi \rightarrow \pi^*$ transition moves to a *lower* energy. From Table XIII we see that the proposed $n \rightarrow \pi^*$ transitions of $(C_6H_5)_3SiC(O)C_6H_5$ and $(C_6H_5)_3GeC(O)C_6H_5$ are blue shifted by increasing solvent polarity and in the same order as the solvent effect on the $n \rightarrow \pi^*$ transition of $(C_6H_5)_3C(O)C_6H_5$ [2]. As expected, the $\pi \rightarrow \pi^*$ 250 mμ transitions of

TABLE XIII

SOLVENT EFFECT ON FIRST TRANSITION ENERGY OF $(C_6H_5)_3C(O)C_6H_5$

Compound	λ_{max} (cm^{-1})		
	Hexane	Ethanol	Acetonitrile
$(C_6H_5)_3CC(O)C_6H_5$	29,650	30,350	30,830
$(C_6H_5)_3SiC(O)C_6H_5$	23,550	23,860	23,880
$(C_6H_5)_3GeC(O)C_6H_5$	23,920	24,310	24,350

$(C_6H_5)_3MC(O)C_6H_5$ move to lower energy with a change in solvent from hexane to ethanol. Further, Gaussian analysis [2] of the band structure of the $n \rightarrow \pi^*$ transition confirms the idea that the maxima are vibrational components of the same transition. The components are separated by about 1000 cm^{-1}, which corresponds to the expected C—O stretching frequency in the excited state.

In spite of the large effect on the $n \rightarrow \pi^*$ transition energy if a Group IV metal is α substituted for carbon, the effect of the transition energy of changing the substituent on the metal or even the metal itself is comparatively small. Thus, we have $(C_6H_5)_3SiC(O)CH_3$ with λ_{max} of 363, 376 and 392 mμ and $(C_6H_5)_3GeC(O)CH_3$ with λ_{max} of 352, 365, and 380 mμ, or $C_6H_5C(O)$-$Si(CH_3)_3$ with λ_{max} of 402 and 413 mμ and $C_6H_5C(O)Si(C_6H_5)_3$ with λ_{max} of 403 and 417 mμ. Substituent changes of X in the benzoyl phenyl of $XC_6H_4C(O)MR_3$ have a larger effect, of course. The $n \rightarrow \pi^*$ transition of *para*-substituted phenyl silyl [102] and phenyl germyl ketones [69] is moved to lower energies by electron withdrawing groups and to a higher energy by electron-donating substituents. Normally the correlation of electronic transition energies with Hammett substituent constants is not particularly good since the substituent changes not only bonding but also antibonding molecular orbital energies. However, since a substituent X in the *para* position of $XC_6H_4C(O)MR_3$ should have little effect on the nonbonding oxygen electrons, we could hope that the $n \rightarrow \pi^*$ transition energy would correlate with σ_{para}. The expectation is that the Hammett σ constant will reflect the perturbation of the substituent on the antibonding π^* orbital. Brook *et al.* found [22] that a linear correlation of the $n \rightarrow \pi^*$ wavelength maximum with Hammett σ or $\sigma^{(-)}$ was fair for *para*-substituted benzoylsilanes. A much better correlation is found if linear energy units are used, however, as may be seen from Fig. 31 in which the $n \rightarrow \pi^*$ transition energy in units of reciprocal centimeters times 10^{-3} is plotted against the Hammett σ_{para}, except for NO_2 where $\sigma^{(-)}$ is used. The only point which deviates significantly from the plot is that of the *tert*-butyl group for which a rationale may be

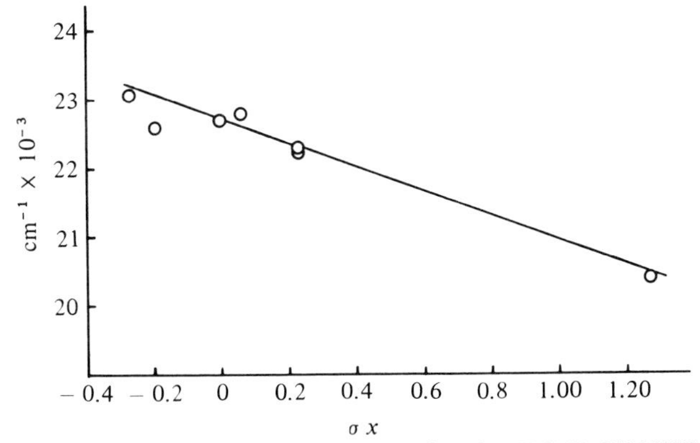

FIG. 31. The $n \rightarrow \pi^*$ transition energy (cm^{-1}) of p-$XC_6H_4C(O)Si(C_6H_5)_3$ vs. σ_X (σ^- where X is NO_2).

extended. It is often suggested that part of the substituent effect of the *tert*-butyl group is steric hindrance to transition state solvation; however, since the only effect on the energy of antibonding π orbital by *tert*-butyl should be an inductive one, it can easily be argued that the appropriate σ constant to be used here would be an inductive $\sigma_I = \frac{2}{3}\sigma_{meta}$ in which case *tert*-butyl no longer deviates significantly from the correlation.

The chromophore M—C=O is an example of the M—A—B—C West [138] Class II chromophore with π bonding at atom A, but lone pairs only on atoms B or C. A simple energy level diagram showing the expected inductive and d orbital resonance effects of a metalloid substituent on the

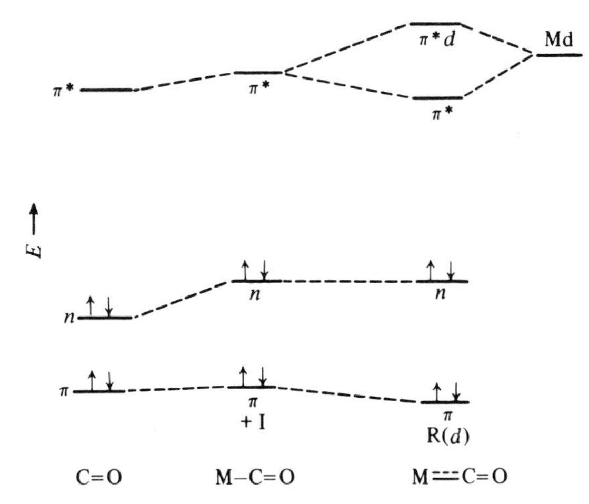

FIG. 32. A qualitative molecular orbital diagram showing the effect of a metal atom and its d orbitals on an M—C=X system such as C=O.

molecular orbital energies of a carbonyl group is given in Fig. 32. The inductive effect of R_3M on C=O is to raise all π, n, and π^* orbitals in energy but to increase the nonbonding level the most. Thus, the inductive effect itself would be expected to lower the $n \rightarrow \pi^*$ transition energy. Further interaction with the vacant metal n d, or $(n + 1)$ p, orbital should leave the nonbonding electron essentially unchanged and lower both the π^* and π orbitals in energy, with the greatest effect on the antibonding orbital.

The question involves whether the red shift of $n \rightarrow \pi^*$ transition of R_3M—C(O)—R ketones, where M is Si, Ge, or Sn, is due to primarily inductive, to a *d* orbital effect, or to a combination of the two. Some time ago Harnish and West [69] suggested the importance of *d* orbital interaction with the antibonding π^* orbital, whereas Orgel suggested [103] the importance of inductive effects.

The experimental evidence now available all seems to point to the importance of the inductive effect. Yates and Agolini [139] found that the order of base strength of the ketones $(C_6H_5)_3MC(O)C_6H_5$ was in the order M = Si > Ge ≫ C. This coupled with the fact that the carbonyl stretching frequencies of $R_3MC(O)R'$ are 40–70 cm^{-1} lower than the corresponding carbon compound emphasize the predominant importance of the inductive effect as expressed by the contributing valence bond structure **93** on the *ground state* properties. Note that *d* orbital interaction with the bonding π orbital, as expressed in valence bond structure **94**, would account for the decreased carbonyl stretching frequency but would also *decrease* the basicity of the oxygen lone pair electrons. Since we expect that the importance of *d–p* π bonding should decrease in the order Si > Ge > Sn, the absence of a substantial blue shift in the $n \rightarrow \pi^*$ transition [104] of $(C_6H_5)_3MC(O)CH_3$ in the order Sn > Ge > Si also led Peddle to suggest the primary importance of the metal inductive effect on the transition energy.

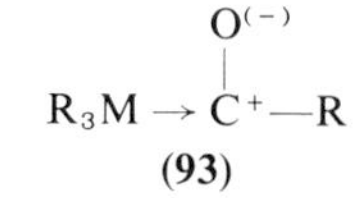

$$R_3M \rightarrow \overset{O^{(-)}}{\overset{|}{C^+}}—R \qquad \textbf{(93)}$$

$$R_3\overset{(-)}{M}{=}\overset{O^+}{\overset{|}{C}}—R \qquad \textbf{(94)}$$

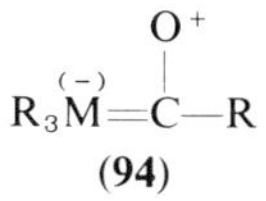

Important evidence for the *absence* of silicon *d–p* π interaction with the carbonyl chromophore has recently been claimed by Brook and co-workers [21]. The evidence rests on the important point that the delocalized antibonding *d–d* π molecular orbital of polyalkylsilanes, $R(SiR_2)_n$—R, is thought to interact strongly with the C=C chromophore. (Evidence for delocalized molecular orbitals in polysilanes which can be constructed from either *d* or *p–d* hybrids was given in the introduction of this section and in Section III-G.) Brook *et al.* note that the spectrum of $[(CH_3)_3Si](CH_3)_2SiCH{=}CH_2$ (λ_{max} 223 mμ) is substantially red shifted compared to $(CH_3)_3SiCH{=}CH_2$ (λ_{max} 202 mμ). The first transition of $[(CH_3)_3Si](CH_3)_2SiCH{=}CH_2$ is

$\sigma \rightarrow \pi^*$, however, whereas that of $(CH_3)_3SiCH{=}CH_2$ is most probably $\pi \rightarrow \pi^*$, and the red shift of the vinylsilane spectrum should be compared with that of $CH_3[Si(CH_3)_2]_2CH_3$ (λ_{max} 197.1–199.5 mμ). The red shift of the spectrum of $CH_3[SiR_2]_nR'$ on substitution by vinyl or phenyl is a general one and has the possible explanation that the group antibonding "d_π" molecular orbital, which would of course be lower in energy than the single vacant metal *d* atomic orbital, interacts strongly with a C=C π molecular orbital. The important result obtained by Brook *et al.* is that the $[(CH_3)_3SiSi(CH_3)_2]$ or $[(C_6H_5)_3SiSi(C_6H_5)_2]$ groups have *no* greater effect on the $n \rightarrow \pi^*$ transitions of the —C(O)CH_3 or —C(O)C_6H_5 chromophore than R_3Si—. For example, the absorption maximum of $(CH_3)_3SiC(O)CH_3$ (λ_{max} 372 mμ) is about the same as $(CH_3)_3SiSi(CH_3)_2C(O)CH_3$ (λ_{max} 370 mμ). The absence then of evidence for extended *d* orbital conjugation in Si—Si—C=O systems would almost certainly preclude any important role for Si, or any Group IV metal *d* orbital *d–p* π interaction for the monosilyl ketones.

Molecular orbital treatments seemed to support the conclusion by Brook. West [138] had estimated the π^* orbital of C=C to be about 0.8 eV below that of the Si 3*d* orbital. Proceeding in a similar fashion and using West's estimate of 1.95 eV for the energy of the Si 3*d* atomic orbital, one finds from the ionization potential of $H_2C{=}O$ (10.85 eV) and the $\pi \rightarrow \pi^*$ transition energy of $CH_3C(O)CH_3$ (1540 Å, 8.0 eV) [86] that the antibonding orbital of a carbonyl is estimated to be at least 1 eV lower than the 3*d* orbital. A different assignment of the $\pi \rightarrow \pi^*$ transition energy to say the 185 mμ absorption of formaldehyde estimates the 3*d*–π^* energy difference to be closer to 3 eV [2]. The conclusion was therefore that silicon *d* orbital conjugation must be expected to be much less in Si—C=O than it was in Si—C=C where it was already minimal.

Yates, Agolini, and co-workers [2] have carried out several interesting molecular orbital calculations on the α-silyl and α-germyl ketones as follows: (1) a simple Hückel calculation using the heteratom parameters of Curtis and Allred [37], where $\alpha_M = \alpha_O + h_M\beta^\circ$, $h_{Si} = -1.20$, and $h_{Ge} = -1.05$, and resonance integrals $\beta_{CM} = k_{CM}\beta$, where $k_{C-Si} = 0.45$, $k_{C-Ge} = 0.30$; (2) a simple Hückel molecular orbital calculation which incorporated both the ω self-consistent charge technique and the more recent self-consistent bond order approach of Boyd and Singer [18], using the expression $\beta_{rs} = \beta_{rs}^\circ \exp[0.55\,\rho_{rs} - 0.3666]$, where ρ_{rs} is the bond order; (3) an extended Hückel calculation of the type developed by Hoffmann [71] in which the valence state ionization potential of silicon was chosen to maximize silicon *d* orbital participation as much as possible.

The models chosen for the calculations were C—O, Si—C—O, Ge—C—O, C_6H_5—C—O, C_6H_5—C(Si)O, and $C_6H_5C(Ge)O$. The Hückel calculations show a decrease in the energy of π^* orbital on metal substitution and the

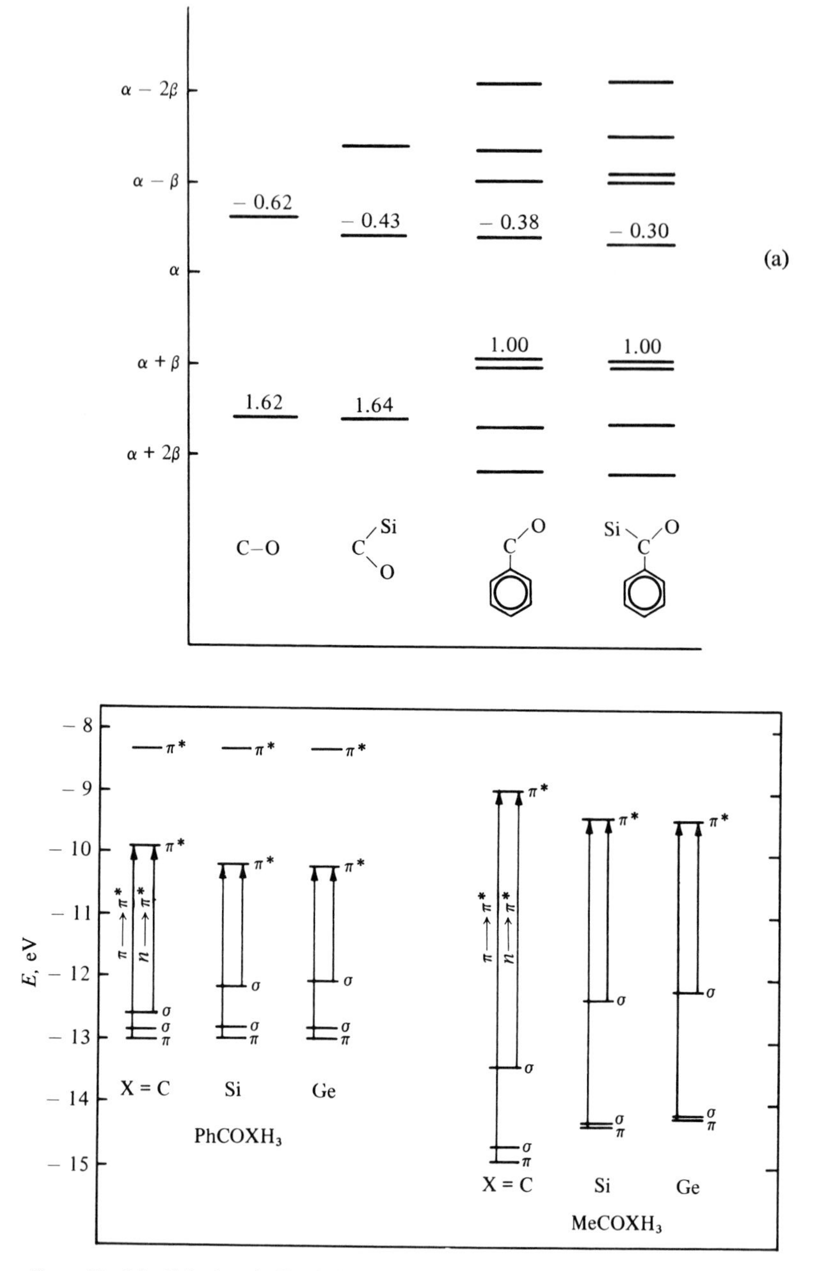

FIG. 33. (a) Calculated Hückel molecular orbital energies for C—O, Si—C—O, C_6H_5—C—O, and $C_6H_5C(O)Si$ [2]. (b) Schematic representation of extended Hückel energies of highest filled and lowest vacant molecular orbitals for $C_6H_5C(O)XH_3$ and $CH_3C(O)XH_3$ [2].

overall result can be summarized in Fig. 33. The correlation of calculated transition energies with the observed $\pi \rightarrow \pi^*$ transition energies is good, provided the 185 mμ aliphatic ketone transition is assigned to $\pi \rightarrow \pi^*$. The $n \rightarrow \pi^*$ transitions fall into a family of two lines differing by about 3000 cm^{-1} ($\sim$0.4 eV). That is, the effective coulomb integral of the nonbonded oxygen orbital of $(R_3C)_2C{=}O$ is some 0.4 eV lower in energy than that of $R_3MC(O)CR_3$. From this we can conclude that in the Hückel approximation the inductive effect is sufficient to account for one-third to one-half of the experimental red shift.

The results of the extended Hückel calculations on $C_6H_5C(O)MH_3$ (M is C, Si, Ge) which included all of the valence shell orbitals and therefore both π and σ results are especially interesting and are presented in Fig. 33b. First, the major effect of the metal is to increase the energy of the nonbonding electrons (of σ symmetry) substantially. Second, contributing to the red shift is a smaller lowering of the π^* orbital *but the surprising result is that the same lowering of the π^* level is obtained to within less than 0.1 eV, even if the d orbitals are omitted from the calculation.* The calculated transition intensities also increase independently of whether *d* orbitals are included.

More recently Bock *et al.* [15a] have put together from a variety of experimental measurements a detailed formulation of the relative changes in π, π^*, and *n* orbitals of $RC(O)M(CH_3)_3$, when M is changed from C to Si. First, measurement of the vertical ionization potential of the carbonyl oxygen nonbonded electrons in $C_6H_5C(O)CH_2Si(CH_3)_3$ (7.96 eV), $C_6H_5C(O)Si(CH_3)_3$ (8.20 eV), and $C_6H_5C(O)C(CH_3)_3$ (9.04 eV) shows the inductive effect on the oxygen lone pair by replacing C with Si to be an increase of 0.84 eV in the oxygen lone pair energy. Next, measurement of the half-wave reduction potentials of $C_6H_5C(O)C(CH_3)_3$ (−1.57 eV) and $C_6H_5C(O)SI(CH_3)_3$ (−1.30 eV) demonstrate a lowering of the π^* orbital of the trimethyl silyl ketone relative to the *tert*-butyl phenyl ketone of 0.27 eV. A small but significant *d*-orbital interaction of Si even with a π bonding orbital is demonstrated by the relative charge transfer transition energies for the tetracyanoethylene π complexes of 2-naphthyl methyl ketone (λ_{max} 19,700 cm^{-1}) and 2-naphthyl trimethylsilyl ketone (λ_{max} 21,700 cm^{-1}). It can probably be safely assumed that this transition represents charge transfer for the highest filled π orbital of the ketone. These experimental results are combined in Fig. 34 from which one derives the conclusion that roughly 25% of the $n \rightarrow \pi^*$ transition red shift of silyl ketones relative to alkyl ketones is due to a *d–p* π^* bonding interaction.

The conflict between the conclusions of Bock *et al.* [15a], Brook *et al.* [21], and Yates and Agolini [139] requires some comment. The failure of the extended Hückel calculations would not be too surprising; in fact it is more surprising if they work. More important is the absence of a large red shift

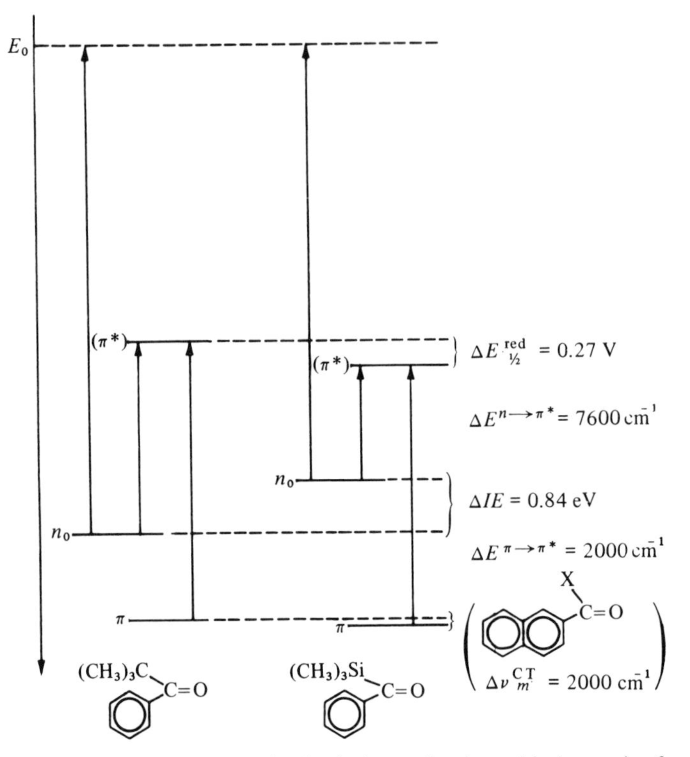

FIG. 34. Experimentally determined relative molecular orbital energies for RC(O)C-$(CH_3)_3$ and RC(O)Si$(CH_3)_3$ [15a].

in the $n \rightarrow \pi^*$ transition of Si—Si—C═O relative to Si—C═O if *d*-orbital effects are important for the monosilyl ketone. One explanation which seems to account for all of the data for polysilanes, α,ω-vinyl polysilanes, and -silyl ketones is as follows: the red shift in the spectrum of CH_2═CH—$(SiR_2)_n$—CH═CH_2 is not a "molecular orbital effect" but must be explained in terms of configuration interaction between olefin $\pi \rightarrow \pi^*$ locally excited states and polysilane locally excited states. In the case of the Si—Si—C═O systems the $n \rightarrow \pi^*$ and —Si—Si— excited states differ by too great an energy for configuration interaction to be important even if the states were of the appropriate symmetry. If this is indeed the case, then we would expect a red shift of the $\pi \rightarrow \pi^*$ transition of $(CH_3)_3$SiSi$(CH_3)_2$—C(O)C_6H_5 relative to R—C(O)C_6H_5 comparable to but smaller than that found for the corresponding styrenes; or better perhaps, since R—C(O)C_6H_5 is a cross-conjugated system, a red shift in the $\pi \rightarrow \pi^*$ transition a simple ketone Si—Si—C(O)-alkyl should be looked for. The apparent absence of significant configuration interaction in the spectrum of $(C_6H_5)_3$SiSi$(C_6H_5)_2$—

$C(O)C_6H_5$ (λ_{max} 244 mμ) between the $(C_6H_5)_3SiSi(C_6H_5)_2$— transition (λ_{max} 245 mμ) and the R—$C(O)C_6H_5$ transition (λ_{max} 247 mμ) is perhaps not a fair test because the 245 mμ transition excited state of $(C_6H_5)_3Si$—$Si(C_6H_5)_3$ may be largely benzene ^1L_a transition excited states and therefore interact only weakly with the $C_6H_5C(O)$ $\pi \rightarrow \pi^*$ state because of intervening Si.

An alternative hypothesis is that lone pair bonding of the oxygen with the *d* orbitals of the second silicon in the Si—Si—C═O system by lowering the energy of the oxygen lone pair fortuitously cancels a similar decrease in the π^* energy by *d–d–p* π^* bonding.

The $n \rightarrow \pi^*$ transitions of $(C_6H_5)_3SnC(O)CH_3$, $(C_6H_5)_3SnC(O)C(CH_3)_3$, and $(C_6H_5)_3SnC(O)C_6H_5$ have been compared with those of the corresponding Ge, Si, and C derivatives. The most intense peak of the $n \rightarrow \pi^*$ transition multiplet of the silyl aliphatic ketones is 1 mμ to the red of the stannyl aliphatic ketone and Peddle suggests this result may reflect a *d–p* π^* interaction of Si which is lacking in the $(C_6H_5)_3MC(O)C_6H_5$ series where the energy matching between the π^* and Si *d* orbitals is poorer (λ_{max} $(C_6H_5)_3SiC(O)C_6H_5$ 424 mμ; λ_{max} $(C_6H_5)_3SnC(O)C_6H_5$ 435 mμ) [104a].

We have earlier noted that allyl derivatives of the Group IV metals absorb at longer wavelengths than the vinyl compounds. Possible explanations for this involved metal *d* orbitals and/or hyperconjugation of the form $R_3M^+CH_2$═CH—$^{(-)}CH_2$. By contrast the $\pi \rightarrow \pi^*$ transition [24] of $(C_6H_5)_3SiCH_2C(O)C_6H_5$ (λ_{max} 243 mμ) has a smaller red shift compared to $(C_6H_5)_3CCH_2C(O)C_6H_5$ (λ_{max} 241 mμ) than does $(C_6H_5)_3Si(CH_2)_2C(O)C_6H_5$ (λ_{max} 242 mμ) relative to $(C_6H_5)_3C(CH_2)_2C(O)C_6H_5$ (λ_{max} 235 mμ). It would be better if we had data for the $\pi \rightarrow \pi^*$ transitions of some of the aliphatic ketones R_3M—CH_2—C(O)(alkyl), but certainly the mechanism operative in the spectra of the allyl- and benzylsilanes does not appear operative here. Since hyperconjugation of the type $R_3M^+CH_2$═C—$O^{(-)}$, should be at least as effective as $R_3M^+CH_2$═C—$C^{(-)}$, the implication is that a part of the red shift of the chromophore M—CH_2—C═CH relative to M—C═CH_2 requires some other explanation such as *d*-orbital involvement, which is absent in its effect on the carbonyl chromophore (see Addendum).

Adopting an earlier suggestion of Brook *et. al* [26], Musker and Ashby have interpreted the greater intensity of the $n \rightarrow \pi^*$ transition of β-silyl ketones such as $(CH_3)_3SiCH_2C(O)CH_3$ as evidence of *d–p* π bonding of the silicon *d* orbitals with the oxygen lone pair electrons which are closer in energy to *d* orbitals than the π electrons [92, 93]. We have already discussed the evidence for this type of bonding in the spectra of phenyl silyl ethers. It is suggested that metal *d* orbital perturbation destroys the symmetry of the nonbonding orbital and allows the previously symmetry forbidden $n \rightarrow \pi^*$ transition greater intensity. This is in keeping with the observed blue shift of the $n \rightarrow \pi^*$ transition of $R_3MCH_2C(O)R'$ relative to $R_3CCH_2C(O)R'$.

The $n \rightarrow \pi^*$ transition of $(CH_3)_3SiCH{=}NC_6H_{11}$, another example of West's Class II chromophore, is found [119] at 285 mμ (ϵ 70). The $n \rightarrow \pi^*$ transition of $C_6H_5CH{=}NC_6H_5$ is near 360 mμ [73], but this is not a fair comparison because of the extended conjugation provided by the phenyl nitrogen. The $n \rightarrow \pi^*$ transitions of alkylketimines $R_2C{=}N{-}R$ are reportedly [136] near 230 mμ; therefore, the effect of α-silyl substitution on the $n \rightarrow \pi^*$ transition of C=N–R appears comparable to its effect on the C=O chromophore.

A third possible example of a Class II chromophore is the trimethyltin-substituted diazomethane (**95**). Although evidence was not definitive, Lappert and Lorberth preferred the alternative structure **96**. The compound has a very strong absorption at 280 mμ with a weaker one near 396 mμ. Structure **96** is isoelectronic with trimethylstannylazide (**97**), λ_{max} 230 mμ (ϵ 313). This 230 mμ absorption hides two transitions, a $\pi_y \rightarrow \pi_x{}^*$ and a $n_{sp^2} \rightarrow \pi^*$ transition. Since the change in the Coulomb integral on substitution of —CH for N_2 should *raise* both the "nonbonding" $\pi_y{}^*$ and the π_x orbitals, with the largest effect on $\pi_x{}^*$, we would predict a small blue shift in the $\pi \rightarrow \pi^*$ transition energies. Compare, for example, the spectrum of pyridine (λ_{max} 252 mμ) with benzene (λ_{max} 254 mμ). The $n_{sp^2} \rightarrow \pi^*$ transition would show a larger *blue* shift, which leads one to believe that the ultra-

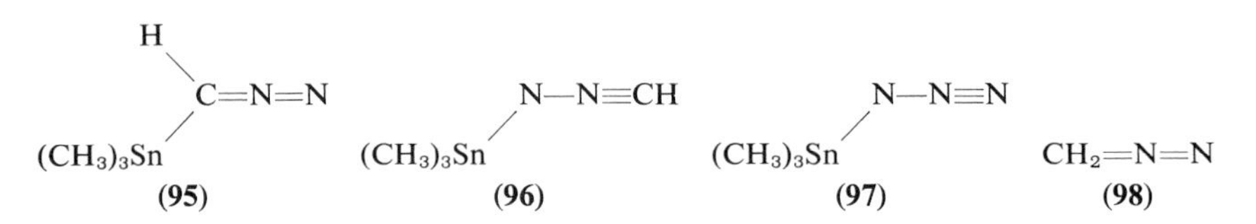

violet spectrum of the compound prepared by Lappert and Lorberth may fit structure **95** better. Diazomethane [90] (**98**), in addition to a very weak broad band near 415 mμ ($\epsilon \leq 10$), has a stronger transition near 220 mμ (ϵ 8000) which is probably one of the allowed $\pi \rightarrow \pi^*$ transitions. It is at least not unreasonable that the combined inductive effect and *d–p* π overlap of the Sn may shift this transition to 280 mμ in the compound $(CH_3)_3SnCHN_2$.

Before leaving the subject we would like to mention a secondary but perhaps important effects which would contribute to the red shift of $R_3MC(R){=}X$— chromophores and which to our knowledge has not been considered before in this respect. Molecular orbital calculations [117] indicate significant interaction between the lone pair nominal "sp^2" orbitals of R—N=N—R to form "bonding" and "antibonding" orbitals. It has also been suggested [13, 87] that the $n \rightarrow \pi^*$ transition of ketones contains a significant contribution of electron density from C—R σ orbitals in the nonbonding molecular orbital. These results are not too surprising if we realize that the overlap between the two sp^2 orbitals is about 50% of that of two parallel p orbitals. The suggestion we would like to bring forth is that in the ketones $R_3M—C(O)R$, for example, as the metal carbon bond σ energy decreases and approaches more closely that of the oxygen lone pair electrons (either in sp^2 or p atomic orbitals), it raises the energy of the nonbonding molecular orbital containing the electron lone pair, resulting in an additional red shift superimposed on the larger inductive red shift. The extent of this effect might be obtained experimentally from the spectra of compounds such as *cis*- and *trans*-$R_3MC(R)=N—R$.

E. Metal Esters and Azides

The third general class of chromophores considered by West in his qualitative molecular orbital approach was the system MA—B=C with lone pair electrons on atom A but π bonds only between B and C.

Organometallic carboxylate esters provide the most common example of this type of chromophore but few data are available. Since the lowest $n \rightarrow \pi^*$ transition of $R—CO_2R'$ is from the carbonyl oxygen, substitution of a metalloid is predicted by simple molecular orbital terms to have little effect on this or the $\pi \rightarrow \pi^*$ transition. Gornorvicz and Ryan [61] have provided some evidence that this analysis is correct. The ultraviolet spectra of $CH_3C(S)OSi(CH_3)_3$ (λ_{max} 245, ϵ 4570) and $(CH_3C(S)O)_2—Si(CH_3)_2$ (λ_{max} 243 mμ, ϵ 4400) are about the same as $CH_3C(S)OC_2H_5$ (λ_{max} 241 mμ, ϵ 9600).

Somewhat more attention [32] has been given the azido derivatives of the Group IV organometallic compounds R_3MN_3. The azide ion itself, N_3^-, has a weak absorption at λ_{max} 220 mμ (ϵ 430) and strongly rising end absorption to 190 mμ. Following the assignments of Closson and Gray, the molecular orbital energy level diagram is given qualitatively for N_3^- by Fig. 35 in which each π orbital is doubly degenerate. The lowest energy transition then is $\pi_g \rightarrow \pi_u^*$ which is quadruply degenerate in this simple approximation in the same fashion as the first $\pi \rightarrow \pi^*$ transition of benzene, discussed in Chapter I. After configuration interaction, the possible excited states are of symmetry ${}^1\Sigma_u{}^+$, ${}^1\Sigma_u{}^-$, and ${}^1\Delta_u$. The ${}^1\Sigma_g{}^+ \rightarrow \Sigma_u{}^+$ is the only allowed transition, corresponding to the nonzero transition dipole moment coupling of the $\pi \rightarrow \pi^*$ transitions of each set of perpendicular π orbitals. The weak transition of N_3^- at 220 mμ can be assigned to the ${}^1\Sigma_g{}^+ \rightarrow {}^1\Delta_u$ state which is doubly degenerate and corresponds roughly to promoting an electron from one π orbital to the perpendicular π^* orbital. This transition will be lower in energy because of more favorable electron spacial correlation terms (see Section I-A).

In the spectra of the R_3MN_3 compounds (M is C, Si, Ge, Sn, Pb) the de-degeneracy of ${}^1\Sigma_g{}^+ \rightarrow {}^1\Delta_u$ transition is removed and for alkyl azides two transitions are observed at about 216 mμ and 287 mμ which were respectively

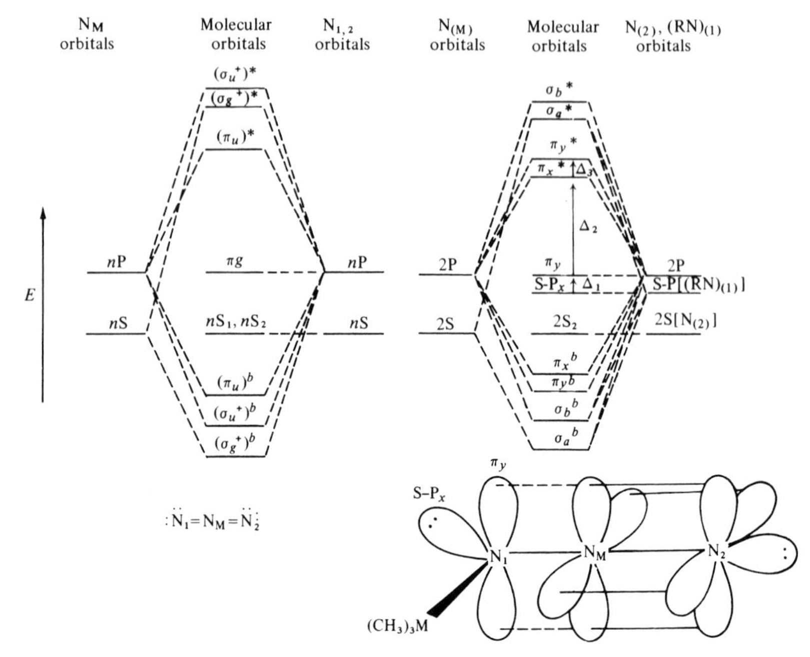

FIG. 35a. Simple molecular orbital correlation diagram of N_3— and RN_3—, *excluding d* orbitals.

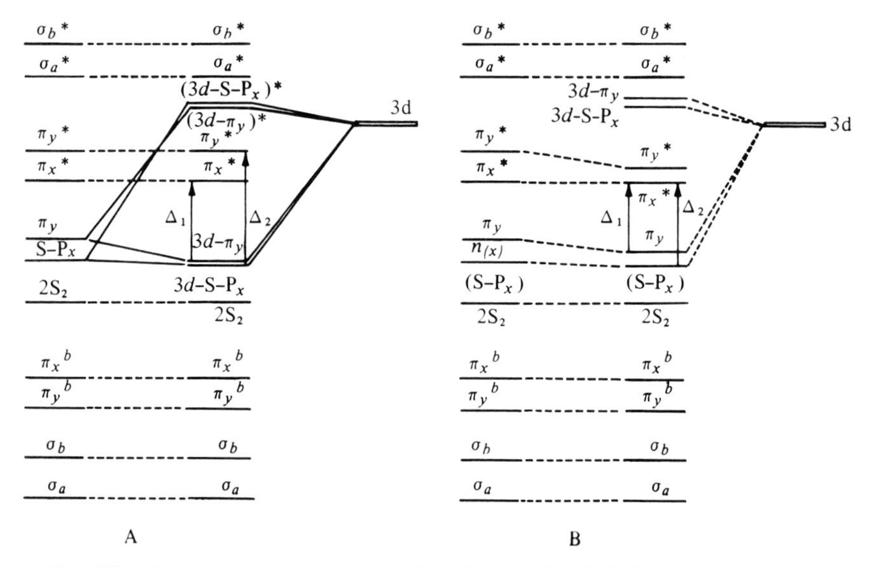

FIG. 35b. A: A simple molecular orbital diagram for R_3MN_3 with predominant d orbital interaction with the p_y lone pair. B: A simple molecular orbital diagram for R_3MN_3 with predominant d orbital interaction with π_y* molecular orbital.

assigned by Thayer and West [133] to the $n_{sp^2}(x) \rightarrow \pi_y$ and $\pi_y \rightarrow \pi_x{}^*$ transitions of structure **99** by Closson and Gray. We would however like to offer an alternative assignment of the 216 mμ transition based on additional data by Thayer and West.

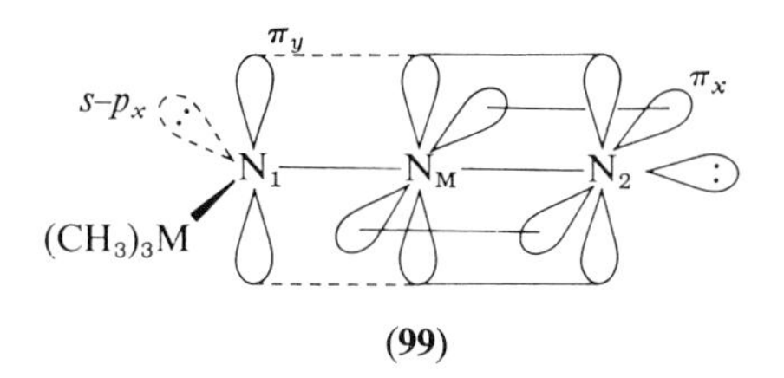

(99)

The simple molecular orbital energies of RN_3 are qualitatively given in Fig. 35. Note that this correlation diagram predicts that both the $\pi_y \rightarrow \pi_y{}^*$ and the $n_{sp^2} \rightarrow \pi_x{}^*$ transitions should be at lower energy than the $n_{sp^2} \rightarrow \pi_y{}^*$ transition. An explanation for this type of failure of the simple molecular orbital diagram based on singlet–triplet transition energy separation and spatial overlap of the molecular orbitals involved has been outlined by Jaffe and was discussed in Chapter I. While these considerations remain relevant with regard to the perpendicular $\pi_y \rightarrow \pi_x{}^*$ type transition compared to the "$\pi_y \rightarrow \pi_y{}^*$" type transition in the azide ion, or alkyl azide, it is more difficult to argue that singlet $n_{sp^2}(x) \rightarrow \pi_x{}^*$ should be at a higher energy than singlet $n_{sp^2}(x) \rightarrow \pi_y{}^*$ on these grounds. The procedure used by Closson and Gray to estimate the energy of the $n_{sp^2}(x) \rightarrow \pi_y$ might be used as well to assign the transition to $n_{sp^2}(x) \rightarrow \pi_x$. Indeed it is not unreasonable that the 2160 Å band hides *both* transitions. Basically, therefore, Closson and Gray appear simply to have made the reasonable assumption that the observed transitions of the alkyl azide are correlated with the observed degenerate transition of the azide ion. However the $n_{sp^2}(x) \rightarrow \pi_x{}^*$ and $n_{sp^2}(x) \rightarrow \pi_y{}^*$ transitions are of different symmetry and there is no reason why they should not cross in a correlation diagram between the states of RN_3 and $N_3^{(-)}$ in which the $n_{sp^2}(x) \rightarrow \pi_x{}^*$ occurs at lower energy.

Thayer and West [133] have measured the ultraviolet spectra of a number Group IV organometallic azides and the results of some of these are given in Table XIV. The most significant result of substituting silicon or germanium for carbon in $(CH_3)_3MN_3$ is a 0.5 eV *blue* shift of the transition assigned by Closson and Gray to $\pi_y \rightarrow \pi_x$ and a similar but smaller blue shift of the long wavelength transition. Because there seems no way to invoke an inductive mechanism to give a blue shift of the spectrum, Thayer and West made the effective assumption that the highest filled orbital of π_y symmetry was essentially a localized $2p$ pair of electrons on the substituted nitrogen. With this assumption, the observed blue shift may be accounted for by dative back bonding of nitrogen lone pair electrons through the metal d orbitals which would increase the $n \rightarrow \pi^*$ transition energy as summarized by the molecular orbital diagram of Fig. 35. By assuming the ground state structure $R_3M—\ddot{N}—N$ the metal d orbitals lower the $P(\pi_y)$ and $n_{sp^2}(x)$ orbital energies but leaves the $\pi_y{}^*$ orbital essentially unchanged. According to Thayer and West, in the absence of any d orbital participation as in lead and tin azides, the inductive effect of the metal raises the energy of the π_y and $n_{sp^2}(x)$ orbitals, causing a red shift of the $sp_x \rightarrow \pi_y$ transition which buries the longer wavelength $\pi_y \rightarrow \pi_x$ maximum. No explanation for the greater intensity of the $(CH_3)_3PbN_3$ spectrum could be offered.

Aside from a general intuitive disbelief that the $\pi_y{}^*$ orbital of structure **99** will not have a sufficiently large N, mixing coefficient such that any d orbital interaction with $\pi_y{}^*$ would be greater than with π_y, we suggest that the *absence*, on changing solvent from isooctane to hydrogen bonding solvents such as methanol, of a substantial blue shift in the observed transitions of alkyl azides or $(CH_3)_3PbN_3$, in view of the strong hydrogen bonding to be expected for a structure such as $R—N^{(-)}—N^+{\equiv}N$ precludes such an assumption. Closson and Gray also suggest that the p_y orbital is sufficiently involved in the π_y molecular orbital that it is not available for hydrogen bonding.

We suggest rather that the way out of these difficulties is to assign the short wavelength transition of the azides to $n_{sp^2}(x) \rightarrow \pi_x{}^*$ (under C_s symmetry $^1A' \rightarrow {}^1A'$ rather than $n_{sp^2}(x) \rightarrow \pi_y{}^*$ ($^1A' \rightarrow {}^1A''$). Reasonable coefficients of mixing of N_1 into the "nonbonding" π_y molecular orbital and $\pi_y{}^*$ will then allow lowering of the $\pi_y{}^*$, π_y, and $n_{sp}(x)$ orbitals through d orbital bonding and an increase in the $n \rightarrow \pi_x{}^*$ and $\pi_y \rightarrow \pi_x{}^*$ transition energies. In the absence of d orbital involvement inductive electron release will have its greatest effect on the nonbonding electrons and a smaller effect on π_y with the result that $n_{sp^2} \rightarrow \pi_x$ red shift will be much greater than that for the $\pi_y \rightarrow \pi_x{}^*$ transition. This approach is summarized in Fig. 35. An explanation for greater intensity of $(CH_3)_3PbN_3$ can now also be offered because as the metal nitrogen bond becomes more ionic and the spectrum more like the azide ion, there will be increased p_x character in the lone pair orbital, leading to better overlap with the $\pi_x{}^*$ orbital and therefore a more intense transition.

Isocyanates and isothiocyanates are isoelectronic with the azides, but are linear. Where *tert*-butylazide has two absorption maxima at 216 mμ (ϵ 500) and 288 mμ (ϵ 23) in its spectrum, *tert*-butylisothiocyanate possesses a single maximum at 248 mμ (log ϵ 3.05). This transition is blue shifted [134] to 238 mμ in the spectra of *all* $(CH_3)_3MNCS$, where M is Si, Ge, or Sn, which strongly supports the assignment of this transition to $n \rightarrow \pi^*$ from the sulfur lone pair electrons. The blue shift of the trimethyl organometallic isothiocyanates is apparently the result of an inductive effect of the $(CH_3)_3M$ group on the π^* orbital. It is easily seen from structure **99** for the azide ion where π_g is nodal at N_M, that replacement of $N_{(M)}$ and $N_{(2)}$ by elements each less electronegative than nitrogen must lead to a large blue shift of the azide ${}^1\Sigma_g{}^+ \rightarrow \Delta_u$ ($\pi_g \rightarrow \pi_u{}^*$) transitions. The ${}^1\Sigma_g{}^+ \rightarrow \Delta_u$ ($n_{sp^2} \rightarrow \pi$, $\pi_y \rightarrow \pi_x$) may be buried therefore in the band assigned to sulfur $n \rightarrow \pi^*$ transition.

A band of unknown origin in the spectra of trimethyltin and trimethyllead isocyanate at 275 mμ may represent the ${}^1\Sigma_g{}^+ \rightarrow \Delta_u$ transition red shifted by the greater electronegativity of oxygen.

The ultraviolet spectrum of the compound reported [84] as $(CH_3)_3SnNNCH$ [λ_{max} 280 (vs), 396 (w) mμ] does not seem to fit well with that of the isoelectronic azide and its structure was considered as $(CH_3)_3SnCHN_2$ in the previous section.

F. *N*-Ketimine- and Azoorganometallic Compounds

The first examples of an *N*-organometallic ketimine of the general structure **100** were synthesized by Kruger [79] and co-workers. These compounds such as *N*-(trimethylsilyl) benzophenoneimine (**100**) are found to have a weak $n \rightarrow \pi^*$ transition near 364 mμ in addition to a much stronger $\pi \rightarrow \pi^*$ transition near 244 mμ.

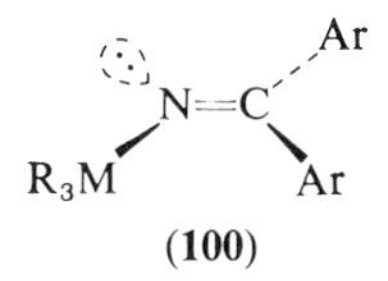

(100)

The *N*-metal ketimines and the trialkylsilylazobenzenes ($R_3SiN{=}NC_6H_5$; R is CH_3, C_2H_5, C_3H_7) whose spectra [80] have also been reported by Kruger and Wannagat, represent the last class of simple chromophores considered by West. In Class IV chromophores, $R_3M—A—B$, there is both π bonding at A and a lone pair of electrons. The inductive effect of the metal substituent on $—A{=}B$ is to raise the energy of the *n* electrons somewhat more than the π and π^* orbitals, which would also be raised in energy. The introduction of the metal *d* orbital will now act to *lower* the π^*, *n*, and π orbital energies in accord with the general principle that the closer the unperturbed orbital is in energy to the metal *d* orbital, the greater the net decrease in energy of the resultant molecular orbital will be. Again all of this may be summarized in a molecular orbital diagram such as Fig. 36. Perhaps it is worth reminding the reader again that since the *n* and π orbitals have different symmetry, different metal *d* orbitals will be used for overlap. The azo compounds are found to exist in both *cis* and *trans* isomers, and as indicated by structure **100** the ketimines are also most probably nonlinear. Apparently then in these compounds the nitrogens' approximate sp^2 hybridization and geometry

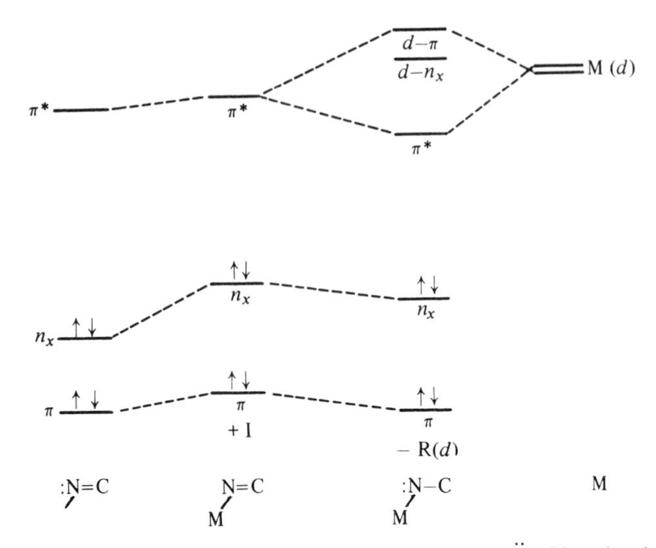

FIG. 36. A qualitative molecular orbital diagram for $R_3M—\ddot{X}=Y$ molecules such as $R_3M—\ddot{N}=CR_2$.

will not favor a large overlap of the metal *d* orbitals and the nitrogen lone pair, so that dative *d–p* π bonding should be small in these molecules.

Chan and Rochow [30] have now reported the ultraviolet spectra of some 36 $R_2C=N—MR_3$ compounds, most of which are benzophenoneketimine derivatives. Some of the data of Chan and Rochow are gathered in Table XIV, and forewarned with the knowledge that the large red shift of the

TABLE XIV

ELECTRONIC TRANSITIONS IN GROUP IV METAL *N*-KETIMINES AND AZIDES

Compound	λ_{max} $n \rightarrow \pi^*$ (mμ) (ϵ)	λ_{max} $\pi \rightarrow \pi^*$ (mμ) (ϵ)
$(C_6H_5)_2C=NSi(CH_3)_3$	364 (77.8)	243 (1.63 × 10^4)
$(C_6H_5)_2C=NSi(nC_4H_9)_3$	370 (93.5)	252 (1.64 × 10^4)
$(C_6H_5)_2C=NGe(CH_3)_3$	347 (86.7)	243 (1.59 × 10^4)
$(C_6H_5)_2C=NSn(CH_3)_3$	361 (94.3)	233 (1.38 × 10^4)
$(C_6H_5)_2C=NSi(C_6H_5)_3$	366 (157)	259 (2.38 × 10^4)
$(C_6H_5)_2C=NSi(CH_2C_6H_5)_3$	368 (114)	254 (1.86 × 10^4)
$(C_6H_5)_2C=NSi(CH_3)_2(OC_4H_9)$	365 (96.2)	250 (1.61 × 10^4)
$(CH_3)_3CN_3$	288 (33)	216 (500)
$(CH_3)_3SiN_3$	255 (19)	212 (260)
$(CH_3)_3GeN_3$	255 (23)	211 (252)
$(CH_3)_3SnN_3$	235 (313)	
$(CH_3)_3PbN_3$	249 (2700)	

$n \rightarrow \pi^*$ of $R_3M—C(O)R$ derivatives with respect to R—C(O)R is primarily an inductive effect, we can hope to rationalize in a consistent fashion the observed substituent effects on transition energy here.

First, we find that the energies of the $n \rightarrow \pi^*$ transition of the series $(C_6H_5)_2C{=}N—M(C_6H_5)_3$ decrease in the order C > Pb > Ge > Sn > Si, which is also the order of decreasing electronegativities as favored by Cotton and Wilkinson [33]. These results and the observation that substitution of strong π electron donors for CH_3 such as OC_4H_9 or $—N(C_2H_5)_2$ do not appreciably change $n \rightarrow \pi^*$ transition energy could be taken to rule out significant nitrogen *n–d* π bonding. Strong π electron donors competing with the *n* electrons of the ketimine nitrogen for metal *d* orbitals would be expected to lead to a red shift of the $n \rightarrow \pi^*$ transition by decreasing *n–d* π bonding, an effect which, if present at all, should be most important for silicon.

If *n–d* π bonding is negligible and if *d–p* π interaction between metal *d* and π^* orbitals is more important than inductive effects, the $n \rightarrow \pi^*$ transition energies should be in the order C $\geq$ Pb > Sn > Ge $\gg$ Si. The observation then of substantially lower $n \rightarrow \pi^*$ transition energies in $R_2C{=}NMR_3'$ where M is Sn rather than Ge must rule out a predominant role for metal *d* orbital for any other than possibly silylketimines.

The role of *d–p* π bonding should be made more significant by replacing methyls of $(C_6H_5)_2C{=}NSi(CH_3)_3$ with inductive electron withdrawal, which would decrease the *d* orbital energy and lead to an increased red shift of the $n \rightarrow \pi^*$ transition. In fact, as may be seen from Table XIV a blue shift is observed (compare, for example, λ_{max} $(C_6H_5)_2C{=}NSi(CH_3)_3$, 364 m$\mu$ with λ_{max} $(C_6H_5)_2C{=}NSi(CH_3)_2CHCl_2$, 360 m$\mu$). Further, Chan and Rochow find a linear relationship between Taft's inductive σ^* of the silicon substituent R and the $n \rightarrow \pi^*$ transition energy.

The conclusion is that provided the Cotton and Wilkinson electronegativity series is accepted, the available data on the $n \rightarrow \pi^*$ transition of Group IV metal organometallic ketimines offers no evidence of important *d* orbital interaction by the metal and the inductive effect of the metal must be of predominant importance. But, if the Allred-Rochow electronegativities are invoked, that is, C > Ge > Si > Sn > Pb, to obtain the observed order of transitions, it is necessary to invoke major *d* orbital interaction by Si and Ge and even significant participation by the Sn *d* orbitals in which the resultant red shifts of the $n \rightarrow \pi^*$ transition are the net effect of *d* orbital and inductive effects.

Chan and Rochow suggest that the enhanced intensity of $n \rightarrow \pi^*$ transition by C_4H_9O and $(C_2H_5)_2N$ substituents may provide evidence of *d–p* π bonding between silicon and nitrogen although it is recognized that steric factors may also be important. It is claimed that the intensity of the $n \rightarrow \pi^*$ transition will increase if the nitrogen is less involved in d–p π bonding with silicon and that such bonding is reduced by competitive bonding from the $(C_2H_5)_2\ddot{N}$— group (compare $(C_6H_5)_2C{=}NSi(CH_3)_3$, $\epsilon = 77.8$, with $(C_6H_5)_2C{=}NSi(CH_3)_2$-$[N(C_2H_5)_2]$, $\epsilon = 114$; but note also $(C_6H_5)_2C{=}NSi(CH_2C_6H_5)_3$, $\epsilon = 114$, and $(C_6H_5)_2C{=}NSi(\text{iso-}C_3H_7)_3$, $\epsilon = 101.7$).

The low energy of the $n \rightarrow \pi^*$ absorption of $(CF_3)_2C{=}NSi(CH_3)_3$ (λ_{max} 351 mμ) is sufficiently unusual to deserve comment. Chan and Rochow have suggested [30] that F hyperconjugation of type **101** may make an important contribution to the excited state. In ground states of molecules the possibility of fluorine hyperconjugation appears to have been pretty well discredited in favor of the simple inductive effect. It is still possible, however, that such hyperconjugation could be important in the excited state of the molecule, i.e., $F_3{\equiv}C$— antibonding orbitals may result in a lowering of the π^* molecular orbital.

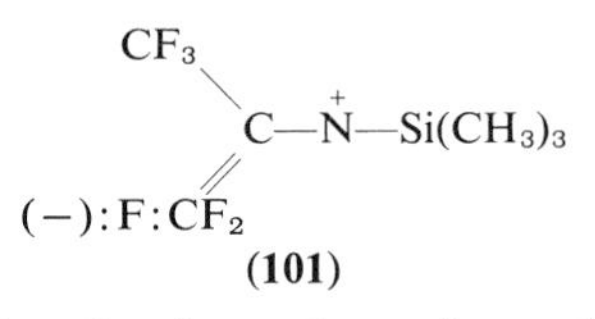

(101)

The weak, long wavelength absorption of azoalkanes near 358 mμ and azobenzene 420 mμ is assigned to the $n \rightarrow \pi^*$ transition. Robin, Hart, and Kuebler [117] find however that there is considerable interaction between the nonbonding electron orbitals and that the nitrogen lone pair electrons are in strongly bonding and antibonding molecular orbitals rather than being localized on one nitrogen.

The ultraviolet spectra [80] of the trialkylsilylazobenzenes R_3Si—N=N—C_6H_5 show a large red shift of this $n \rightarrow \pi^*$ transition to λ_{max} 575 mμ (R is CH_3) and λ_{max} 587 mμ (R is n-C_3H_7). The $\pi \rightarrow \pi^*$ transition of azoisopropane has been assigned to a transition at 1640 Å (7.56 eV) (Robin *et al.* [117] calculate a $\pi \rightarrow \pi^*$ transition energy nearer 11 eV), and the ionization potential of HN=NH has been reported as 9.85 eV. The electron affinity of the silicon 3d orbital has been estimated at about 1.95 eV. From the ionization potential of HN=NH and $\pi \rightarrow \pi^*$ transition energy (7.56 eV), the antibonding π^* orbital of —N=N— is estimated to be only 0.5 eV or less in energy below the silicon 3d orbitals.

The substitution of Si for C has a smaller red shift effect (0.79 eV) on the $n \rightarrow \pi^*$ transition of the —N=N—R chromophore than it did on the $n \rightarrow \pi^*$ transitions of $(C_2H_5)_2C{=}N$— (1.27 eV) or the silyl ketones for which a red shift of 0.8–1.1 eV was found. If d–p π^* bonding is important in R_2M—N=N—R compounds, there must also be important d–n bonding to explain the absence of a substantially larger red shift in the $n \rightarrow \pi^*$ transition. The results can be explained by an inductive effect alone.

G. The Metal–Metal Bond Chromophore

It is by now well known, due largely to the research of Gilman, Kumada, and their co-workers, that catenated metal–metal bonds of the Group IV metals absorb as chromophores in the near ultraviolet. The absorption maximum of $(CH_3)_3SiSi(CH_3)_3$ at 190 mμ (6.5 eV) (ϵ 8500–7230) moves in the compound $(CH_3)_3Si[Si(CH_3)_2]_3Si(CH_3)_3$ to 250 mμ (5.0 eV) (ϵ 18,400), and as may be seen from Table XV, a similar result is obtained for the alkyl

TABLE XV

Electronic Transitions in $R(R_2M)_nR$

Compound	λ_{max}, mμ (ϵ)
$CH_3[(CH_3)_2Si]_2CH_3$	190 (18,000)
$CH_3[(CH_3)_2Si]_3CH_3$	215 (9020)
$CH_3[(CH_3)_2Si]_4CH_3$	186 (40,000), 235 (14,700)
$CH_3[(CH_3)_2Si]_5CH_3$	250 (18,470)
$CH_3[(CH_3)_2Si]_6CH_3$	220 (sh) (14,000), 260 (21,100)
$CH_3[(CH_3)_2Si]_{10}CH_3$	215 (sh) (28,200), 230 (sh) (21,000); 255 (sh) (24,600), 279 (42,700)
$CH_3[(CH_3)_2Ge]_2CH_3$	190 (sh) (20,000)
$CH_3[(CH_3)_2Ge]_3CH_3$	190 (sh) (35,000), 215 (10,000)
$CH_3[(CH_3)_2Ge]_4CH_3$	185 (sh) (55,000), 230 (15,000)
$CH_3[(CH_3)_2Ge]_5CH_3$	190 (sh) (80,000), 210 (35,000), 240 (20,000)
$CH_3[(CH_3)_2Sn]_2CH_3$	182 (40,000), 210 (25,000)
$CH_3[(CH_3)_2Sn]_6CH_3$	205 (55,000), 246 (23,000)
$CH_3CH_2[(CH_3CH_2)_2Sn]_6—CH_2CH_3$	325 (28,000)
$(C_6H_{11})_3PbPb(C_6H_{11})_3$	253 (40,000), 280 (sh), 315 (sh) (20,000)

derivatives of Ge, Sn, and Pb. In the series $R(R_2M)_nR$, when R is alkyl, where comparison is possible, we find that for a given value of n the transition energy of the first absorption maximum increases in the order Pb < Sn < Si ≤ Ge ≪ C, which is in the order of the metal–metal bond energies except for the interchange of Si and Ge. The question then raised is are these transitions of $R(R_2M)_nR$, where M is Si, Ge, Sn, and Pb, different in any fundamental way from the electronic transitions of the saturated hydrocarbons, where M is C?

The first transition of ethane $\sigma \rightarrow \sigma^*$, is found near 132 m$\mu$ (9.4 eV). In pentane, this transition has moved to about 149 mμ, or a red shift relative to ethane of the $\sigma \rightarrow \sigma^*$ transition by 1.1 eV, which is quite comparable to the 1.4 eV red shift of the corresponding silanes. We may ask ourselves further is the difference in σ bond energy between $(CH_3)_3SiSi(CH_3)_3$ and ethane sufficient to account for the difference in their first electronic transition energies, assuming the first transition is from the M—M bond? In each case, in the approximation of simple molecular orbital theory, the energy of the first $\sigma_{M-M} \rightarrow \sigma^*_{M-M}$ transition of R_3MMR_3 should be $2\beta_{M-M}$, where β_{M-M} represents the energy of the M—M bond. If we equate β with the bond energies, for $\beta_{C-C} = 3.6$ eV and $\beta_{Si-Si} = 3.0$ eV [39], the electronic transitions energies should be in the ratio $2\beta_{C-C}/2\beta_{Si-Si} = 1.2$. This compares reasonably well with the observed ratio of λ_{max} (eV) $CH_3CH_3/(CH_3)_3SiSi(CH_3)_3$ equals 1.4. (The use of ratios here is dictated by the well-known fact that there are large differences between so-called ground state β's, used in calculating bond energies, etc., and the spectroscopic β's required in calculating electronic transition energies.) Since there is not much reason to classify the transitions of the alkylpolysilanes, etc., as fundamentally different from the transitions of the alkanes, it would seem pertinent at this point to examine the nature of the first $\sigma \rightarrow \sigma^*$ transition in the alkanes.

The various approaches to a quantum mechanical treatment of saturated hydrocarbons was reviewed by Klopman [77] in 1963. Among the more successful simple molecular orbital methods is the Sandorfy "*C*" approximation [124] in which only the carbon skeleton is included in the calculation. The formulation of the problem is similar to the simple Hückel method for π electrons in that the molecular orbitals are described as linear combinations of sp^3 atomic orbitals neglecting overlap. Where the resonance integrals for overlapping sp^3 orbitals on adjacent atoms is given in units of β, the resonance integral of sp^3 orbitals on the same atom is expressed by $k\beta$. The integral β as usual is evaluated from either spectroscopic or thermodynamic data for the C—C bond and the parameter m is chosen to give the best fit with experimental data. For calculating ground state energies the values $\beta = 38.866$ kcal and $k = 0.355$ have been suggested. The Sandorfy "*C*" approximation as used by Fukui *et al.* [48] has been found to give good agreement between calculated and observed ionization potentials. An attempt to apply the Pariser-Parr-Pople SCF method to alkanes and their electronic transitions has been made by Katagiri and Sandorfy [75] with the following somewhat preliminary conclusions. The first calculated excitation in methane, ethane, and propane is from a C—H rather than C—C orbital, although in butane and higher hydrocarbons the transition is from a C—C σ orbital. Experimentally a fairly linear relationship is observed between the first transition of a linear hydrocarbon and its ionization potential, and the Pariser-Parr-Pople SCF calculated ionization potentials are also found to be linearly related to the calculated transition energies.

Experimentally the definitive work on the electronic transitions on the alkanes has very recently been done by Raymonda and Simpson [115] who measured the vacuum ultraviolet spectra of a large variety of normal, branched chain, and cyclic hydrocarbons. These results appear to rule out assignment of the lowest energy transition to $\sigma_{C-H} \rightarrow \sigma^*$ but may be rationalized if the lowest energy transition is presumed to be $\sigma_{C-C} \rightarrow \sigma^*$. The approach of Raymonda and Simpson retains the idea of the localized σ and σ^* orbitals. If we consider only the C—C skeleton, the first $\sigma \rightarrow \sigma^*$ transition of a linear hydrocarbon with nC—C bonds is n-fold degenerate. If, however, we introduce configuration interaction in which the degeneracy is split in this case by the electrostatic interaction energy between transition densities, the term level diagram for $\sigma_{C-C} \rightarrow \sigma^*_{C-C}$ transitions is given by Fig. 37. This indepen-

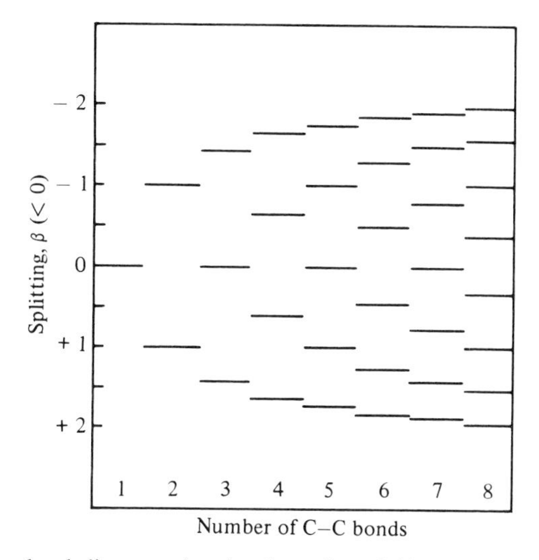

FIG. 37. Energy level diagram showing how the n-fold degenerate first excited state of an n-bond alkane is split by the electrostatic interaction between adjacent transition densities in units of the interaction integral β [115].

dent systems model thus predicts that the electronic transitions of $n + 1$ will "bracket" those of n in the hydrocarbon series $CH_3(CH_2)_nH$ and that the lowest transition energy will approach a limit of 2β below transition energy of ethane. Here the integral β is the interaction resonance integral between excited states and should not be confused with other β.) As may be seen from Fig. 38, the topological mapping of the $\sigma \rightarrow \sigma^*$ transitions predicted by the independent systems model finds good agreement with spectra. The predictions regarding branched chain and cyclic hydrocarbons will be of particular interest to us later, therefore the term level diagram for these are given in Figs. 39 and 40. The first electronic transitions of 1-methylpropane and 2,3-dimethylpropane are predicted to be doubly and triply degenerate, respectively, at the same energy as that of propane. In fact transitions symmetrically above and below the $\sigma \rightarrow \sigma^*$ propane frequency

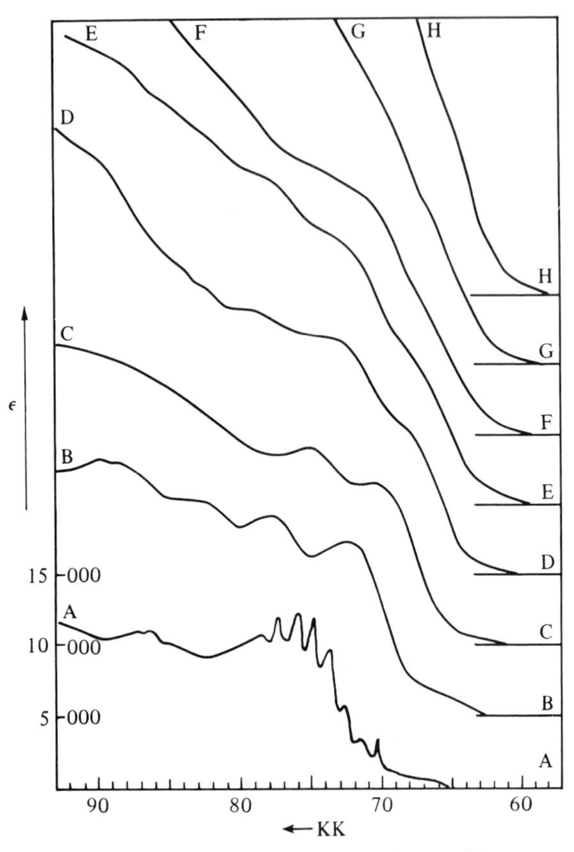

FIG. 38. Spectra of some linear alkanes: (A) ethane; (B) propane; (C) butane; (D) pentane; (E) hexane; (F) heptane; (G) octane; (H) nonane [115].

are attributed to Jahn-Teller splitting of these degenerate excited states. The spectrum of 2,2,3,3-tetramethylbutane (hexamethylhexane) is predicted and found to have a transition to the *red* of butane. In a number of highly branched alkanes an additional lower energy transition is found which Raymonda and Simpson assigned to a transition in the system C—C—C in which the electron is promoted from the σ level of one bond to the σ^* level of the adjacent bond in a charge transfer type of transition. The important point, however, is that all of the branched chain hydrocarbons have electronic transitions to the *red* of the *longest* C—C nonbranched unit contained within the hydrocarbon, although these transitions sometimes appear only as shoulders.

The energy level diagram according to Raymonda and Simpson for the

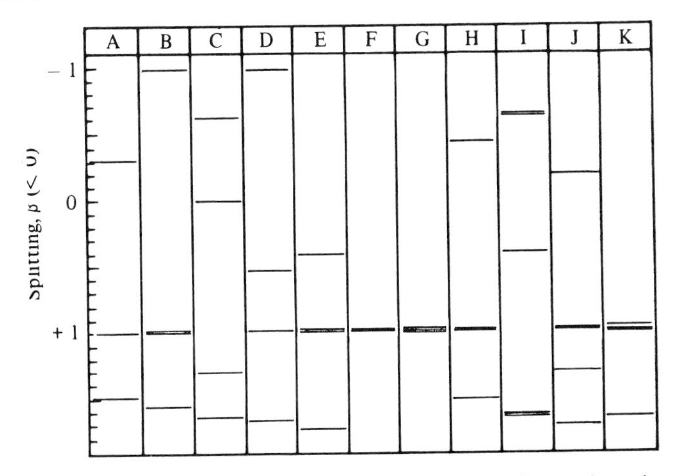

FIG. 39. Energy level diagram for some branched chain alkanes in units of $\beta < 0$: (A) isopertane; (B) 2,3-dimethylbutane; (C) 3-methylpentane; (D) 2-methylpentane; (E) 2,4-dimethylpentane; (F) isobutane; (G) neopentane; (H) 2,2-dimethylbutane; (I) 3-ethylpentane; (J) 2,2,4-trimethylpentane; (K) hexamethylethane.

cyclic hydrocarbons is given in Fig. 40. The predictions here, which are also in agreement with the spectral results, are that because the first $\sigma \rightarrow \sigma^*$ transition is found to be forbidden, the first allowed transition of the cyclic hydrocarbons will be to the blue of the normal alkane, and the transition

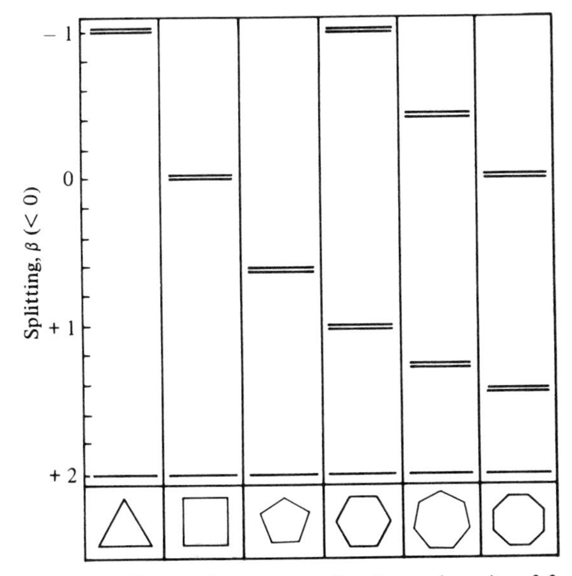

FIG. 40. Energy level diagram for some cyclic alkanes in units of $\beta < 0$ [115].

energy should decrease in the order cyclobutane > cyclopentane > cyclohexane > cycloheptane.

Now that we have a somewhat better appreciation of the electronic transitions in alkanes we can return to the problem of the nature of the transition in the alkyl polysilanes.

Harada *et al.* [68a] note that if the far UV spectrum of $(CH_3)_3SiSi(CH_3)_3$ (λ_{max} 139, 163, 192 mμ) were shifted down into the UV one would have a spectrum similar to that of higher linear polysilanes, suggesting that the maxima for the linear polysilanes have their origin within one Si—Si bond.

The ultraviolet transitions of the alkyl-substituted linear polysilanes and polygermanes have been the subject thus far of two simple molecular orbital calculations. Assigning the transitions to $\sigma \rightarrow \sigma^*$, P. P. Shorygin and coworkers claim that the spectra of the silanes and germanes may be satisfactorily described by a Sandorfy "*C*" calculation using the parameters for $\beta_{M-M} = -3.4$ eV, $m = 0.27$. However since the actual calculated transition energies were not given, we cannot comment on the result.

Several workers [111] have more or less simultaneously realized that one might describe the polysilane transitions as promoting an electron from a σ orbital to a delocalized molecular orbital made up of a linear combination of $3d$ orbitals (or since there is some evidence that the $4s$ and $4p$ energies may be comparable to $3d$, a linear combination of $4s$, $4p$, or hybrid orbitals). The result is that as the silicon chain lengthens a correlation diagram such as given by Fig. 41 is obtained in which the energy of the *lowest* vacant

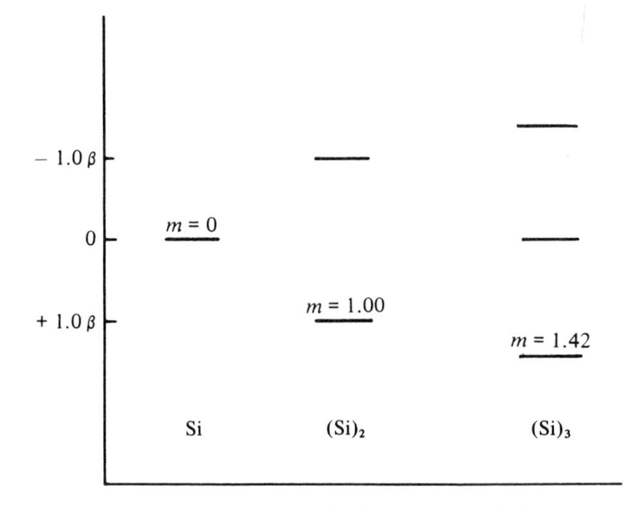

FIG. 41. Simple molecular orbital energies ($E = \pm m\beta$) relative to $\alpha_{Si} = 0$ for a linear combination of n silicon d orbitals.

orbital relative to the α_{Si} of the atomic orbital is $m\beta_{Si-Si}$, and m is the familiar Hückel energy coefficient given for the *lowest* molecular orbital of n atoms by the expression (13),

$$m = 2 \cos \pi/(n + 1) \tag{13}$$

Since the absolute value of $|m|$ is the same for both the lowest and highest molecular orbital for acyclic systems, we do not have to worry about whether the atomic orbitals have p- or d-type symmetry. It is easily seen then that the transition energy E_t is given by an expression such as (14),

$$E_t = \alpha_{Si} + m\beta^* - E_\sigma = (\alpha_{Si} - E_\sigma) - m\beta^* \tag{14}$$

If the transition takes place from a highest filled orbital of nearly constant energy, a plot of transition energies against m should be a straight line of slope β^*. Such a plot is given in Fig. 42 for the transition energies of

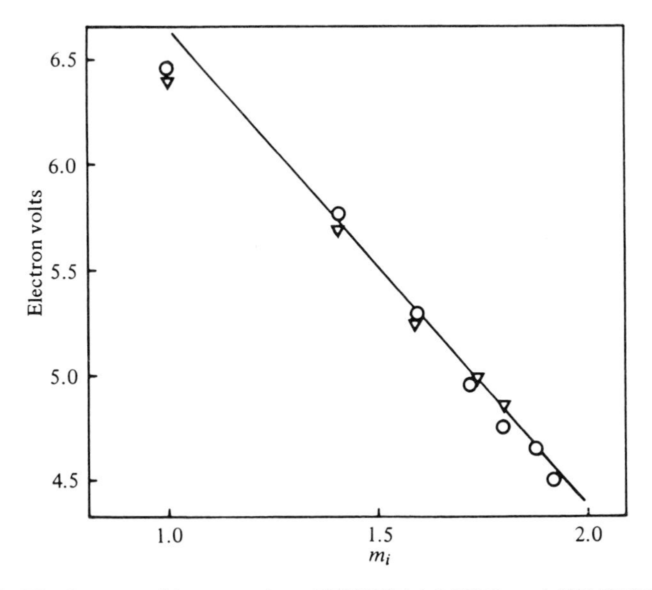

FIG. 42. The first transition energies of $H[Si(CH_3)_2]_nH$ (▽) and $CH_3[Si(CH_3)_2]_nCH_3$ (○) vs. the energy, m, of the lowest vacant d molecular orbital.

$H{-}[Si(CH_3)_2]_n{-}H$ and $CH_3{+}Si(CH_3)_2{+}_nCH_3$, where n = 2–6, 8, 10, and ignoring the disilanes for the moment, the observed correlation is quite good; it also has been found [111] that with an appropriate choice of parameters, phenyl and vinyl silanes may also be fitted to the line. However, the experimental resonance integral β_{Si-Si} obtained (2.8 eV) seems quite high considering the more diffuse character of $3d$ (or $4p$) orbitals compared to carbon $3p$ where the C—C resonance integral β°_{C-C} is usually taken to be about 2.34 eV. Note now that a linear correlation of transition energy with m would also be obtained if the energy of the highest filled σ orbital were also a linear function of m such as Eq. (15); E_t is then given by Eq. (16).

$$E_\sigma = \alpha'_{Si} - m\beta_\sigma' \tag{15}$$

$$E_t = (\alpha^*_{Si} - \alpha'_{Si}) + m(\beta_\pi{}^* + \beta_\sigma') \tag{16}$$

Although we do not have the data for silanes, the reader may be justifiably surprised to find that the ionization potentials, i.e., E_σ, of the normal hydrocarbons with the exception of ethane give an excellent linear relationship [132] with m as predicted by Eq. (15) for which $\beta_\sigma' = 1.7$ eV. Furthermore,

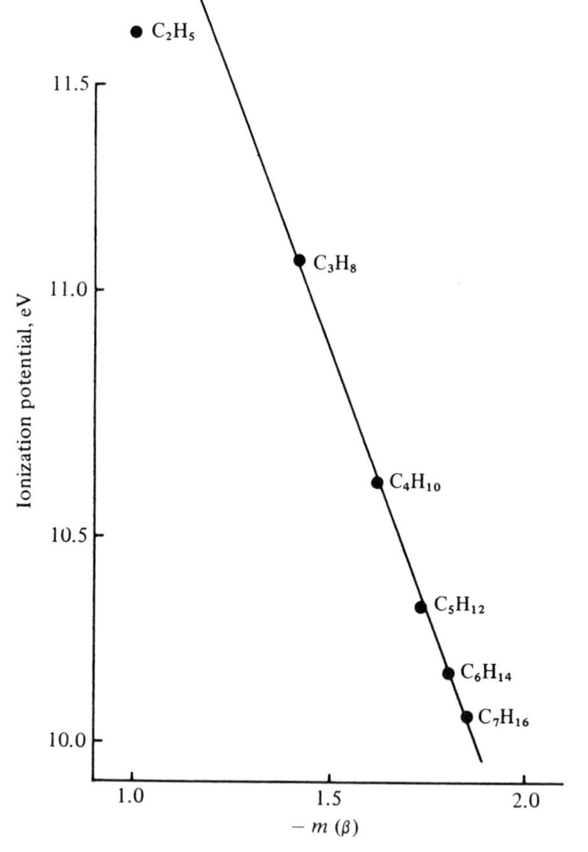

FIG. 43. The ionization potentials of linear hydrocarbons H—$(CH_2)_n$—H vs. the energy m in units of β for highest filled molecular orbital of model H—$(\ddot{X})_n$—H.

it should be noted that if we apply the correction for E_σ of disilane different from the E_σ of $R(SiR_2)_nR$, $n > 2$, obtained as the deviation of ethane from the linear I_pVs/m relationship of Fig. 43, the disilanes fall much closer to the correlation line of Fig. 42. It would be nice to have the ionization potential data for the silanes themselves, but it has been observed that the transition energies of the polysilanes give a linear correlation with the ionization potentials of the corresponding hydrocarbons [109]. From the more general Eq. (16), the resonance integral $\beta^*_{\text{Si-Si}}$ is estimated to be approximately $2.8 - 1.7$ or ~ 1.1 eV.

By now it will come as no surprise that, as may be seen from Fig. 44, a plot of the first $\sigma \rightarrow \sigma^*$ transition energies of the normal hydrocarbons

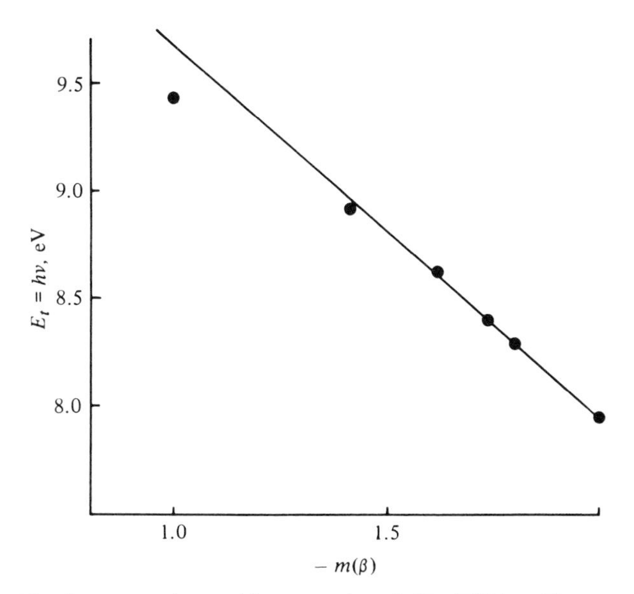

FIG. 44. The first $\sigma \rightarrow \sigma^*$ transition energies of $H-(CH_2)_n-H$ vs. energy, m, of simple highest filled molecular orbital of $H-(\ddot{X})_n-H$.

$H(CH_2)_nH$, where $n = 2, 3, 4, 5, 6$, and ∞, *also* gives a linear relationship but with $\beta = 1.7$ eV.

Applying the relationship that $\beta(\text{experimental}) = \beta^* + \beta_\sigma$ from Eq. (16) we reach the conclusion that $\beta^* \sim 1.7 - 1.7 \sim 0.0$, from which we would conclude that the $\sigma \rightarrow \sigma^*$ transition in the normal alkanes was from a "delocalized ground state molecular orbital" to an excited state of relatively constant energy in which the promoted electron involved finds itself in a highly localized orbital. This result conforms to a model, suggested by Stevenson [130] to account for the observed linear relationship between the

Hückel *m* on ionization potential, in which the —CH_2— unit is regarded as a pseudo-π-electron donor —$\ddot{X}$—. Extending this model to an interpretation of the normal hydrocarbon first absorption maximum, amounts to describing the transition as $\pi_\sigma \rightarrow \sigma^*$, where the σ^* orbital derives from the localized $-(-\ddot{X}-\ddot{X}-)-$ σ bond.

As a result, we would like to tentatively suggest that unless the ionization potentials of the linear silanes prove to be much more sensitive to chain length than the hydrocarbons, the result that $\beta_{Si}^* \cong 1.1$ eV whereas $\beta_C^* \sim 0.0$ is evidence of an excited state electron delocalization mechanism present in the polysilanes but absent in hydrocarbons, that is, a lowest energy delocalized antibonding molecular orbital with large mixing coefficients for the Si 3*d* (or 4*s* and *p* plus hybrids) atomic orbitals. This proposal is also in keeping with the result that for the compounds $CH_3[M(CH_3)_2]_nCH_3$, where $n = 4, 5$, the transition energies of germanium are greater than those of silicon and at least according to the data of Shorygin *et al.* [126] for $n = 6$, the transition energy of Sn is less than that of the Si compound. These results are consistent with our present concepts of the relative importance of *d* orbital participation in the series Si, Ge, Sn, and Pb. The hypothesis of molecular orbitals composed of purely *d* orbitals is, of course, very naive and only serves as a sort of first-order approximation, since it must be recognized that the *d* atomic and σ molecular orbitals are not orthogonal and we can in no way invoke any kind of σ–π separability. Therefore, for linear silanes, germanes, etc., the near ultraviolet absorption and decreasing transition energy with chain length of the alkyl-substituted Group IV organometallic compounds appears to be primarily the result of lower metal–metal σ bond energy and increasing polarizability of the metal atom and σ bond framework. Superimposed on this, at least for silicon, is important delocalized "*d*" orbital interaction in the vacant σ^* orbitals (see Addendum).

In important contrast to alkyl-substituted alkanes which show transitions at longer wavelengths than the unsubstituted alkanes, side chain silyl-substituted *alkyl*silanes show substantial blue shifts of the first transition. For example, although permethyltrisilane (**102**) and -tetrasilane (**103**) have

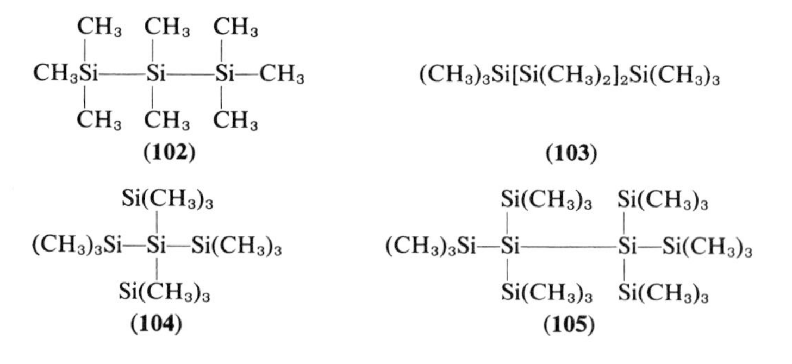

spectral absorption maxima at 215 and 235 mμ, respectively, the branched chain silanes **104** and **105** have no absorption maxima above 210 mμ. The phenylsilanes appear to be much less sensitive to side chain substitution. If we compare compounds **106** (λ_{max} 243 mμ) and **107** (λ_{max} 243 mμ) with the related side-chain-branched silanes **108** (λ_{max} 241 mμ) and **109** (λ_{max} 242 mμ), there is a much smaller blue shift.

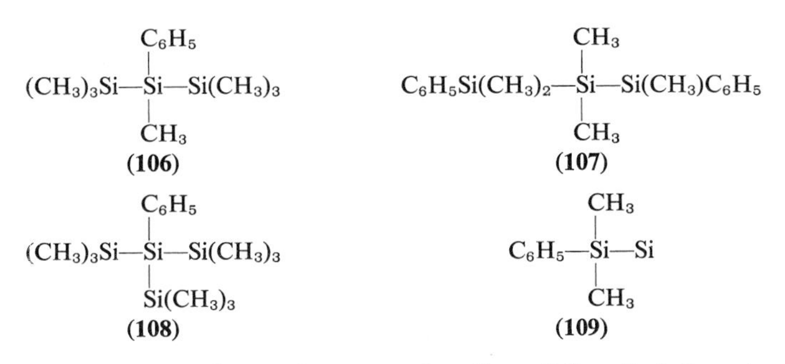

Let us return to our first-order approximation of the alkylsilane transition as promoting the electron to a delocalized molecular orbital described as a linear combination of 3*d* or higher atomic orbitals. If the atomic orbitals used were pure *s* or *p* orbitals, then the correlation of transition energy with *m* in Fig. 41 would predict a large red shift of the first transition of a linear silane on substitution at an internal site by a silyl group because, as we have tried to indicate in Fig. 45, the substituent atomic *s* or *p* orbital and resultant molecular orbital over the chain will be bonding. On the other hand, because of the symmetry of the atomic *d* orbital a molecular orbital which is a linear combination of *d* orbitals has a nodal plane at each silicon perpendicular to the plane of the Si—Si—Si bond. Thus, the *d* orbital of the substituent Si(4) directed towards Si(2) in Fig. 45 and the molecular orbital of atoms 1, 2, and 3 are orthogonal (nonbonding) and the substituent will not have the effect of lowering the energy of the lowest vacant molecular orbital of the $(Si)_n$ framework. In this fashion at least, we can offer a reasonable, if speculative, explanation of the *absence* of a large red shift in the spectra of branched chain silanes which will preserve our first approximation of the excited state, *provided the lowest vacant molecular orbital is taken as a linear combination of the silicon* 3*d orbitals, and not* 4*s or* 4*p*.

The large *blue* shift observed for structures **104** and **105**, however, requires some additional explanation.

At least part, if not all, of the blue shift can be attributed to the fact that the replacement of a methyl or hydrogen by the very bulky trimethylsilyl group must introduce substantial steric strain in the ground state. Moreover, since we are removing an electron from a bonding σ orbital and placing it into an antibonding orbital, the equilibrium bond distances will be larger

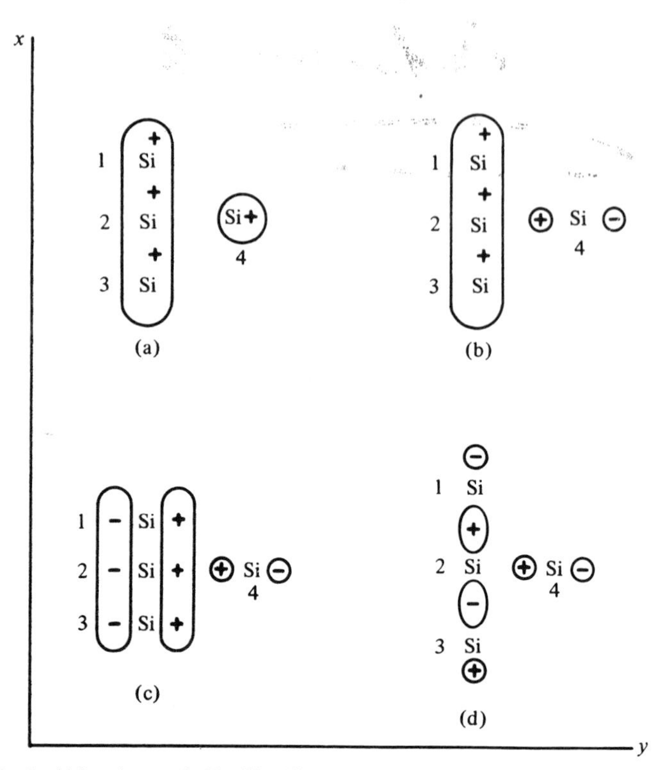

FIG. 45. Combinations of Si—Si—Si molecular orbitals from atomic s, p, or d orbitals with the fourth Si, where the p or d orbital is at chain Si—2 (top view).

in the excited state. As a result of the greater equilibrium bond distances and the Franck-Condon nature of the transition, the excited state finds itself in a "compressed" conformation. Steric strain in this case should be greater in the excited state than the ground state and lead to a blue shift of the transition maximum. (The O–O transition will however be at a lower energy.) The 243 mμ absorptions of the arylsilanes **108** and **109** probably involve a large contribution from the ^1L_a $\pi \rightarrow \pi^*$ transition of the benzene ring and would therefore be much less sensitive to steric effects of this sort. An additional contribution to the blue shift would result from steric hindrance to solvation of the excited state.

The ultraviolet spectra [29, 50, 126] of the cyclic permethylpolysilanes $[Si(CH_3)_2]_n$ in Fig. 46, $n = 5$ [λ_{max} 210 (sh), $\epsilon > 10{,}000$; 261 mμ, ϵ 1100; 272 mμ, ϵ 970] and $n = 6$ [λ_{max} 195 (sh) mμ, ϵ 45,000; 232 mμ, ϵ 5800; 255 (sh), ϵ 2000] and $n = 7$ [λ_{max} 217 (sh) mμ; 242 mμ, ϵ 2100] agree with prediction of the independent systems model of Raymonda and Simpson discussed earlier for alkanes in that the first *allowed* transition of the cyclic

$[Si(CH_3)_2]_n$ appears at much shorter wavelength than $H-(-Si(CH_3)_2-)-_nH$ since the first transition of the planar cyclic molecules is forbidden according to the Raymonda and Simpson independent systems model. In the simple Hückel molecular orbital model considered earlier which represented the lowest vacant molecular orbital as a linear combination of d orbitals and the highest filled orbital as a linear combination of pseudo-p π orbitals [each —$Si(CH_3)_2$— acting as a lone pair electron donor] the lowest vacant molecular orbitals for the five-, six-, and seven-membered rings respectively are given as $1.62\beta^*$, $2.00\beta^*$, and $1.8\beta^*$. The first transition of a D_{6h} cyclohexasilane is from a B_{2g} to A_{1u} molecular orbital and is therefore ${}^1A_{1g} \rightarrow B_{2u}$ which is not symmetry allowed. The first allowed transition is to the degenerate $1.00\beta^*$ orbitals. The five- and seven-membered rings involve transitions between degenerate orbitals which give rise to allowed transitions. For cyclopentasilane after configuration interaction the transitions are $A_1' \rightarrow E_1'(x, y)$ polarized and $A_1' \rightarrow E_2'$ forbidden. The observed order of transition energies expected on the basis of this model then are for $[Si(CH_3)_2]_n$, $n = 6 > 5 > 7$, which at least is not contradicted for $n = 6, 5$ if we assign the shoulders at 195 and 210 mμ to a part of the first allowed transitions of these molecules. The predicted and observed order of transition energies in the cyclic alkanes on the other hand is in the order cyclobutane > cyclopentane > cyclohexane > cycloheptane. (Preliminary ionization potentials by Pitt [109] indicate that the observed order may be due to ground state ring strain.)

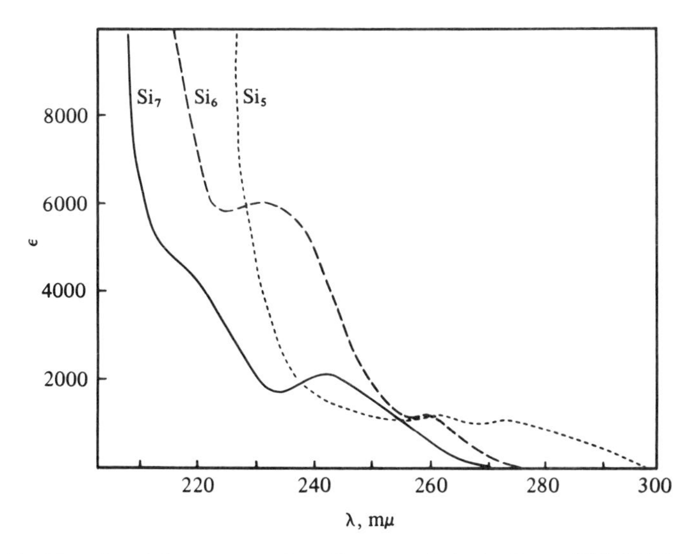

FIG. 46. The ultraviolet spectra of cyclic permethylpolysilanes $[Si(CH_3)_2]_n$: (·····) n = S; (----) n = 6; (——) n = 7 (R. West [138]).

H. Substituted Metal–Metal Chromophores

Historically even before it was discovered that the metal–metal bonds themselves were chromophores, interest in the spectra of catenated silanes, germanes, etc. was stimulated by the work of several groups [51, 66, 101] reporting the spectra of phenyl-substituted derivatives of these compounds. By now a variety of phenyl-, vinyl-, and halogen-substituted cyclic and acyclic derivatives of Group IV metal catenates have had their ultraviolet spectra recorded and some of these are listed in Table XIV.

Replacement of terminal CH_3 by $CH_2{=}CH—$ in $CH_3[Si(CH_3)_2]_4CH_3$ results [52] in a small red shift, 0.2 eV, of the absorption maximum to λ_{max} 243.5 mμ. It seems significant that the effect of two vinyl groups on the spectrum of $CH_3[Si(CH_3)_2]_4CH_3$ is less than that of two $(CH_3)_3Si—$ since the absorption maximum of $CH_3[Si(CH_3)_2]_6CH_3$ is found at 260 mμ. In the absence of additional strong transitions to the red of 243 mμ in the spectrum of $CH_2{=}CH(SiCH_3)_4CH{=}CH_2$, *this transition and the related transitions in other vinyl polysilanes must correlate with the* $\sigma \rightarrow \pi_\sigma{}^*$ *transition of the alkyl polysilanes and not the* $\pi \rightarrow \pi^*$ *transition of the olefin.* In terms of locally excited states the effect of vinyl substitution on the spectrum of a polysilane may be described in terms of configuration interaction between the C═C $\pi \rightarrow \pi^*$ transition [which in the absence of configuration interaction should occur near 180 mμ (6.9 eV)] and the $-\!\!(\!-Si(Me)_2-\!)\!\!-_n$ transition. In the disilane $(CH_3)_3SiSi(CH_3)_2—R$, the metal–metal bond transition occurs near 6.3 eV and the stronger configuration interaction, resulting from the nearness in energy of the two states may be regarded as moving the lowest energy transition to 225 mμ, a red shift of some 1.4 eV.

A qualitative molecular orbital description of the transition of divinyltetramethyldisilane is also possible as is given in Fig. 47. Based on the observation that the transition energy of $(CH_3)_3SiSi(CH_3)_3$ is less than that of $RCH{=}CH_2$ and the assumption that the highest Si—Si σ orbital is higher in energy than the C═C π orbital, we have the two possibilities presented as a and b in Fig. 47. In the single configuration approximation Fig. 47a represents the lowest energy transition as remaining localized in the Si—Si framework whereas in Fig. 47b the lowest energy configuration has considerable charge transfer character from Si—Si to C═C. After configuration interaction such distinctions would become considerably less important, but the essential feature of electron promotion from the highest filled metal–metal σ bond molecular orbital would be retained.

Armstrong and Perkins [4] have carried out antisymmetrized free electron molecular orbital calculations for the $\pi \rightarrow \pi^*$ transitions of $CH_2{=}CH$-$(SiMe_2)_2CH{=}CH_2$ and $CH_2{=}CH(SiMe_2)_4CM{=}CH_2$ with end walls of 25–30 eV and a central barrier of 8.4–10.2 eV. Good agreement was obtained

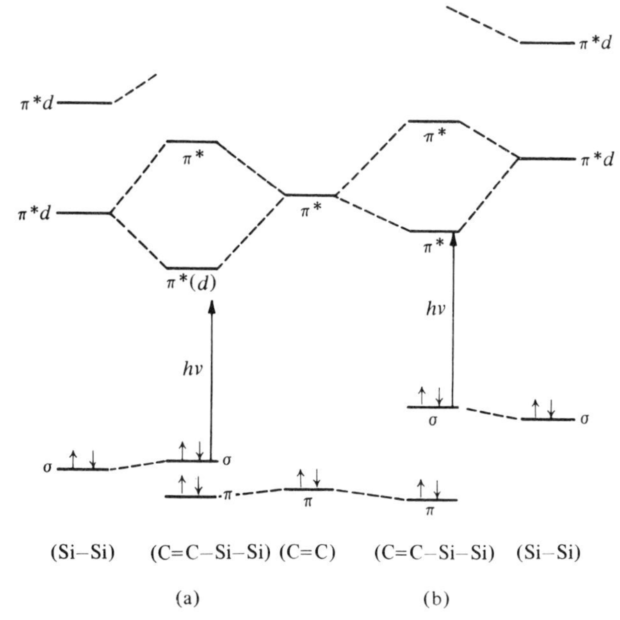

FIG. 47. A qualitative molecular orbital energy level correlation diagram for $CH_2{=}CH_2$, $(CH_3)_3SiSi(CH_3)_3$ and $CH_2{=}CH{-}Si(Me_2){-}Si(Me_2){-}CH{=}CH_2$ with $E_{\sigma,Si-Si} > \pi_{C=C}$ and (a) $\pi^*_{C-C} > \pi^*_{Si-Si}$ or (b) $\pi^*_{C-C} < \pi^*_{Si-Si}$.

between the calculated transition energies and the observed maxima at 225 and 245 mμ, respectively. However the $\sigma \rightarrow \pi^*$ transition was not considered and as indicated in the preceding paragraphs, the agreement obtained may not be with the appropriate transition.

The ultraviolet spectra of the phenyl-substituted polysilanes, polygermanes, etc., are naturally complicated by the transitions of the phenyl rings themselves. In a few instances such as the spectrum of $(C_6H_5)_3SiSi(CH_3)_3$, where it is not buried by more intense transitions, the typical ^1L_b transition may be discerned in the neighborhood of 265–280 mμ. The general features and behavior of the spectra of $(C_6H_5)(MR_2)_nC_6H_5$, etc., where M is Si, Ge, Sn, or Pb, correspond generally to those of the saturated derivatives (but with a few important differences) in that (1) the order of transition energies of the first allowed transition in the series $(C_6H_5)_6M_2$ increase in the order Ge > Si > Sn ≫ Pb; (2) for a given metal M in a series such as $C_6H_5(MR_2)_nC_6H_5$ there is a decrease in transition energy with increase in n; (3) the cyclic derivatives such as $[(C_6H_5)_2M]_n$ absorb at shorter wavelengths than the corresponding acyclic molecules but with transition energies in order $n = 6 \approx 5 > 4$; (4) as was mentioned earlier, branched chain the phenylsilanes appear to absorb at slightly shorter wavelengths than their linear analogs but are not as sensitive to side chain branching as the permethylpolysilanes. It was suggested that this might be the result of considerable $\pi \rightarrow \pi^*$ character in the transition of the phenyl derivatives as a result of configuration interaction with the ^1L_a benzene transition which would reduce the importance of steric strain in the excited state.

In fact since the ^1L_a transition of benzene, or "center of gravity" of states from the degenerate $\pi \rightarrow \pi^*$ transition, is near 6.0 eV compared with a transition energy of 6.3 eV for $(CH_3)_3SiSi(CH_3)_3$, it would not be unreasonable to assign the 240 mμ transition of the 1,2-diphenyldisilanes to a red shifted ^1L_a transition of the aromatic ring. Opposing such an assignment is the absence in longer-chain homologs, $C_6H_5(SiR_2)_nC_6H_5$, of a transition which could be assigned to promotion of an electron from the metal–metal bond chromophore, and the effect of ring substituents on the transition energy. A CH_3O group shifts the ^1L_a transition of benzene to the red by 14 mμ (0.8 eV) but shifts the absorption maximum of $[p\text{-}XC_6H_4Si(CH_3)_2]_2$ (at 238 mμ where X is H) by only 4 mμ (0.08 eV); whereas, the $p\text{-}(CH_3)_3Si$ substituent, which has a very small effect, if any, on the ^1L_a transition of benzene, produces a red shift of 8–9 mμ in the spectrum of the diphenyldisilane.

It is unlikely that a single configuration will provide a completely adequate description of the first intense transition in the spectra of the aryl polysilanes, aryl polygermanes, etc., but an assignment to intramolecular charge transfer from the highest filled —$(MR_2)_n$—R σ orbital to aryl ring lowest vacant π^* has much to recommend it. The unexpectedly large effect of the $p\text{-}(CH_3)_3Si$ group is explained on the basis of its ability to stabilize structures such as **110** by *d* orbital π bonding. Substituent effects of groups such as CH_3O, Cl, etc.,

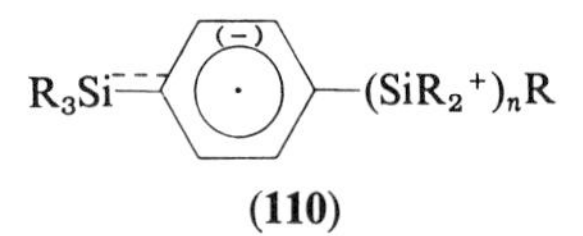

(110)

which are inductively electron withdrawing but π-electron donating, are complicated by the fact that although electron donation decreases the electron affinity of the ring expected to produce a blue shift in charge transfer transition, the same effect is produced by moving the ^1L_a transition to the red, which increases the configuration interaction between the ^1L_a and charge transfer excited states. The charge transfer character of the first allowed transition in the series C_6H_5—$(MR_2)_n$—R′ will be greatest for $n = 2$, as n increases the mixing coefficient for the phenyl orbitals in the lowest vacant molecular orbital would become smaller and the charge transfer character of the transition decrease. These features are illustrated in Fig. 48 which is a molecular orbital correlation diagram constructed on the assumption that

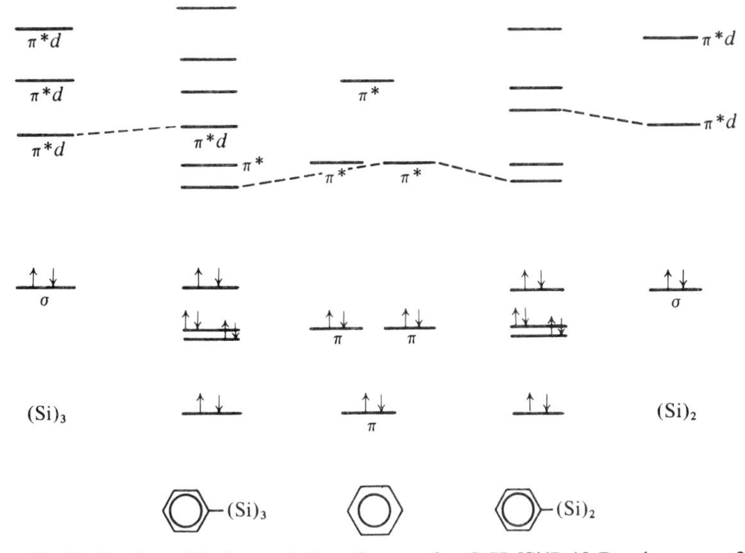

FIG. 48. Molecular orbital correlation diagram for $C_6H_5[Si(R_2)]_nR$, where $n = 2, 3$.

difference in energy between benzene π and π^* orbitals is less than that between highest filled and lowest vacant $-(-MR_2-)-_n$ ($n = 2$) orbitals and that the energy of highest $-(-MR_2-)-$ σ is greater than the phenyl π orbital. In any case it should be remembered that configuration interaction with ^1L_a transition is expected to be important.

In the spectra of pentachlorobenzene and pentachlorotoluene [7] the ^1L_a transition is found at 219 mμ (ϵ 21 $\times$ 10^3) and the ^1L_b transition near 290 mμ. Introduction of an alkyl group stearically larger than methyl results in increased intensity and a red shift of both the L_a and L_b transitions [7]. Gilman and Morris [57] have reported the ultraviolet absorption maxima of the compounds $C_6Cl_5-(-Si(CH_3)_2-)-_nC_6Cl_5$ ($n = 1$–6) and observe no new transition whose energy depends on n. The ^1L_a transition of C_6HCl_5 is *blue* shift 3–6 mμ in the spectra of the perchlorophenylsilanes and possesses an unusually large ϵ which increases from 9.1×10^4 for $n = 2$ to 1.1×10^5 for $n = 6$. Since the spectra themselves were not presented, we cannot be certain; nonetheless, it seems safe to assume that the transitions characteristic of the $-(-Si(CH_3)_2-)-_n$ chromophore are buried beneath the more intense ^1L_a transition of the C_6Cl_5 ring and account for the increasing apparent intensity of this absorption with n. Absorption maxima for $C_6H_5[Si(CH_3)_2]_3C_6F_5$ are reported [57] at 208 mμ (ϵ 21,300), 221 mμ (sh) and 239 mμ (ϵ 14,700).

Creemers and Noltes [35] in reporting the spectra of a number of mixed Ge and Sn compounds, such as **111–114**, note that the difference in absorption maximum of compound **111** (λ_{max} 242 mμ) as compared to $(C_6H_5)_3SnGe(C_6H_5)_3$ (λ_{max} 243 mμ) is substantially less than that of $(C_6H_5)_3SiSi(CH_3)_3$ (λ_{max} 236 mμ) and $(C_6H_5)_3SiSi(C_6H_5)_3$ (λ_{max} 246 mμ), and suggest this may reflect better metal–phenyl orbital overlap in the case of silicon. Substitution of Sn for Ge results in blue shifts of the transitions. Branched chain derivatives such as **113** (λ_{max} 269 mμ) do not appear to absorb at any shorter wavelength than their straight chain analogs **114** (λ_{max} 252 mμ).

$(C_2H_5)_3SnGe(C_6H_5)_3$ **(111)**

$(C_6H_5)_3Ge{-}Sn(C_2H_5)_2{-}Ge(C_6H_5)_2{-}Sn(C_2H_5)_2{-}Ge(C_6H_5)_3$ **(112)**

$(C_6H_5)_3Ge{-}Sn(C_2H_5)(Ge(C_6H_5)_3){-}Ge(C_6H_5)_3$ (Sn bearing C_2H_5 above and $Ge(C_6H_5)_3$ below) **(113)**

$(C_6H_5)_3Ge{-}Sn(C_2H_5)_2Ge(C_6H_5)_3$ **(114)**

The ultraviolet spectra of compounds of the general series $[(C_6H_5)_3M]M'$, where M is Sn or Pb and M′ is Ge, Sn, or Pb, are reported by Drenth and co-workers [40]. These compounds exhibit a principal intense transition in the 276–358 mμ region of 60 to 80 × 10^3. In addition, however, where M or M′ is Pb a second transition of lower intensity (ϵ 20 to 30 × 10^3) is usually found at a longer wavelength. It can also be argued that the pronounced shoulder near 310 mμ (ϵ 20,000) in the spectrum of hexacyclohexyldilead represents a similar "extra" transition in the dilead derivatives. Drenth and co-workers have suggested the possibility of ascribing this additional transition to an interaction between the filled lead 5*d* orbital (or appropriate hybrids) and the vacant orbitals, such as Pb 4*f* or Sn 5*d*, of the adjacent metal. This additional transition in the spectra of the lead compounds could then be ascribed to a transition between the resultant bonding and antibonding molecular orbitals.

The α and/or ω substitution of the permethylpolysilanes by halogen has little effect on the transition energy but does (with one or two exceptions) lead to an increase in the intensity of the absorption (see Addendum).

REFERENCES

1. K. A. Andrianov, V. I. Sidorov, V. P. Bazou, and L. M. Khanashvili, *Izv. Akad. Nauk SSSR Ser. Khim.* **1966**, 179.
2. F. Agolini, S. K. Lemenko, I. Csizmadia, and K. Yates, *Spectrochim. Acta* **24A**, 169 (1968).
3. H. Alt, H. Bock, F. Gerson, and J. Heinzer, *Angew. Chem. Intern. Ed. Engl.* **6**, 942 (1967).

4. D. R. Armstrong and P. G. Perkins, *Theoret. Chim. Acta* **5**, 69 (1966).
5. W. H. Atwell, D. R. Weyenberg, and H. Gilman, *J. Org. Chem.* **32**, 885 (1967).
6. R. E. Bailey and R. West, *J. Organometal. Chem.* (*Amsterdam*) **4**, 430 (1965).
7. M. Ballester and J. Castaner, *J. Am. Chem. Soc.* **82**, 4259 (1960).
8. R. Baney and R. Krager, *Inorg. Chem.* **3**, 1657 (1964).
9. S. H. Bauer and C. F. Aten, *J. Chem. Phys.* **39**, 1254 (1963).
10. J. A. Bedford, J. R. Bolton, and R. H. Prince, *Trans. Faraday Soc.* **59**, 53 (1963).
11. S. Bell and A. D. Walsh, *Trans. Faraday Soc.* **62**, 3005 (1966).
12. R. A. Benkeser, C. E. De Boer, R. E. Robinson, and D. M. Sauve, *J. Am. Chem. Soc.* **78**, 682 (1956).
13. R. S. Berry, *J. Chem. Phys.* **38**, 1934 (1963).
14. H. Bock and H. Alt, *Angew. Chem. Intern. Ed. Engl.* **6**, 943 (1967).
15. H. Bock and H. Alt, *Chem. Commun.* **1967**, 1299.
15a. H. Bock, H. Alt, and H. Seidl, *J. Am. Chem. Soc.* **91**, 355 (1969).
16. K. Bowden and E. A. Braude, *J. Chem. Soc.* **1952**, 1068.
17. K. Bowden, E. A. Braude, and E. R. H. Jones, *J. Chem. Soc.* **1946**, 948.
18. G. V. Boyd and N. V. Singer, *Tetrahedron* **22**, 3383 (1966).
19. E. H. Braye, W. Hubel, and I. Caplier, *J. Am. Chem. Soc.* **83**, 4406 (1961).
20. A. G. Brook, *J. Am. Chem. Soc.* **79**, 4373 (1957).
21. A. G. Brook, D. G. Anderson, J. M. Duff, P. F. Jones, and D. M. MacRae. *J. Am. Chem. Soc.* **90**, 1076 (1968).
22. A. G. Brook, K. Kivisk, and G. E. LeGrow, *Can. J. Chem.* **43**, 1175 (1965).
23. A. G. Brook and G. J. D. Peddle, *J. Organometal. Chem.* (*Amsterdam*) **5**, 106 (1966).
24. A. G. Brook and J. B. Pierce, *Can. J. Chem.* **42**, 298 (1964).
25. A. G. Brook and J. B. Pierce, *J. Org. Chem.* **30**, 2566 (1965).
26. A. G. Brook, M. A. Quigley, G. J. D. Peddle, N. V. Schwartz, and C. M. Warner, *J. Am. Chem. Soc.* **82**, 5102 (1960).
27. D. A. Brown and N. J. Fitzpatrick, *J. Chem. Soc. A* **1967**, 316.
28. J. F. Brown and P. I. Prescott, *J. Am. Chem. Soc.* **86**, 1402 (1964).
29. E. Carberry and R. West, *J. Organometal. Chem.* (*Amsterdam*) **6**, 582 (1966).
30. L. H. Chan and E. G. Rochow, *J. Organometal. Chem.* (*Amsterdam*) **9**, 231 (1967).
31. J. Chatt and A. Williams, *J. Chem. Soc.* **1954**, 4403.
32. W. D. Closson and H. B. Gray, *J. Am. Chem. Soc.* **85**, 290 (1963).
33. F. A. Cotton and G. Wilkinson, "Advanced Inorganic Chemistry." Wiley (Interscience), New York, 1962.
34. Donald P. Craig, A. Maccoll, R. S. Nyholm, L. E. Orgel, and L. E. Sutton, *J. Chem. Soc.* **1954**, 332.
35. H. M. J. C. Creemers and J. G. Noltes, *J. Organometal. Chem.* (*Amsterdam*) **7**, 237 (1967).
36. C. W. N. Cumper, A. Melnikoff, and A. I. Vogel, *J. Chem. Soc. A* **1966**, 242.
37. M. Curtis and A. L. Allred, *J. Am. Chem. Soc.* **87**, 2554 (1965).
38. L. C. Cusachs and J. R. Linn Jr., *J. Chem. Phys.* **46**, 2919 (1967).
39. I. M. T. Davidson and I. L. Stephenson, *J. Organometal. Chem.* (*Amsterdam*) **7**, P-24 (1967); *Chem. Commun.* **1966**, 747.
40. W. Drenth, M. J. Janssen, G. J. M. Van Der Kerk, and J. A. Vliegenthart, *J. Organometal. Chem.* (*Amsterdam*) **2**, 265 (1964).
41. W. Drenth, L. C. Willemsens, and G. J. M. Van Der Kerk, *J. Organometal. Chem.* (*Amsterdam*) **2**, 279 (1964).
42a. C. Eaborn, "Organosilicon Compounds." Butterworth, London and Washington D.C., 1960.

42b. C. Eaborn, *J. Chem. Soc.* **1953**, 3148.
42c. C. Eaborn, *J. Chem. Soc.* **1953**, 4154.
42d. C. Eaborn and R. Shaw, *J. Chem. Soc.* **1954**, 2027.
42e. C. Eaborn and R. Shaw, *J. Chem. Soc.* **1955**, 3306.
42f. C. Eaborn and S. H. Parker, *J. Chem. Soc.* **1954**, 939.
42g. R. Eastmond and D. R. Walton, *Chem. Commun.* **1968**, 204.
43. E. A. V. Ebsworth, *Chem. Commun.* **1966**, 530.
44. E. A. V. Ebsworth and S. G. Frankiss, *J. Chem. Soc.* **1963**, 661.
45. Yu. P. Egorov and R. A. Loktionova, *Teoriya i Eksperim. Khim. Akad. Nauk Ukr. SSR* **1**, 160 (1965).
46. F. W. G. Fearon and H. Gilman, *J. Organometal. Chem.* (*Amsterdam*) **10**, 409 (1967).
47. H. B. Fritz and K. E. Scharzhans, *J. Organometal. Chem.* (*Amsterdam*) **5**, 181 (1966).
48. K. Fukui, K. Kato, and T. Yonezawa, *Bull. Chem. Soc. Japan*, **33**, 1197 (1960).
49. J. M. Gaidis and R. West, *J. Am. Chem. Soc.* **86**, 5699 (1964).
50. H. Gilman and W. H. Atwell, *J. Organometal. Chem.* (*Amsterdam*) **4**, 176 (1965).
51. H. Gilman, W. H. Atwell, and G. L. Schwebke, *Chem. Ind.* (*London*) **1964**, 1063.
52. H. Gilman, W. H. Atwell, and G. L. Schwebke, *J. Organometal. Chem.* (*Amsterdam*) **2**, 369 (1964).
53. H. Gilman, W. Atwell, P. K. Sen, and C. L. Smith, *J. Organometal. Chem.* (*Amsterdam*) **4**, 163 (1966).
54. H. Gilman and D. R. Chapman, *J. Organometal. Chem.* (*Amsterdam*) **5**, 392 (1966).
55. H. Gilman and G. E. Dunn, *J. Am. Chem. Soc.* **72**, 2178 (1950).
56. H. Gilman and R. Harrell, *J. Organometal. Chem.* (*Amsterdam*) **9**, 67 (1967).
57. H. Gilman and P. J. Morris, *J. Organometal. Chem.* (*Amsterdam*) **6**, 102 (1966).
58. H. Gilman and C. L. Smith, *J. Organometal. Chem.* (*Amsterdam*) **6**, 665 (1966).
59. I. P. Goldshtein, E. N. Gurzanova, N. N. Zemlyanski, O. P. Syutkina, F. M. Panov, and K. A. Kocheskov, *Izv. Akad. Nauk SSSR Ser. Khim.* **1967,** 2201.
60. L. Goodman, A. Konstam, and L. H. Sommer, *J. Am. Chem. Soc.* **87**, 1013 (1965).
61. G. A. Gornowicz and J. W. Ryan, *J. Org. Chem.* **31**, 3439 (1966).
62. G. N. Gorshkova, M. A. Chubarova, A. M. Sladkov, L. K. Luneva, and V. I. Kasatochkin, *Zh. Fiz. Khim.* **40**, 1433 (1966).
63. B. G. Gowenlock and K. J. Morgan, *Spectrochim. Acta* **17**, 370 (1961).
64. M. Gverdtsiteli, T. P. Doksopulo, M. M. Menteshashvidi, and I. Abkhazava, *Soobshch. Akad. Nauk Gruz.* **40**, 333 (1965).
65. D. N. Hague and R. H. Prince, *Proc. Chem. Soc.* **1962**, 300.
66. D. N. Hague and R. H. Prince, *Chem. Ind.* (*London*) **1964**, 1492.
67. D. N. Hague and R. H. Prince, *J. Chem. Soc.* **1965**, 4690.
68. I. Haiduc, I. Haiduc, and H. Gilman, *J. Organometal. Chem.* (*Amsterdam*) **11,** 459 (1968).
68a. Y. Harada, J. N. Murrell, and H. Sheena, *Chem. Phys. Letters* **1**, 595 (1968).
69. D. F. Harnish and R. West, *Inorg. Chem.* **2**, 1082 (1963).
70. E. Hennge and H. Renter, *Naturwissenschaften* **49**, 513 (1962).
71. R. Hoffmann, *J. Chem. Phys.* **39**, 1397 (1963).
72. G. Husk and R. West, *J. Am. Chem. Soc.* **87**, 3993 (1965).
73. H. H. Jaffe and M. Orchin, "Theory and Applications of Ultraviolet Spectroscopy." Wiley, New York, 1962.

74. F. Johnson, R. S. Gohlke, and W. Nasutavicus, *J. Organometal. Chem.* (*Amsterdam*) **3**, 233 (1965).
75. S. Katagiri and C. Sandorfy, *Theoret. Chim. Acta* **4**, 203 (1966).
76. L. I. Katzin, *J. Chem. Phys.* **23**, 2055 (1955).
77. G. Klopman, *Tetrahedron Suppl.* **19**, 2, 111 (1963).
78. H. P. Koch, *J. Chem. Soc.*, **1949**, 387.
79. C. Kruger, E. G. Rochow, and U. Wannagat, *Chem. Ber.* **96**, 2132 (1963).
80. C. Kruger and U. Wannagat, *Z. Anorg. Allgem. Chem.* **326**, 288, 296 (1964).
81. J. A. Ladd, *Spectrochim. Acta* **22**, 1157 (1966).
82. S. R. LaPaglia, *J. Mol. Spectry.* **7**, 427 (1961).
83. S. R. LaPaglia, *Spectrochim. Acta* **18**, 1295 (1962).
84. M. F. Lappert and J. Lorberth, *Chem. Commun.* 1967, 836.
85. J. Marrot, J. Maire, and J. Cassan, *Compt. Rend.* **260**, 3931 (1965).
86. S. F. Mason, *Quart. Rev.* **15**, 287 (1961).
87. S. F. Mason, *Mol. Phys.* **5**, 3431 (1962).
88. F. A. Matsen, *J. Am. Chem. Soc.* **72**, 5243 (1950).
89. G. Milazzo, *Gazz. Chim. Ital.* **71**, 73 (1941).
90. E. Müller, H. Haiss, and W. Rundel, *Chem. Ber.* **93**, 1541 (1960).
91. R. S. Mulliken, *J. Chem. Phys.* **3**, 506 (1935).
92. W. K. Musker and R. W. Ashby, *J. Org. Chem.* **31**, 4237 (1966).
93. W. K. Musker and G. Larson, *J. Organometal. Chem.* (*Amsterdam*) **6**, 627 (1966).
94. W. K. Musker and G. B. Savitsky, *J. Phys. Chem.* **71**, 431 (1967).
95. D. V. Muslin, N. S. Vasileiskaya, M. L. Khidekel, and G. A. Razuvaev, *Izv. Akad. Nauk SSSR Ser. Khim.* **1966**, 181.
96. S. Nagakura and J. Tonaka, *J. Chem. Phys.* **22**, 236 (1954).
97. J. Nagy and P. Hencsei, *J. Organometal. Chem.* (*Amsterdam*) **9**, 57 (1967).
98. J. Nagy, J. Reffy, A. Kuszmann-Borbely, and K. P. Becker, *J. Organometal. Chem.* (*Amsterdam*) **7**, 393 (1967).
99. W. P. Neumann and K. Konig, *Ann. Chem.* **677**, 1 (1964).
100. W. P. Neumann and K. Kuhlein, *Tetrahedron Letters* **23**, 1541 (1963).
101. W. P. Neumann and J. Pedain, *Ann. Chem.* **672**, 34 (1964).
102. A. Nicholson and A. L. Allred, *Inorg. Chem.* **4**, 1747 (1965).
103. L. E. Orgel, *in* "Volatile Silicon Compounds" (E. A. V. Ebsworth, ed.) p. 81. Pergamon Press, Oxford, 1963.
104. G. J. D. Peddle, *J. Organometal. Chem.* (*Amsterdam*) **5**, 486 (1966).
104a. G. J. D. Peddle, *J. Organometal. Chem.* (*Amsterdam*) **14**, 139 (1968).
105. D. Peters, *J. Chem. Soc.* **1959**, 1761.
106. V. A. Petukhov, V. F. Mironov, and A. L. Kravchenko, *Izv. Akad. Nauk SSSR Ser. Khim.* **1966**, 156.
107. V. A. Petukhov, V. F. Mironov, and P. P. Shorygin, *Izv. Akad. Nauk SSSR Ser. Khim.* **1964**, 2203.
108. L. W. Pickett and E. Sheffield, *J. Am. Chem. Soc.* **68**, 217 (1946).
109. C. G. Pitt, private communication, Triangle Research, Durham, North Carolina, 1968.
110. C. G. Pitt and M. S. Fowler, *J. Am. Chem. Soc.* **89**, 6793 (1967).
111. C. G. Pitt, L. L. Jones, and B. G. Ramsey, *J. Am. Chem. Soc.* **89**, 5471 (1967).
112. J. R. Platt, *J. Chem. Phys.* **19**, 263 (1951).
113. C. N. R. Rao, *J. Sci. Ind. Res.* (*India*) **17b**, 56 (1958).
114. C. N. R. Rao, J. Ramachandran, and A. Balasubramanian, *Can. J. Chem.* **39**, 171 (1961).

115. J. W. Raymonda and W. T. Simpson, *J. Chem. Phys.* **47**, 430 (1967).
116. V. O. Reikhsfel'd, I. E. Saratov, and L. Gubanova, *Zh. Obshch. Khim.* **35**, 2014 (1965).
117. M. B. Robin, R. Hart and N. A. Keubler, *J. Am. Chem. Soc.* **89**, 1564 (1967); *J. Chem. Phys.* **44**, 1803 (1966).
118. R. Rudman, W. C. Hamilton, S. Novick, and T. Goldfarb, *J. Am. Chem. Soc.* **89**, 5157 (1967).
119. T. Saegusa, Y. Ito, S. Kobayashi, and K. Hirota, *J. Am. Chem. Soc.* **89**, 2240 (1967).
120. H. Sakurai and M. Kumada, *Bull. Chem. Soc.* (*Japan*) **37**, 1894 (1964).
121. H. Sakurai, H. Yamamori, and M. Kumada, *Bull. Chem. Soc. Japan* **38**, 2024 (1965).
122. H. Sakurai, K. Tominaga, and M. Kumada, *Bull. Chem. Soc. Japan* **39**, 1280 (1966).
123. H. Sakurai, H. Yamamori, and M. Kumada, *Chem. Commun.* **1968**, 198.
124. C. Sandorfy, *Can. J. Chem.* **33**, 1337 (1955).
125. I. E. Saratov and V. O. Reikhsfel'd, *Zh. Obshch. Khim.* **36**, 1069 (1966).
125a. C. E. Scott and C. C. Price, *J. Am. Chem. Soc.* **81**, 2670 (1959).
126. P. P. Shorygin, V. A. Petukhov, O. M. Nefedov, S. P. Kolesnikov, and V. I. Shiryaev, *Teoriya i Experim. Khim. Akad. Nauk Ukr. SSR* **2**, 190 (1966).
127. V. I. Sidorov, V. P. Bazov, L. M. Khanashvili, and K. A. Adrianov, *Izv. Akad. Nauk SSSR Ser. Khim.* **1966**, 192.
128. N. O. V. Sonntag, S. Linder, E. I. Becker, and P. E. Spoerri, *J. Am. Chem. Soc.* **75**, 2284 (1953).
129. W. C. Steele and F. G. A. Stone, *J. Am. Chem. Soc.* **84**, 3599 (1962).
130. D. P. Stevenson, quoted in Platt [112].
131. O. M. Steward and O. R. Pierce, *J. Am. Chem. Soc.* **83**, 4932 (1961).
132. A. Streitwieser, Jr., "Molecular Orbital Theory for Organic Chemist," p. 198. Wiley, New York, 1961.
133. J. S. Thayer and R. West, *Inorg. Chem.* **3**, 889 (1964).
134. J. S. Thayer and R. West, *Advan. Organometal. Chem.* **5**, 180 (1967).
135. M. E. Volpin, V. G. Dulgova, Y. T. Struckhov, N. K. Bokig, and D. N. Kursanov, *J. Organometal. Chem.* (*Amsterdam*) **8**, 87 (1967).
136. R. Waack and M. A. Doran, *J. Phys. Chem.* **67**, 148 (1963).
137. R. Waack and M. A. Doran, *Chem. Ind.* (*London*) **1965**, 563.
138. R. West, *J. Organometal. Chem.* (*Amsterdam*) **3**, 314 (1965).
139. K. Yates and F. Agolini, *Can. J. Chem.* **44**, 2229 (1966).

CHAPTER V

Organometallic Derivatives of Group V and Tellurium

ORGANOMETALLIC DERIVATIVES OF GROUP V

A. Alkyl Derivatives

The methylamines possess [78] a long wavelength transition which is relatively insensitive to the number of methyl groups (λ_{max} CH_3NH_2, 215 mμ; λ_{max} $(CH_3)_3N$, 227 mμ), and therefore may be assigned to an $n \rightarrow 3s$ transition. A shorter wavelength transition moves rapidly with increasing alkyl substitution to a longer wavelength and receives the assignment of $n \rightarrow \sigma^*$ (λ_{max} CH_3NH_2, 174 mμ; λ_{max} $(CH_3)_3N$, 199 mμ), so that in the case of triethylamine the $n \rightarrow \sigma^*$ transition has buried the less intense $n \rightarrow 3s$ absorption.

An analogous situation to that of the amines appears to be obtained for the alkylarsines [11] and phosphines [42]. The spectrum of CH_3PH_2 has maxima at 187 (ϵ 1500), 196 (ϵ 60), and 210 mμ (ϵ 130); and that of $(CH_3)_2PH$ has a maximum at 189 mμ (ϵ 6300). The absorption band of $(CH_3)_3P$ with a maximum at 201 mμ is distinctly different, being very broad with an ϵ of 18,800, from which it would appear that the $n \rightarrow \sigma^*$ has overtaken the $n \rightarrow 4s$ transition. A very early report [79] that alkylphosphines absorbed in the 220–250 mμ region is incorrect.

The alkylcyclopentaphosphines (**115** and **116**) exhibit ultraviolet transitions [43, 56] at significantly longer wavelengths than acyclic models such as $CF_3PHP(CF_3)PHCF_3$, with λ_{max} 208 (ϵ 7600), 224 (sh) mμ (ϵ 5500). In the case of cyclotetraphosphine (**115**), where R is CF_3, λ_{max} of 221 (ϵ 3030), 239 (ϵ 3250), and 259 (ϵ 3080) mμ are obtained and where R is $(C_2H_5)_2$ CH or another aliphatic alkyl, λ_{max} of 275, 282, and 290 mμ are obtained. The cyclopentaphosphine spectra have a single broad absorption band at a longer wavelength with maxima, where R is CF_3, at 240 mμ (ϵ 5970) and 298 mμ, where R is CH_3, with considerable tailing towards the visible.

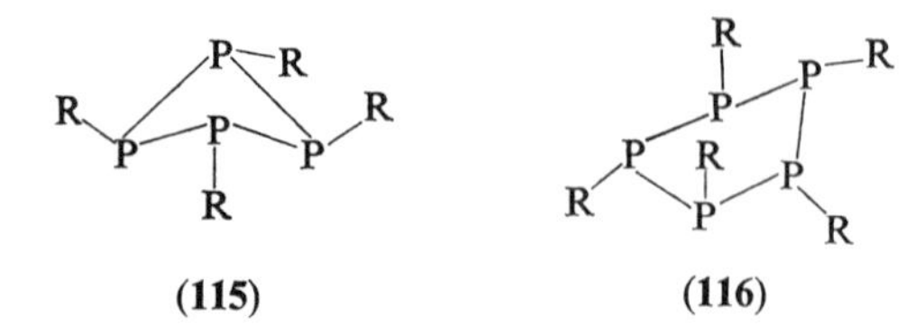

(115) (116)

The transitions of the cyclophosphines have been interpreted [56] in terms of aromatic delocalization of the phosphorus lone pair electrons through overlap of the lone pair atomic orbitals with phosphorus vacant *d* (or appropriate hybrid) atomic orbitals. The ground state of the molecule may then be depicted [43] by resonance hybrids of structures **117a–117b** and the electronic transitions become analogous to those in benzene and other aromatic compounds.

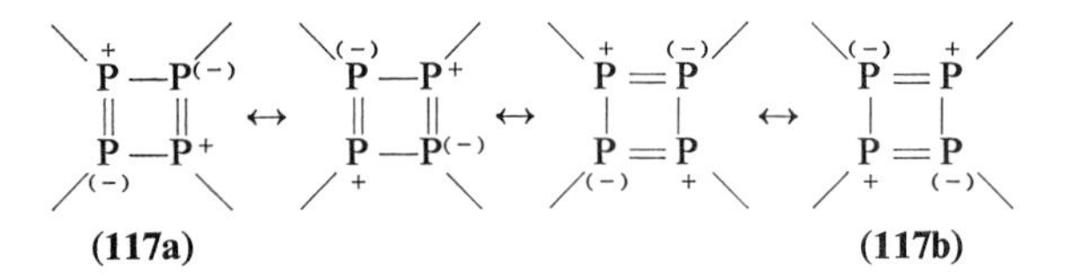

(117a) (117b)

Support for the importance of such ground state *d–p* π bonding can be found [43] in the much smaller pK_a's of the alkyltetracyclophosphines ($pK_a = 1$) compared with tertiary phosphines such as $(CH_3)_3P$ ($pK_a = 8.6$) and the lower activation energy [53] for inversion of configuration in asymmetric diphosphines RR'PPRR' as compared to monophosphines.

Free electron molecular orbital calculations were carried out by Mahler and Burg [56] which predicted longest wavelength transition energies of 260 mμ for the four-membered ring and 230 mμ for the five-membered ring. As may be seen from the free electron orbitals indicated in Fig. 49, the tetracyclophosphine is predicted to be a diradical, which it is not. The obvious rationale is that the degeneracy of the orbitals is removed by the nonplanarity of the ring. This removal of degeneracy, however, introduces another transition at a much lower energy than that predicted at 260 mμ, which if the free electron model is correct, might be found in the near infrared. Where the free electron model predicts a transition energy of 230 mμ for the methyl pentamer, the transition is actually found near 300 mμ; this difference could be rationalized as the expected result of configuration interaction between the degenerate excited states as is observed in the simple Hückel molecular orbital prediction of the benzene spectrum.

Cowley and White [20] found good agreement between calculated and

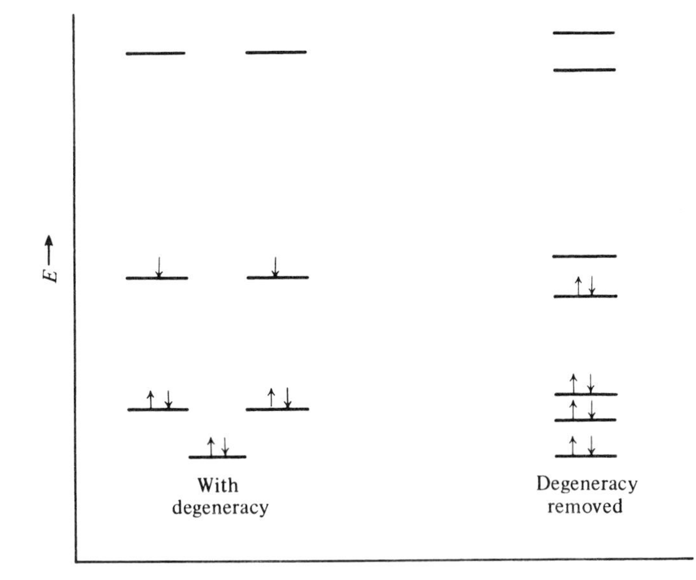

FIG. 49. Molecular orbitals according to free electron model for $(RP)_4$.

observed spectra of $(CF_3P)_4$ using the extended Hückel method only if the phosphorus 3*d* orbitals were included in the calculations.

A number of other "explanations" of the absorption of the cyclic phosphines are also possible, however. For example, if only the phosphorus lone pair orbitals are used to create delocalized molecular orbitals, the energy of the highest filled orbital of $CF_3PHP(CF_3)PHCF_3$ is 0.4β higher than that of $CF_3PHPHCF_3$. Assuming then promotion of the electron to a σ^* level by analogy with the monophosphine, the difference between the 204 and 224 mμ transitions leads to a 1.2 eV value of β, with the resulting predicted transition energies for the cyclotetraphosphines and cyclopentaphosphine in the neighborhood of 260 and 230 mμ, respectively. Again the cyclopentaphosphine transition would originate from degenerate orbitals and configuration interaction could move the transition toward the red. Whereas the inductive effect of substituting CF_3 for CH_3 in $(RP)_4$ or $(RP)_5$ could legitimately be expected to have only a small effect on an "aromatic" transition of the type proposed by Mahler and Burg, since the first-order inductive effect would be to lower all phosphorus molecular orbitals equally, a transition to $—CF_3\ \sigma^*$ should be at significantly higher energy than that to $—CH_3\ \sigma^*$. For example, $(CF_3)_3P$ has no transition [56] above 200 mμ.

As still another alternative or perhaps in addition to the above, by analogy with the spectra of the polysilanes (see Chapter IV), we might postulate electronic excitation to vacant delocalized 3*d* (or 4*s* or 4*p*) molecular orbitals with only a minimum of *p–d* π bonding or electron delocalization by *d* orbitals in the ground state.

We do not wish to imply by the above arguments that the essential features of the Mahler-Burg description of the cyclophosphine transitions are necessarily incorrect, but merely to point out what we believe to be equally possible interpretations, even though in a sense this "muddies the water" without shedding additional light. Certainly other explanations may be offered for the lower basicity of the cyclic phosphines, such as greater *s* character in the cyclic lone pair electron orbitals than in the monophosphines. Therefore, it seems best in this context to consider as many reasonable assignments as possible.

The ultraviolet spectrum of $(CF_3As)_4$ has been reported by Cowley *et al.* [19] to have maxima at 197 mμ (ϵ uncertain), and 224 mμ (ϵ 3900). The shorter wavelength absorption compared with $(CF_3P)_4$ was taken to suggest a more restricted delocalization of lone pair electrons, presumably because of decreased *d–p* π bonding. It may also reflect of course a decrease in ground state *p–p* π interaction or excited state *d–d* bonding.

B. Alkenyl and Alkynyl Derivatives

An earlier report that the spectrum of dibutylvinylphosphine had a maximum only at 182 mμ appears to be superseded by the work of Weiner and Pasternack [81] on the ultraviolet spectra of vinylphosphines and -arsines $(CH_2{=}CH)_nMR_{(3-n)}$. A selection of their data is given in Table XVI. (For early work on vinylarsine chlorides [64] see the Addendum, last page.)

TABLE XVI

Ultraviolet Maxima of Vinyl Phosphines and Arsines

Compound	λ_{max}, mμ ($\epsilon \times 1.0^{-3}$)
$(CH_3)_2AsC_2H_3$	231 (3.07)
$(C_2H_5)_2AsC_2H_3$	235 (2.81)
$(C_2H_5)_2PC_2H_3$	244 (2.71)
$C_2H_5As(C_2H_3)_2$	230 (4.27)
$C_2H_5P(C_2H_3)_2$	237 (5.20)
$(C_2H_3)_3As$	227 (4.98)
$(C_2H_3)_3P$	235 (6.41)

The alkylvinylphosphines and arsines exhibit a single broad absorption above 220 mμ, and a second intense, short wave length transition with a maximum somewhere below 200 mμ. These transitions of the vinylphosphines and -arsines may be compared with the transitions of the enamines [67] **118** (235 mμ) and **119** (227 mμ).

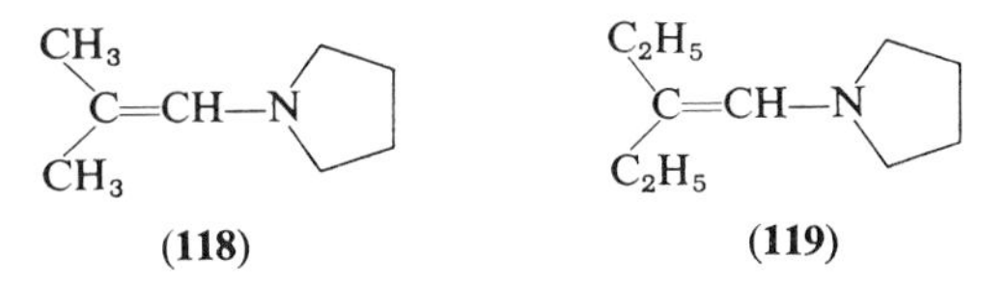

It is of interest at this point to evaluate the relative energies of the N, P, and As lone pair electrons. Although there is a large difference between the first ionization potential of N (14.5 eV) and P (10.9 eV) or As (10.5 eV), a large part of the difference disappears if we compare $(CH_3)_3N$ (9.3 eV) and $(CH_3)_3As$ (8.3 eV) ionization potentials.

Weiner and Pasternack [81] have pointed out that it is unreasonable to regard the transitions of the vinylamines, vinylphosphines, and vinylarsines as the olefin $\pi \rightarrow \pi^*$ transition which has been shifted toward the red, by the *p–p* π interaction of a lone pair orbital with a *higher* energy filled olefin π orbital, as indicated in Fig. 50a. In view of the fact that N is known to be

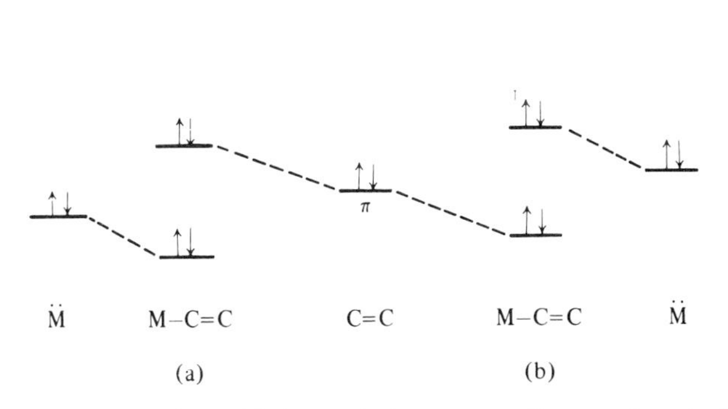

FIG. 50. Simple molecular orbital correlation energy diagram for vinyl Group V derivatives: (a) lone pair energy less than π; (b) lone pair energy greater than π.

much better at *p–p* π bonding than P or As (for example, the P and As analogs of pyridine are unknown), the largest red shift of the olefin transition would be expected for the vinylamines and not the vinylphosphines.

As an alternative one can postulate molecular orbitals formulated, as suggested by the ionization potential data, from N, P, and As lone pair orbitals *higher* in energy than the filled π orbital of ethylene (Fig. 50b). In such a case we might legitimately describe the transition as $n \rightarrow \pi^*$ or charge transfer from lone pair N, P, As, etc. to the olefin π^*, orbital since the heteroatom would have the largest mixing coefficient in the highest filled molecular orbital. The amount of charge transfer character should increase in the order As > P > N.

Weiner and Pasternack [81] have assigned the long wavelength transitions of the vinylphosphines and vinylarsines to $n \rightarrow \pi^*$ or charge transfer. With this assignment the higher energy transition of the vinylamines (enamines) relative to the vinylphosphines and vinylarsines becomes understandable as a result of the greater ionization potential of nitrogen. Other data are also in agreement with the charge transfer assignment:

(1) For a substituent R, the transition energy of $CH_2{=}CHMR_2$, where M is P or As, decreases in the order $C_3F_7 > CH_2{=}CH > CH_3 > C_2H_5$, which is in the order of increasing electron availability of M due to inductive electron release by R.

(2) The transition energies are shifted 1–2 mμ to a higher energy with a solvent change of isooctane to methanol. This effect is characteristic of $n \rightarrow \pi^*$ transitions.

(3) The transitions are not found in the spectra of the corresponding phosphine oxides, which simultaneously shows the importance of lone pair electrons on M and the absence of a major role played by P or As *d* orbitals since it would be assumed that *d* orbital interactions would be more important where M bears a formal positive charge.

(4) In view of the similar ionization potentials expected for R_3As and R_3P, the higher transition energy for the vinylarsines as compared to the vinylphosphines may be rationalized in the charge transfer model as the result of a longer C—M bond length where M is As rather than where M is P. On the basis of Eq. (8), page 9, it can be estimated that the Coulomb term e^2/r will result in a blue shift of roughly 0.3 eV for the arsine versus the vinylphosphine. Weiner and Pasternack have suggested that the lower transition energy of the vinylphosphines may reflect the relative importance of d–p π bonding.

Note however also the order [14a] of valence state ionization potentials of X (sp^4, V_3) $\rightarrow X^+$ (sp^3, V_3) for which N (13.4 eV), P (12.22 eV), and As (12.63 eV) and the measured ionization potentials of NH_3 (10.5 eV), PH_3 (10.0 eV) and AsH_3 (10.6 eV).

(5) The substitution of ethyl for methyl leads to an 8 mμ *blue* shift in the ultraviolet spectra of enamines **118** and **119**. This result is readily understood in terms of a decreased electron affinity of the olefin π^* orbital due to the greater inductive effect of C_2H_5 vs. CH_3. In molecular orbital terms the substituted carbon must have a smaller atomic orbital mixing coefficient in the highest filled molecular orbital than the lowest vacant orbital [Eq. (9), p. 16].

C. Charrier and co-workers [17, 18] have reported the ultraviolet spectra of a number of diphenylphosphine acetylenes **120** below and a few alkylphosphine acetylenes **121** along with the spectra a few of the corresponding phosphine oxides and sulfides.

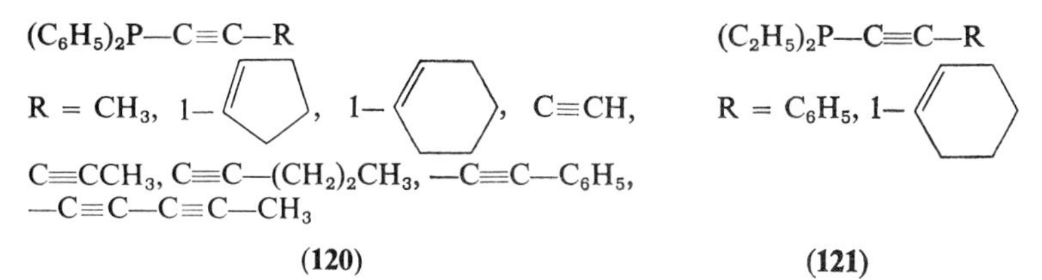

(120) **(121)**

Boglyubov and Petrov [10] report the spectra of nine sulfides of vinylphosphines and acetylenic phosphines (**122–124**), $n = 1$–3.

$$(C_6H_5)_{3-n}\overset{\overset{\displaystyle S}{|}}{P}\text{—}(CH\text{=}CHC_6H_5)_n \qquad (CH_3)_{3-n}\text{—}\overset{\overset{\displaystyle S}{|}}{P}\text{—}(C\equiv C\text{—}CH_3)_n$$

(122) **(123)**

$$(CH_3)_{3-n}\overset{\overset{\displaystyle S}{|}}{P}\text{—}(C\equiv C\text{—}C_6H_5)_n$$

(124)

In all compounds **123–124** as n increases, there is a transition of moderate intensity which moves toward the red. For example, in the case of compound **123**, for $n = 1, 2, 3$, the maxima are found at 221, 234, and 250 mμ. This result was interpreted as evidence d–p π conjugation with the phosphorus atom. The d orbital effect may be somewhat increased by the inclusion of the sulfur d orbitals, just as disilanes have a larger spectroscopic effect than monosilanes on the spectra of phenyl and vinyl derivatives.

The spectra of vinyldi-n-butylphosphine oxide and vinylphosphonic acid derivatives (**125–128**) all have [51] intense absorption maxima in the region 174–178 mμ, indicating the absence of significant d orbital perturbation of the $\pi \rightarrow \pi^*$ transition of a primary olefin. It is interesting to note once again however the homoconjugation effect found in the spectra of the silanes where the maximum of $CH_2\text{=}CHCH_2PO(OC_4H_9)_2$ at 182 mμ is significantly shifted toward the red compared to that of $CH_2\text{=}CHPO(OC_4H_9)_2$ at 177 mμ.

$$CH_2\text{=}CH\text{—}POCl_2 \quad \textbf{(125)} \qquad CH_2\text{=}CHPO[N(CH_3)_2]_2 \quad \textbf{(126)}$$

$$CH_2\text{=}CH\text{—}PO(OC_4H_9)_2 \quad \textbf{(127)} \qquad CH_2\text{=}CHPO(CH_3)(OC_4H_9) \quad \textbf{(128)}$$

It is obvious that the large red shift of the $\pi \rightarrow \pi^*$ transition found in α,β-unsaturated ketones due to the extended conjugation provided by the carbonyl finds no analogy in the vinylphosphine oxides or phosphonic acid derivatives in terms of resonance structures such as **129**. Arguments relative to kinetic data indicating *d* orbital participation for nucleophilic attack at the double bond should be applied to a transition state such as **130** where *d–p* π overlap must be more important than for neutral carbon.

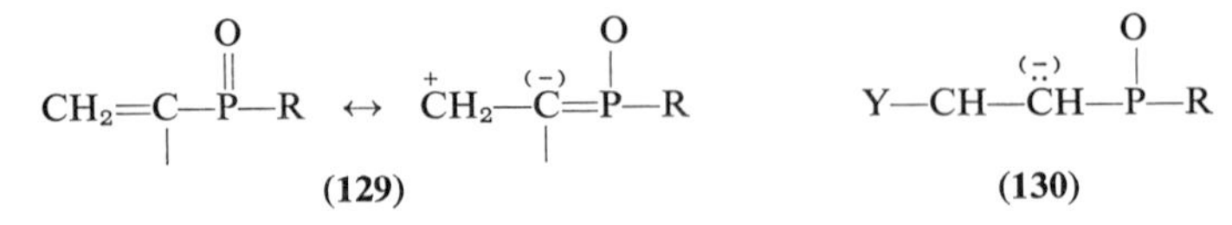

C. Heterocyclic Organometallic Compounds

Although the absorption spectra of pentaphenylphosphole (**131**), λ_{max} 248, 320, 358 mμ, and pentaphenylarsole (**132**), λ_{max} 224, 360 mμ, are complicated by the presence of phenyl group [13], it is perhaps worth noting that the transitions occur at about the same energy. In the solid state both compounds show an emission spectrum maximum at 480 mμ. Interestingly enough, the absorption maxima of the oxide of structure **131**, where R is C_6H_5, are found at 260, 272, and 385 mμ with emission in the solid at 528 mμ.

Since the 320 mμ transition of the pentaphenylphosphole is absent in the oxide and the corresponding hexaphenylsilole, it has been suggested [45] that this absorption may be due to an $n \rightarrow \pi^*$ transition.

The 2,4,6-triphenylphosphabenzene (**133**) absorption [60, 62] maximum in methanol at 278 mμ (ϵ 41,000) may be compared with those of the analogous 2,4,6-triphenylpyridine at 254 mμ (ϵ 49,500) and 317 (ϵ 9390) and 1,3,5-triphenylbenzene, λ_{max} 254 mμ (ϵ 56,000). Substitution of a *p*-CH_3O into the phenyl rings of compound **133** moves the transition to 299 mμ [62]. The lower energy of the 278 mμ band of the triphenylphosphabenzene relative to those of the triphenylpyridine and triphenylbenzene is presumably the result of the lower ionization potential and greater polarizability of phosphorus. It has been observed that the ultraviolet spectrum [26] of 9-phosphaanthracene (**134**), λ_{max} 380, 400, 429 mμ; log ϵ 3.9, except for the

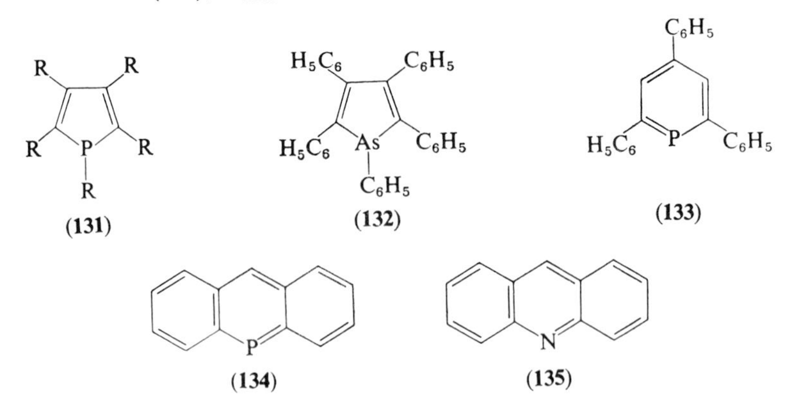

substantial red shift of the spectrum, resembles in structure the spectrum of anthracene more closely than that of acridine (**135**), possibly because of the similar electronegativities of phosphorus and carbon (see Addendum).

The ultraviolet spectrum of compound **136** consists of an intense maximum at 267 mμ (ϵ 18,600) emerging from a large shoulder. This may be compared with a transition energy of 269 mμ (ϵ 4600) for compound **137** or 278 mμ (ϵ 2500) and 217 mμ (ϵ 3000) for *trans*-diphenylphosphinestyrene (**138**). It is suggested [1] that this and NMR data support the hypothesis of delocalization of the 4π electrons of compound **136** over the phosphorus atoms.

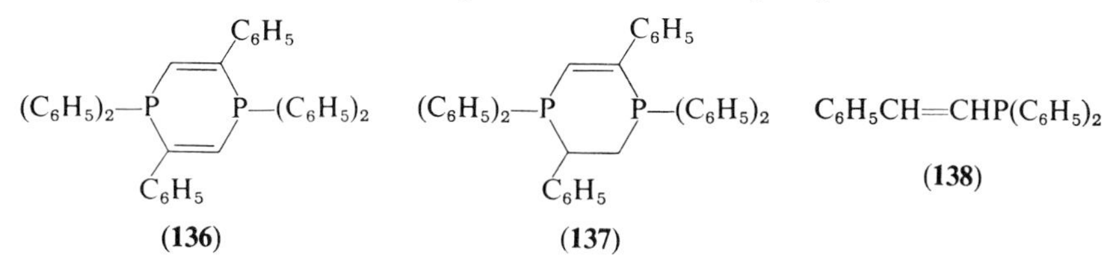

Compound **139** has an absorption maximum in its spectrum at λ_{max} 327 mμ (ϵ 7300) [62]. The phosphine oxide spectrum in methanol shows two absorption maxima at 340 and 432 mμ due to the tautomeric equilibrium between compounds **140** and **141**.

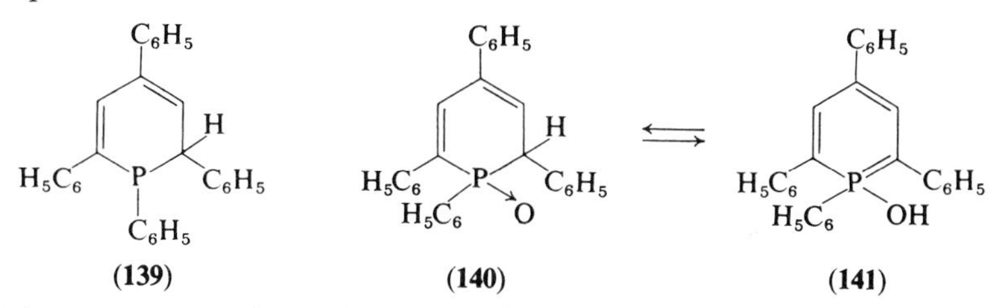

The spectrum of 10-phenyl-9,10-dihydro-9,10-azaphosphaphenanthrene was observed by Dewar and Kubba [28a] to resemble closely that of the corresponding 10-phenyl-10,9-borazarophenanthrene, which was taken to suggest that the azaphosphaphenanthrene and related compounds might be aromatic. However, NMR measurements on azaphosphaphenanthrenes by Gaidis and West [83] fail to show electron delocalization in the ground state.

D. Phosphoranes (Ylides)

The ultraviolet–visible spectra of the phosphoranes or phosphorus ylides with the general formula $R_3P—\ddot{C}R_2$ are of particular interest since they would offer close to optimum conditions for the observation of the effect of *p–d* bonding on the electronic transitions of organophosphorus compounds. There would appear to be ample chemical evidence for the importance of *d–p* bonding in the ground state of phosphoranes [45].

Perhaps the most interesting comparisons are to be made between the spectra of $R_3P{=}CR_2$, the free or solvent separated carbanion $R_2\ddot{C}$-alkyl, the intimate ion pair R_2(alkyl)C:M^+, and the olefin $R_2C{=}CH$. Along these lines comparisons of the following sort can be made.

Compound **142** in ethanol has an absorption maximum [31] at 267 mμ (ϵ 7600) which should probably be assigned to perturbed $\pi \rightarrow \pi^*$ transitions of the benzene ring since the same transition is found through spectra of $(C_6H_5)_3P^+{-}CH_3$ and compound **143**. In addition the phosphorane **143** spectrum has a transition at 290 mμ (log ϵ 4.10), 295 mμ (log ϵ 4.01). It is this last transition of compound **143** which is to be compared with absorption of the initimate ion pair 2-butenyllithium at 291 mμ or the butadiene absorption maximum of 217 mμ.

$[(C_6H_5)_3\overset{+}{P}{-}\ddot{C}H{-}\overset{+}{P}(C_6H_5)_3]^+BF_4^-$
(142)

$[(C_6H_5)_3\overset{+}{P}{-}CH{=}CH{-}\ddot{C}H{-}\overset{+}{P}(C_6H_5)_3]^+Br^{(-)}$
(143)

CH_3–cyclopentylidene(:)–$P(C_6H_5)_3$
(144)

The only reported [63a] transition energy for a simple alkylphosphorane is the absorption maximum of 416.5 mμ (ϵ 2200) for compound **144**. The blue shift of the absorption maximum of compound **142** relative to **144** is probably best understood in terms of the qualitative simple molecular orbital diagram of Figs. 51a and 51b. In the case of compound **144** (Fig. 51b), the

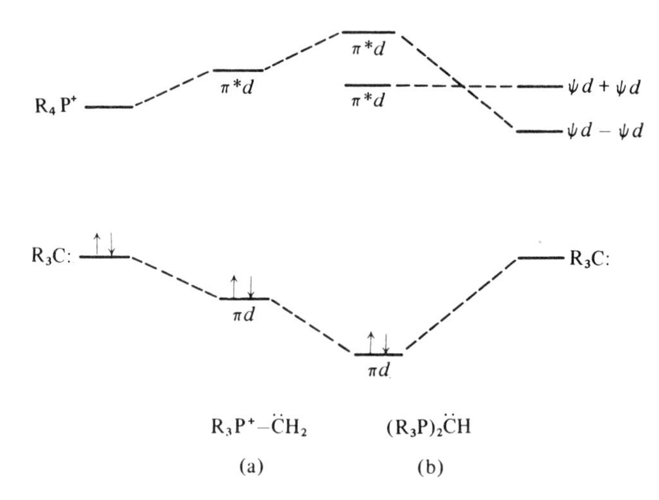

FIG. 51. (a) Qualitative molecular orbitals correlation diagrams for $R_3P^+{-}\ddot{C}H_2^-$. (b) Qualitative molecular orbitals correlation diagrams for $(R_3P)_2\ddot{C}H$.

molecular orbitals are obtained from a linear combination of the symmetry orbitals of the phosphorous taken as $d_{(1)} + d_{(2)}$ and $d_{(1)} - d_{(2)}$. A rough valence bond equivalent is to think of the transition of compound **144** as $R_3P^+—\ddot{C}R_2 \rightarrow R_3\dot{P}—\dot{C}R$ but the transition of compound **142** as $R_3P^+—C{=}PR_3 \rightarrow R_3\dot{P}—C\dot{-}PR_3$.

Similarly we can compare the longest wavelength transition 521 mμ maximum of the 9-fluorenyl solvent separated ion pair, the 9-fluorenyl-lithium intimate ion pair, λ_{max} 415, 435, and 457 mμ in ether, and the 320 mμ transition in the spectra [33] of compound **145** in methylene chloride, with the 382 mμ transition in $CHCl_3$ of 9-triphenylphosphoniumfluorenylide (**146**).

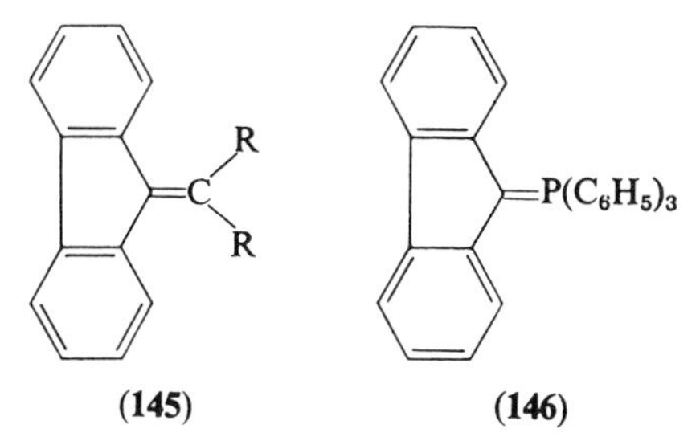

(**145**) (**146**)

The phosphorane **147** raises the intriguing question of possible aromaticity [61], but the transition energies (λ_{max} 362, 429 mμ) seem closer in energy to those of the intimate ion pair **148** (λ_{max} 395 mμ in THF) than those of naphthalene at 275 and 305 mμ, or 1-phenylbutadiene at 280 mμ.

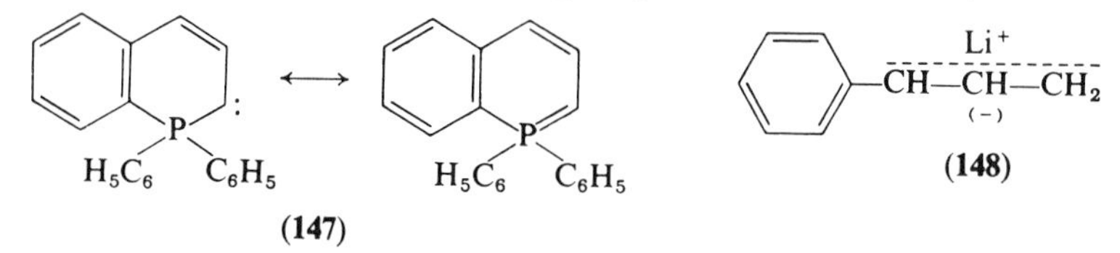

(**147**) (**148**)

From the above and similar [61] comparisons available, it appears that the best model for predicting the transition energy of a phophorane, $R_3P^+—\ddot{C}R_2$, is obviously not the hydrocarbon $R_2C{=}CH_2$ nor free carbanion $R_2\ddot{C}H$, but the intimate ion pair $R_2\ddot{C}M^+$. In fact, at least with the data available, the close correspondence between the transition energies of carbanion alkali metal ion intimate ion pairs may indicate some fundamental similarity, the most obvious one being depicted in structures such as **149** and **150**, for, unlike *p–p* π bonding wherein the electron density is greatest

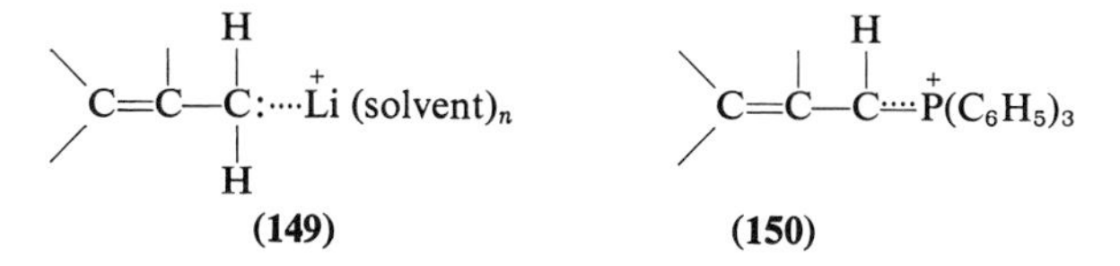

(**149**) (**150**)

midway between the atoms, *p–d* π bonding is essentially a polar bond with the greatest overlap and therefore greatest electron density, in the region of the group contributing the *p* orbital.

Somewhat in line with the above observation is the apparent red shift in the absorption maximum of compound **151** (348, 365, *475* mμ in benzene) compared with that of compound **152** in methanol at 342 mμ [62].

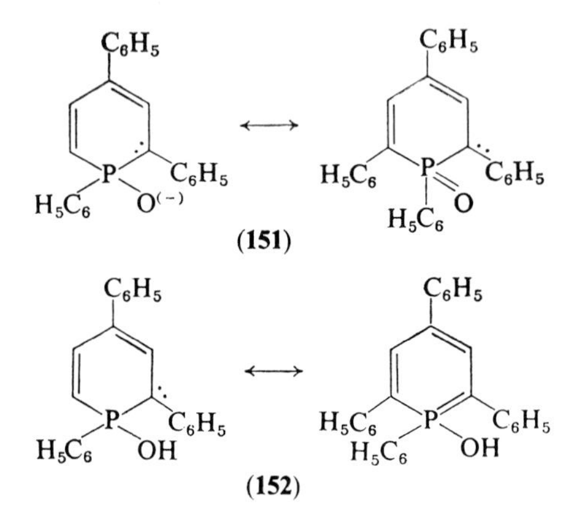

The spectrum of one of the more interesting phosphoranes [69], triphenylphosphoniumcyclopentadienylide, is given in Fig. 52. It is particularly interesting to compare the electronic transition energies of this molecule for which roughly equal contributions of resonance structures **153a** and **b** have been estimated on the basis of the observed dipole moment of 7.0 D, and the transition energies of the dimethylfulvalene (**154a** and **b**) with a dipole moment of 1.44 D, presumably the result of contributing structure **154b**. The spectra of compound **153** in acetonitrile at 250 mμ (ϵ 21,600) and 295

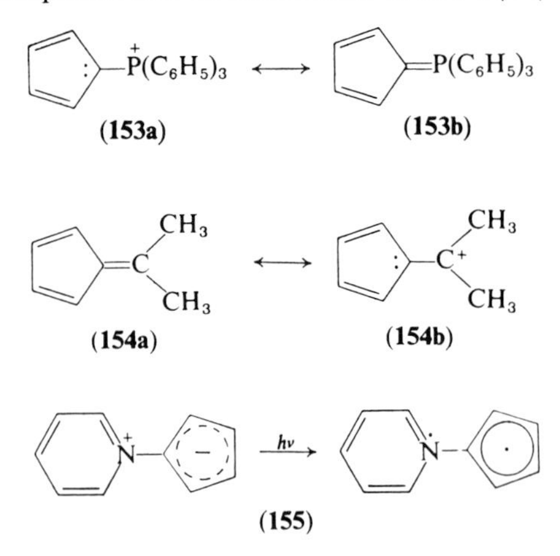

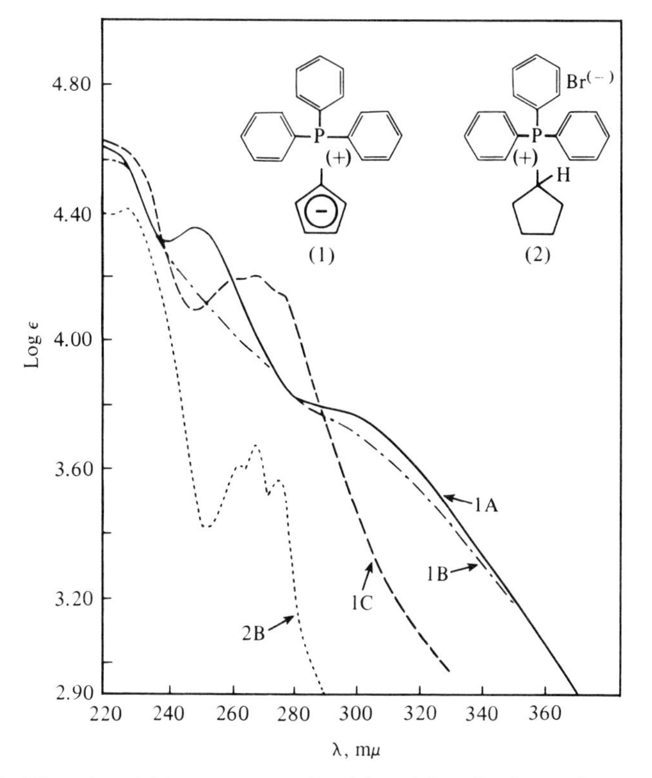

FIG. 52. The ultraviolet spectrum of triphenylphosphoniumcyclopentadienylide: (a) in acetonitrile; (b) in 95% EtOH; (C) in 95% EtOH·HCl [69, p. 68].

mμ (ε 5900) therefore bears a closer correspondence to that of the dimethylfulvalene at 270 mμ (ε 16,000) and 364 mμ (ε 200) than might otherwise have been expected from our earlier discussion. Kosower and Ramsey [52a] have suggested that by analogy with the intramolecular charge transfer transition of pyridiniumcyclopentadienylide (**155**), λ_{max} 495 mμ (ε 1.9×10^4) in CH_3CN, the long wavelength transition of compound **153** might also be designated as an intramolecular charge transfer transition from cyclopentadiene anion to triphenylphosphonium cation moiety.

The transitions at 250 and 295 mμ in compound **153** may be compared with [55] absorptions at 299 mμ (ε 26,500) and 335 mμ (sh) of the analogous arsenic (**156**) or antimony [54] compounds (**157**), λ_{max} 349 mμ (ε 42,000). Although part of the red shift between the absorption maxima of compounds **153** and **156** is undoubtedly the result of ring phenyls, the general red shift from structure **156** to **157** may reflect a decrease in *d* orbital participation in π bonding and a shift of the spectrum of the ylides to that of the free tetraphenylcyclopentadienide anion.

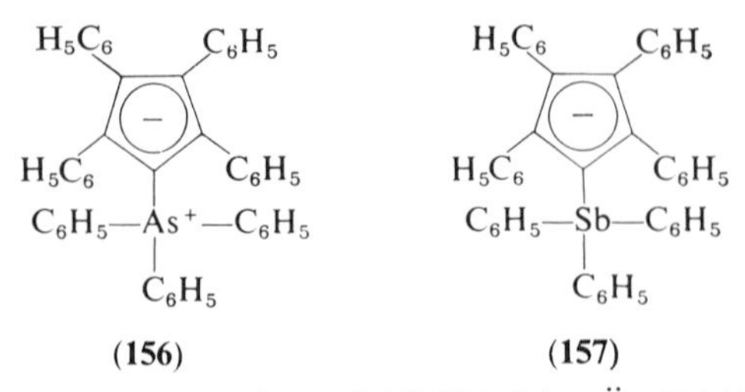

The 325 mμ (ϵ 7000) transition of $(C_6H_5)_3P^+—\ddot{C}—P^+(C_6H_5)_3$ is absent both in the spectrum of $[(C_6H_5)_3P^+]_2\ddot{C}H$ and $(C_6H_5)_3P^+O^-$. A reasonable valence bond description might be given by Eq. (17):

$$(C_6H_5)_3P^+—\ddot{C}—P^+(C_6H_5)_3 \rightarrow (C_6H_5)_3P{=}C{=}P(C_6H_5)_3 \quad (17)$$

and in this aspect the comparison with other four-electron systems such as $CH_2{=}CH—\ddot{C}HLi^+$ [λ_{max} 315 mμ (ϵ 4600)] and $CH_2{=}C{=}CHC_2H_5$ [λ_{max} 227 mμ (ϵ ~ 1000)] seem relevant.

The valence bond description as indicated is somewhat ambiguous, since in molecular orbital terms the observed absorption may correspond to a transition between molecular orbitals taken as linear combinations of the phosphorus "3*d*" orbitals and the carbon lone pair orbitals, or it may, for example, be the result of configuration interactions between charge transfer states in which the acceptor orbital is unspecified but which may be represented by a structure such as $R_3\dot{P}—\dot{C}—P^+R_3$.

Shevchuk and Dombrovski [77], and Grigorenko *et al.* [40] have reported the ultraviolet spectra of 28 compounds of the general structure $ArC(O)\ddot{C}(R)P^+(C_6H_5)_3$ in which Ar ranges in structure from phenyl and other substituted phenyls to α-furyl and threnyl, and R is H and CH_3 [77], or I, and —SCN [40]. The benzoylmethylenetriphenylphosphorane (R is H, Ar is C_6H_5) spectrum shows maxima of 268 mμ (log ϵ 3.84), 274 mμ (log ϵ 3.85), and 318 mμ (log ϵ 4.07). Again the 268–274 mμ absorption appears to be associated with the phenyl transition of the $(C_6H_5)_3P$ group whereas the 317 mμ transition is of interest as peculiar to the phosphorane structure $P^+\ddot{C}(R)C(O)Ar$. Substitution of R as I for H has only a small effect on the transition, moving it a few millimicrons toward the blue, whereas the —SCN group produces a much larger blue shift. In terms of substituent effects in the aryl ring only the *para*-nitro group produced a significant change in moving the transition to 360 mμ, which may indicate some transfer of negative charge into the ring in the excited state.

The betaine (**158**) spectrum [66] shows a dramatic change from the spectrum of 1-phenylbutadiene (λ_{max} 280 mμ) with absorption maxima at 510 mμ (log ϵ 4.25), 375 mμ (log ϵ 3.90), and 267 mμ (log ϵ 3.97) which are shifted toward the *red* with change to *less* polar solvent. The valence bond structure **158b** probably makes the predominant contribution to the excited state (see Addendum).

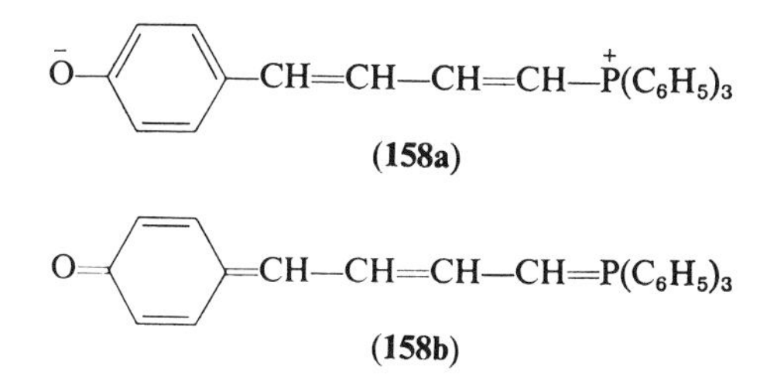

E. The $R_3P—N=Y$ Chromophores

A parallel problem to the general question of the importance of *d–p* π bonding has been one of whether or not *p–d–p* π bonding provided for extended conjugation over the contributing atoms or groups, or whether the use of separate and orthogonal *d* orbitals resulted in localized *p–d* π bonds. One of the best known systems where this problem has generated considerable discussion is the phosphonitrilic or phosphazene system (**159**).

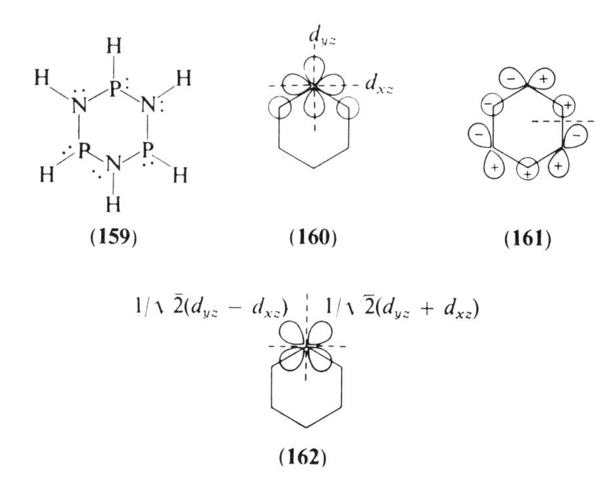

In the simplest approach Craig [22] argued that in the phosphazene system d_{xz}–p_z should be more important than d_{yz}–p_z overlap. Even in the lowest energy molecular orbital (**161**), however, for d_{xz}–p_z overlap we are left with a node between at least one P and N in a six-membered ring, and Dewar [28] and co-workers suggested that symmetry favored d_{yz}–p_z interaction. In a linear or acyclic system, other things being equal, one might conclude the Craig orbitals would be the correct ones since we do not here have the problem of joining the ends. Dewar *et al.* [28a], however, have raised the argument that the greatest *d–p* π bonding will be obtained by the use of equivalent $d_{xy} \pm d_{yz}$ hybrid orbitals, as indicated in structure **162**. Since the new hybrid orbitals are orthogonal, this represents the bonding in the phosphazene series as a set of three molecular orbitals localized on *P–N–P* (**163**). In other words, in the usual case of *p–d–p* π bonding, we should not expect extended conjugation. Since one should also consider the d_{xy} and $d_{x^2-y^2}$ orbitals and include the absence of planarity, in detail the problem becomes one of great complexity as Craig and Paddock [21] pointed out in a more recent paper, and there may be some electron delocalization. The bulk of the ultraviolet spectra data however more strongly support the argument against *p–d–p* π extended conjugation. For example, the ultraviolet spectra of the phosphazenes (**159**), phenyl phosphinimido compounds (**164**), or the phosphorane (**151**) discussed earlier show no evidence of expected aromaticity or extended conjugation involving *d* orbitals.

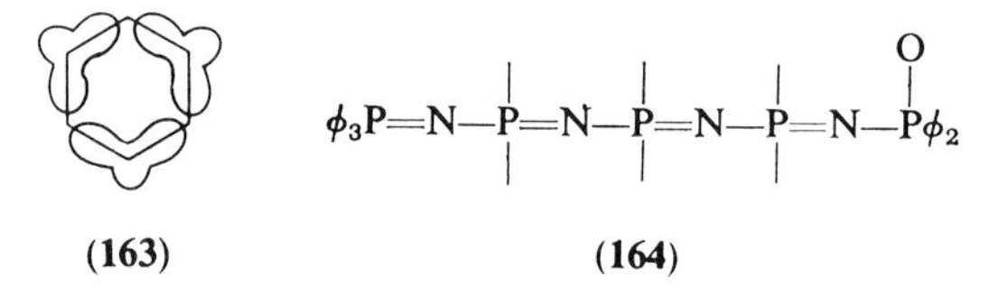

(**163**) (**164**)

The orthogonality of orbitals such as those of Dewar and Lucken does not mean the spectrum will be the simple sum of the spectra of the localized chromophore. A comparable set of "orthogonal" molecular orbitals in $CH_2{=}C{=}CHCH_2CH_3$ [λ_{max} 227 mμ (log ϵ 2.9), 170 (broad) mμ (log ϵ 3.9–4.0)] certainly does not result in a spectrum [15] which is the sum of say the spectra of $CH_2{=}CH_2$ and $CH_2{=}CH{-}CH_2CH_3$.

The quinoidal phosphonium azo-*o*-phenoxides of the general structure **165** possess transitions in the visible which are dramatically shifted toward the red compared to the absorption maxima of other phosphoniumazobenzenes, including the corresponding molecule where R is C_6H_5. For example, where R is *N*-morpholine, and $R^1 = R^2 = R^3 = R^4 = H$, the maxima are 498 m$\mu$ (log ϵ 3.07) and 350 mμ (log ϵ 2.9). This result was interpreted by Reid and Appel [70] as evidence of extended conjugation provided by resonance structures such as **166**, where R is dimethylamino or *N*-morpholine. In fact, we assume the major contributing structure to the ground state to be resonance structure **165a**, and *d* orbital participation to be more important in π^* than π orbitals. Therefore, *p–d* π interaction of the type proposed would raise the energy of π^* more than π orbitals, leading to greater transition energies. Or in slightly different terms *p–d* π bonding by the amino groups does not lengthen the $\ddot{O}^{(-)}{-}C_6H_5{-}N{=}N{-}P^+$ chromophore but may actually shorten it by competitively removing the N=P bonding. This is certainly found to be the case in the spectra of other azophosphonic acid derivatives discussed shortly.

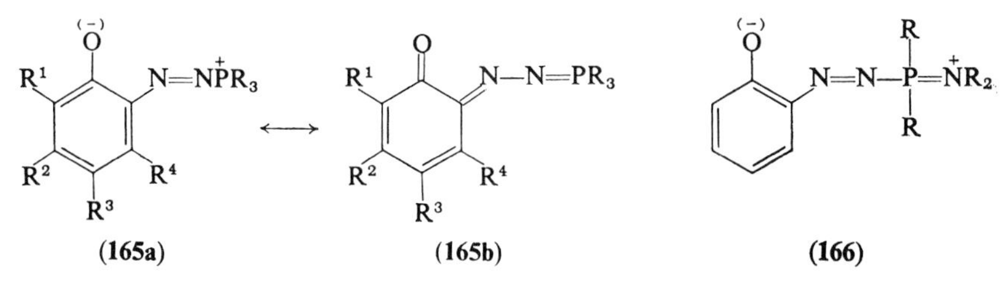

(165a) **(165b)** **(166)**

The results of Reid and Appel [70] seem to require some special explanation such as assigning the long wavelength transition to the allowed $n \rightarrow \pi^*$ of this *cis*-type structure (**167**). It must be emphasized that we know of no other evidence to support this alternative, and others can certainly devise other "possible" special explanations.

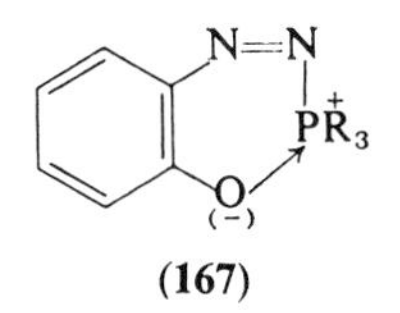

(167)

Bock and co-workers [9] have reported the ultraviolet–visible spectra of azobisphosphonic [8], phenylazophosphonic acid [7] and azophosphinic acid derivatives (**168** and **169**), where X is usually O but in a few cases S, and R is C_6H_5, OC_6H_5, $[N(CH_3)_2]$, OK, or O-alkyl. Before considering

(168) **(169)**

electronic transitions in azophosphorus compounds, however, it is necessary to examine the spectra of azoalkanes.

The first three transitions of [71] the *trans*-azoalkanes occur near 358 mμ (28,000 cm^{-1}), $\epsilon \sim 10$; 190 mμ (52,000 cm^{-1}), ϵ 4000; and 165 mμ (60,000 cm^{-1}), $\epsilon \sim 7 \times 10^3$ (Fig. 53).

It is generally agreed that the first of these transitions is $n_+ \rightarrow \pi^*$ from the nonbonding combination of lone pair atomic orbitals. Earlier assignments of the 190 mμ transition indicated either an $n_+ \rightarrow \sigma^*$ assignment or $n_- \rightarrow \pi^*$ from the bonding combination of the nitrogen lone pair; more recently, however, using Gaussian type orbitals, calculations by Robin *et al.* [71] which include σ electrons explicitly confirm the $n_+ \rightarrow \sigma^*$ assignment for this transition where the lowest σ^* is N—R(NH). In fact, in the *trans* diimide the difference in energy between the n_+ and n_- orbitals is calculated to be an astonishing 7 eV, placing $n_- \rightarrow \pi^*$ 7.0 eV higher in energy than

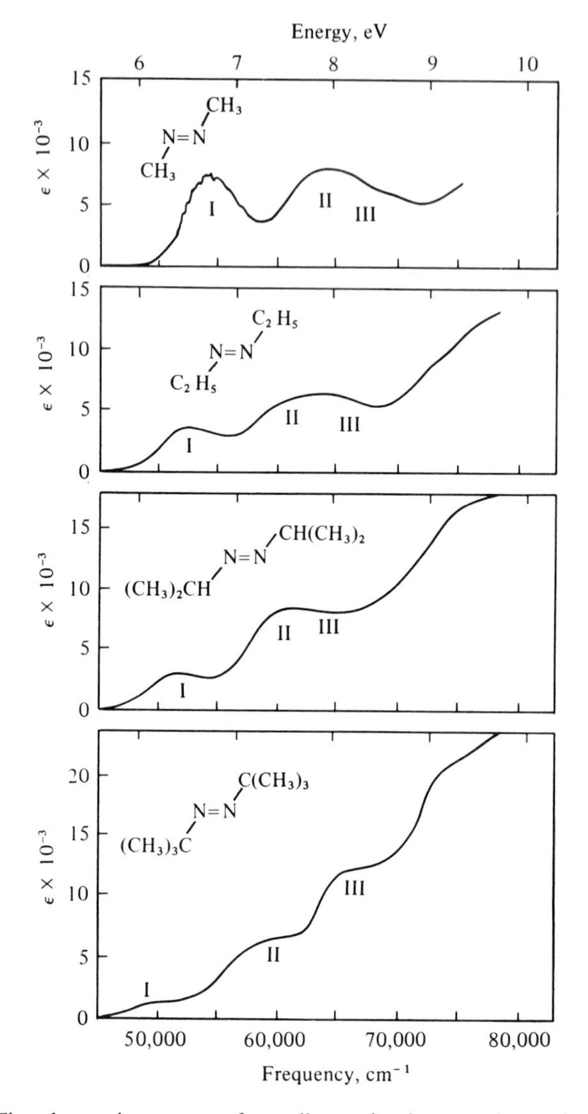

FIG. 53. The electronic spectra of azoalkanes in the gas phase: (a) azomethane; (b) azoethane; (c) azoisopropane; (d) azoisobutane [71].

$n_+ \rightarrow \pi^*$. The 165 mμ transition, which in the past has been regarded as the $\pi \rightarrow \pi^*$ transition analogous to that found in olefins, is also reassigned by Robin and co-workers to an $n_+ \rightarrow \sigma^*$ ($^1A_g \rightarrow {}^1B_u$) transition. The $\pi \rightarrow \pi^*$ transition is calculated to be somewhere in the range of 100 mμ or 12 eV, which can be compared with the lowest $\pi \rightarrow \pi^*$ transition of N_2 at 12 eV in contrast to the 6 eV transition of HC≡CH. As Robin *et al.* [71] point out, the earlier $\pi \rightarrow \pi^*$ assignment of the 165 mμ transition is strongly supported by the similarities between the spectra of *trans*-azobenzene and *trans*-stilbene, similarities which are strange if the assignments of Robin *et al.* [71] are correct.

The spectra of the azobisphosphonic and -phosphinic acid [8] derivatives [R is C_6H_5, OC_6H_5, $N(CH_3)_2$, OK] exhibit a weak ($\epsilon \sim$ 10–20) transition in the region of 570–530 mμ ($\sim$18,000 cm^{-1}) which can be assigned to $n_+ \rightarrow \pi^*$, corresponding to the 358 mμ (28,000 cm^{-1}) band of the azoalkanes (see Fig. 54). In addition, except for where X is S, R is C_6H_5 and X is O, R is OK, there

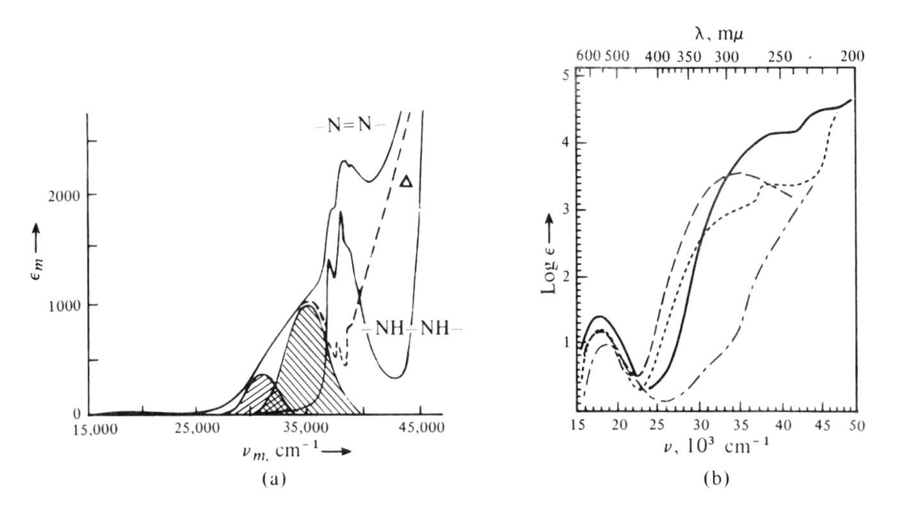

FIG. 54. The ultraviolet spectra of azobisphosphonic acid derivatives: (a) absorption difference (----) of tetraphenyl azobisphosphonate and tetraphenylhydrazino-1,2-bisphosphonate. (b) For Y_2OP—N=N—POY_2, Y is (———) C_6H_5; (----) [$N(CH_3)_2$]; (·····) OC_6H_5; (-·-·-)—OK. [H. Bock, *Angew. Chem.* **4**, 457 (1965).]

is a second absorption band near 285 mμ (35,000 cm^{-1}) which is most prominent where X is O, R is [$N(CH_3)_2$] but is also clearly evident where R is C_6H_5 or C_6H_5O on the side of the localized C_6H_5 absorptions. Gaussian analysis of this second band reveals two transitions at 286 mμ ($\epsilon \sim 10^3$) and 322 mμ ($\epsilon \sim$ 500).

The red shift of the azo $n \rightarrow \pi^*$ transition on substitution of R_3P for alkyl is substantially less than that obtained for silyl (R_3Si) substitution, but greater than that for carbonyl substitution, for example, λ_{max} $C_6H_5O_2CN$=NCO_2-C_6H_5 435 mμ. Since tetracoordinate phosphorus is, if anything, more electronegative than carbon, the approximately 10,000 cm^{-1} (1.24 eV) red shift of the azo $n \rightarrow \pi^*$ transition on replacing alkyl with X is R_2P cannot be the result of a simple inductive effect, but instead requires for explanation substantial interaction between the *d* orbitals of phosphorous and the —N=N— π^* orbital to create a new π^* *d* molecular orbital of lower energy, as indicated in the simple molecular orbital diagram of Fig. 55.

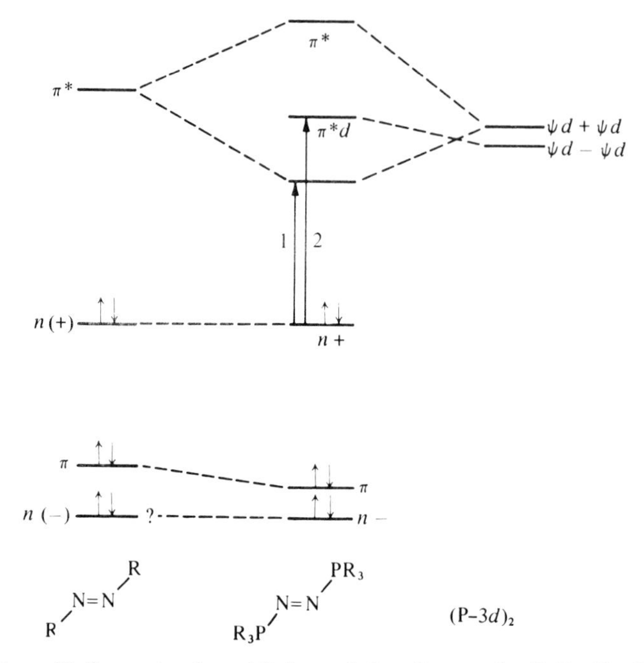

FIG. 55. A qualitative molecular orbital correlation diagram for R_3P—N=N—PR_3.

In Fig. 55 the *d* phosphorous symmetry orbitals, which are the (+) and (−) combinations of the two R_3P^+ individual atomic *d* orbitals, are assumed to be lower in energy than the —N=N— π^* orbital and after combination with the N_2 orbitals result in a set of molecular orbitals having the same symmetry characteristics as those of butadiene. The first $n_+ \rightarrow \pi^*$ transition, which we have already assigned to 530–570 mμ, is $^1A_g \rightarrow {}^1B_g$ and not allowed. The second predicted $n_+ \rightarrow \pi^*$ transition would have the symmetry $^1A_g \rightarrow {}^1A_u$ which is allowed but is perpendicularly polarized and would therefore not be as strong as might otherwise be expected.

The absorption in the region of 285 mμ, or if Gaussian analysis is correct, one of the transitions within this band would appear to be a reasonable assignment for this second $n \rightarrow \pi^*$ transition. There are two arguments against assigning this absorption to $\pi \rightarrow \pi^*$ as assumed by Bock *et al.* [9]. First, after subtracting where necessary the absorption due to the phenyl ^1L_b transitions, the transition lacks the intensity to be expected of a $\pi \rightarrow \pi^*$

transition in a *trans*-type, butadiene-type molecule, especially if the absorption intensity is divided between two transitions. The second objection relates to the observation that unless there is significant lone pair *n–d* bonding, we should expect a smaller red shift of the $\pi \rightarrow \pi^*$ transition than the $n \rightarrow \pi^*$ transition on substituting R_3P for alkyl. In fact the proposed $\pi \rightarrow \pi^*$ red shift is *at least* 25,000 cm^{-1} (assigning $\pi \rightarrow \pi^*$ of R—N=NR to 165 mμ) whereas the $n \rightarrow \pi^*$ shift is only about 10,000 cm^{-1}.

Finally, it is very possible that one or both of the transitions near 285 mμ in the spectra represent $n_+ \rightarrow \sigma^*$ transitions shifted toward the red by overlap of the *d* orbitals of one phosphorus with the backside lobe of the N—P sp^2 bond of the opposite phosphorus. In the absence of any calculations, however, it is not possible to seriously evaluate this idea.

Now that at least some attempt has been made to assign the transitions of R_2XP—N=N—PXR_2, we will attempt to explain the effect of a change in R or X on the spectra. First, where X is O, there is a blue shift in the $n \rightarrow \pi^*$ transition of 1200 cm^{-1} as R is progressively changed from R is C_6H_5, $N(CH_3)_2$, or OC_6H_5 to OK. As Bock *et al.* have pointed out in a review [9] of the spectra of azo compounds, the major change in the π orbitals of P—N—N—P by substituents on P may only be brought about by substituent effects on the electronegativity of P, since as we have said earlier the available evidence precludes direct conjugation of π_p systems through *d* orbitals. In this case increasing *p–d* π donation of electrons in the order $O^{(-)} > OC_6H_5 > C_6H_5$, may be regarded as a *decrease* in the electronegativity and effective atomic number of P, thereby raising the energy of the lowest P—N—N—P *p–d* π^* orbital.

If the assignment of the 285 mμ absorption to the second $n \rightarrow \pi_2^*$ is correct, this transition should exhibit an even larger blue shift since the *p–d* orbital mixing coefficients are larger in this orbital. In the only spectrum which is not complicated by the presence of C_6H_5—P≡ absorption, the first $n \rightarrow \pi^*$ of $(KO)_2OP$—N=N—$PO(OK)_2$ is found at 543 mμ (18,800 cm^{-1}), and there is a weak shoulder at 320 mμ ($\epsilon \sim 5$); however, the 292 mμ transition of $[(CH_3)_2N]PNNP[(CH_3)_2N]_2$ has apparently been shifted toward the blue to the region of 250 mμ, a shift of some 5000 cm^{-1}. There appears to be a smaller blue shift but of similar magnitude in the spectra where R is (OC_6H_5) and R is $(OC_6H_5)(OK)$. These results do not conflict with the proposed $n \rightarrow \pi_2^*$ assignment.

The first $n \rightarrow \pi^*$ transition of the phenylazophosphinic or phenylazophosphonic acid derivatives is found near 500 mμ (20,000 cm^{-1}) which may be compared with 398 mμ for C_6H_5N=NCH_3, and the discussion previously given of this transition in the azodiphosphonic acid derivatives applies here equally well. The second transition, occurring near 290 mμ in spectra of these derivatives even where R is $[N(CH_3)_2]_2$ or O-alkyl, is an order of magnitude

TABLE XVII

ULTRAVIOLET MAXIMA OF R—N=N—R′

R	R′	λ_{max}, mμ (ϵ)
R′	$PO(C_6H_5)_2$	568 (23), (255) (13,000)
R′	$PO[(N(CH_3)_2]_2$	562 (13), (287) (3000)
R′	$PO(OK)_2$	532 (13), (320) (5), (250) (10^2)
R′	$PS(C_6H_5)_2$	538
C_6H_5	$PO(Alkyl)_2$	(513) (77), (286) (11,500)
C_6H_5	$PO(C_6H_5)_2$	501 (93), 288 (9600)
C_6H_5	$PO(OK)_2$	499 (65), 286 (11,300)

($\epsilon \sim 3 \times 10^4$) more intense than in the azodiphosphonic acid spectra (see Table XVII). It seems clear that this represents a *d–p* π red shift of the $\pi \rightarrow \pi^*$ transition found at 260 mμ in the spectra of phenylazoalkanes. The position of this maximum is essentially unchanged on going in series from compound **169** where X is O and R is $[N(CH_3)_2]$ (λ_{max} 288 mμ), to where R is (OK) (λ_{max} 286 mμ), in sharp contrast to the result for the series of compounds **168**.

There is one final interesting observation with regard to the $n \rightarrow \pi^*$ transition of azophosphonic acid derivatives **168** and **169**. Unlike the absorption bands of other derivatives, the $n \rightarrow \pi^*$ band of azophosphonic acid bands are asymmetrical with a sharp rise in absorption on the long wavelength side of the absorption band, and a more gradual decrease in intensity toward higher energy. The extreme version of this sort of asymmetrically shaped band occurs when the transition probability between vibrational levels of ground and excited state is greatest for the O–O transition. In turn, this is true when there is no difference between the equilibrium bond distances in the ground and excited states. Therefore, as Bock *et al.* [9] have suggested, the asymmetry of the $n \rightarrow \pi^*$ band indicates that excitation into the lowest π^* molecular orbital does not cause appreciable lengthening of the N—N bond. This would not be the case if the lowest π^* orbital, which is nonbonding between nitrogens had appreciable nitrogen atomic orbital coefficients. It is however understandable for large P_d mixing coefficients; in the extreme case the $n \rightarrow \pi^*$ would be $n \rightarrow 3d$. Furthermore, bonding of the type $R_2XP{=}N{-}N{=}PXR_3$ which tends to lengthen the —N=N— bond in the ground state will be less important in the excited state where resonance structures such as $R_2X\dot{P}{-}\dot{N}{=}\ddot{N}{-}PXR_2 \leftrightarrow R_2XP{-}\ddot{N}{=}\dot{N}{-}\dot{P}XR_2$ are important.

The spectra of the phosphoniumazofluorenides (**170**) and azodiphenylmethides (**171**) have been treated by Goetz and Juds [35] and representative data are given in Table XVIII. The fluorenylides exhibit four major absorption bands in the regions 380 mμ ($\epsilon \sim 2 \times 10^4$), 300 m$\mu$ ($\epsilon \sim 7 \times 10^3$), 250 m$\mu$ ($\epsilon \sim 5 \times 10^4$), and 220 m$\mu$ ($\epsilon \sim 4 \times 10^4$). Since the latter three

TABLE XVIII

ABSORPTION MAXIMA OF TRIARYLPHOSPHAZINES AND TRIARYLPHOSPHINIMINES

Compound	λ_{max}, mμ (log ϵ)
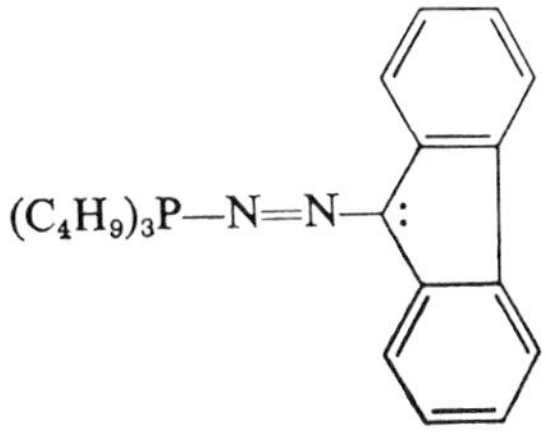 $(C_4H_9)_3P—N=N—$:	369 (4.33), 366 (4.32); 306 (3.82), 295 (3.79); 249 (4.78); 215 (4.23)
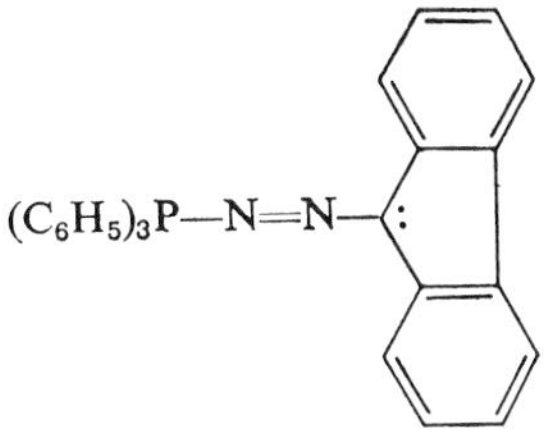 $(C_6H_5)_3P—N=N—$:	388 (4.47), 375 (4.47); 337 (3.91), 308 (3.93); 247 (4.83); 227 (4.64)
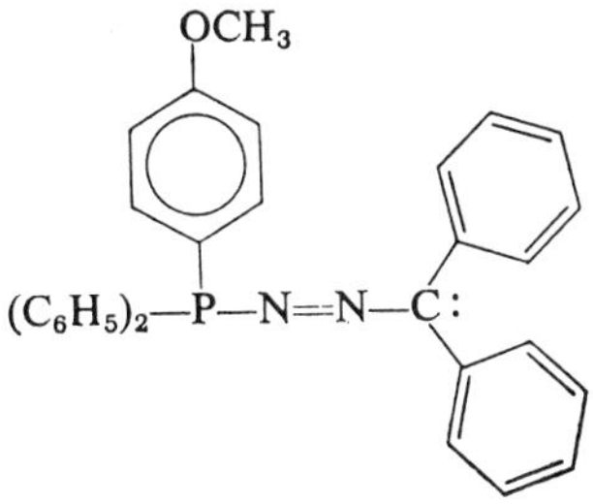 OCH_3 $(C_6H_5)_2—P—N=N—C:$	342 (4.08); 244 (4.83), 240 (4.68)
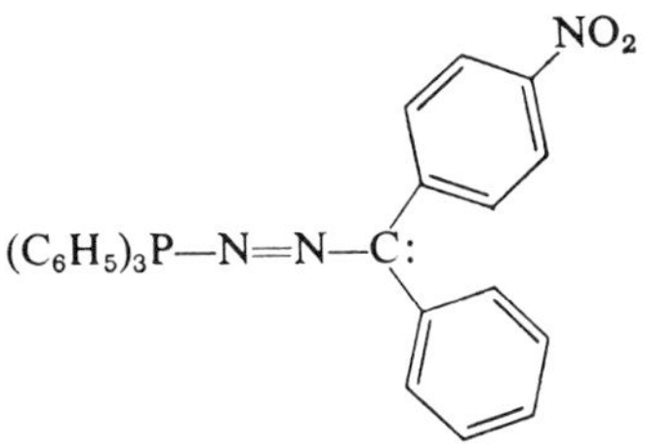 NO_2 $(C_6H_5)_3P—N=N—C:$	422 (4.22); (310) (~3.8)
$(C_6H_5)_3P=NC_6H_5$	270 (sh) (3.6)
p-$ClC_6H_4N=P(C_6H_5)_3$	290 (sh); 254 (4.3)
p-$NO_2C_6H_4N=P(C_6H_5)_3$	384 (4.4); 255 (sh)

absorptions probably correspond to the ^1L_b, ^1L_a, and ^1B transitions of the fluorene chromophore, the 380 mμ transition is of principal interest as being characteristic of the P=N—N= chromophore bonded to fluorene. The spectra of the diphenylmethide series exhibit a similar transition in the region 435–340 mμ in addition to the expected $\pi \rightarrow \pi^*$ transitions of the $(C_6H_5)_2\ddot{C}$— group.

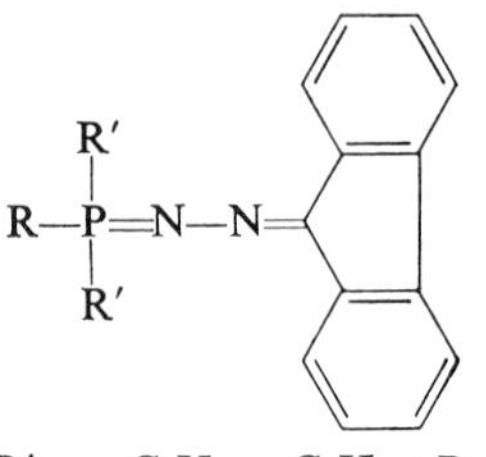

$R = R' = n\text{-}C_4H_9$, $n\text{-}C_5H_{11}$; $R = C_6H_5$, $R' = CH_3$; $R = p\text{-}XC_6H_4$, $R' = C_6H_5$; X = H, Cl, Br, $OCH_3N(CH_3)$

(170)

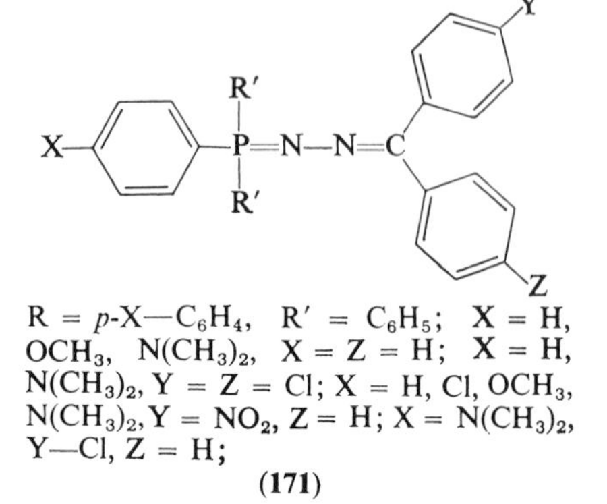

$R = p\text{-}X—C_6H_4$, $R' = C_6H_5$; X = H, OCH_3, $N(CH_3)_2$, X = Z = H; X = H, $N(CH_3)_2$, Y = Z = Cl; X = H, Cl, OCH_3, $N(CH_3)_2$, Y = NO_2, Z = H; X = $N(CH_3)_2$, Y—Cl, Z = H;

(171)

Over the greatest range the change in energy of the longest wavelength transition with change in substituent on the phosphorus phenyl of compound **170** amounts to only 200 cm^{-1} (0.025 eV), from λ_{max} 382 mμ where X is H to λ_{max} 396 mμ where X is $N(CH_3)_2$. In the case of compound **171**, where Y and Z are Cl, the substitution of $N(CH_3)_2$ in the *para* position of a phosphorus phenyl leads to a red shift of this transition of only 380 cm^{-1} (0.05 eV). However, in compound **171**, if Y is NO_2 and Z is H, $p\text{-}N(CH_3)_2$ substitution for X leads to a larger red shift of about 700 cm^{-1} (0.09 eV). Substitution of alkyl groups for aryl groups on P leads to a blue shift of the transition, for example, the absorption of compound **170**, where $R = R' = n\text{-}C_4H_9$, is found at λ_{max} 368 mμ.

The spectrum of compound **171**, where X is H, Y is NO_2, and Z is H (λ_{max} 422 mμ), more closely resembles that of the corresponding $(p\text{-}NO_2C_6H_4)(C_6H_5)CN_2$ (λ_{max} 392 mμ) than the ketazine $(p\text{-}NO_2C_6H_4)(C_6H_5)C{=}N—N{=}C(C_6H_5)(p\text{-}NO_2C_6H_4)$ (λ_{max} 337 mμ). Therefore, it is perhaps worthwhile to attempt to correlate the molecular orbitals of $R_3P{=}N—N{=}CR_2$ with those of CH_2N_2. This is done in Fig. 56.

Considering only the π orbitals in the plane perpendicular to that of the molecule, CH_2N_2 has one bonding π orbital and two antibonding π^* orbitals. These probably give rise to the $\pi \rightarrow \pi^*$ transition at 220 mμ (ϵ 9000) in the spectrum of diazomethane, which is shifted to 288 mμ in the spectrum of diphenyldiazomethane [27]. The effect of R_3P: is to introduce an additional orbital, *d*, plus an extra pair of electrons. If the bonding interaction between the R_3P: *d* orbital and π of N_2CR_2 were very strong, we would expect a blue shift of the lowest $\pi \rightarrow \pi^*$ absorption maximum as is found in the case of $(p\text{-}NO_2C_6H_4)CN_2$ vs. $(p\text{-}NO_2C_6H_4)(C_6H_5)C{=}NN{=}C(C_6H_5)(p\text{-}NO_2C_6H_4)$

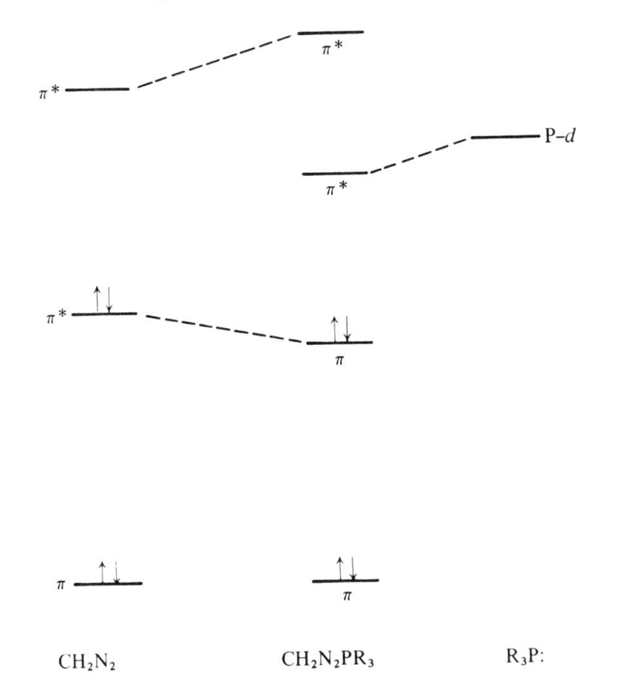

FIG. 56. A suggested qualitative correlation diagram between the molecular orbitals of CH_2N_2 and $CH_2N_2PR_3$.

or as an additional example $CH_2{=}CH—CH_2Li$ vs. $CH_2{=}CH—CH{=}CH_2$. A somewhat weaker interaction, as indicated qualitatively in Fig. 56, gives the observed red shift.

Goetz and Juds [35] have chosen to interpret the moderate red shift of the longest wavelength transition in spectra of the compound series **170** and **171,** with increasing π electron donating ability of substituents on phosphorus, as evidence of weak π–π conjugation between R and —N=N—R′, which raises the energy of highest filled π to a greater extent than π^*. There is also the alternative explanation that *d–p* π bonding by the phosphorus substituents decreases the electronegativity of phosphorus, and by an inductive effect decreases also the electronegativity of the adjoining nitrogen, which could also have the effect of raising the energy of the π orbital relative to π^* to give the observed red shift.

The ultraviolet spectra of the phosphinimines with the general structure **172**, where Y is H and *p*-C≡N, *p*-NO_2, and H, and where Y is *p*-$N(CH_3)_2$, CF_3, X is NO_2, have been studied by Horner and Oediger.

Although Horner and Oediger [44, 44a] concluded that there was evidence in the spectra of those compounds where X is *p*-NO_2 of extended conjugation

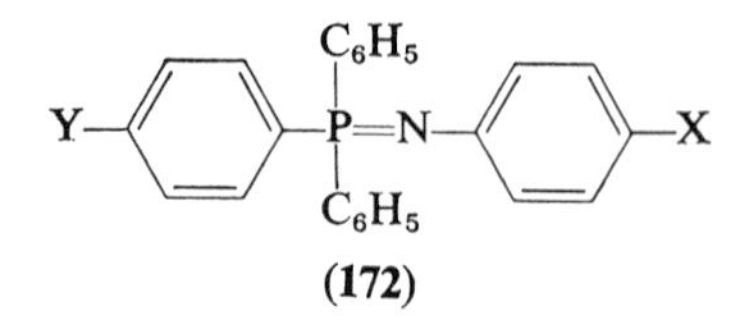

(172)

of the type in structure **172**, Zhmurova and Kirsanov [82] have later observed that the spectra of all the compounds investigated closely approximate a simple sum of the spectra of the $(XC_6H_9)(C_6H_5)_2P$ moiety and the amines $XC_6H_4NH_2$. Therefore there is no good evidence of such extended conjugation. Representative absorption maxima are again to be found in Table XVIII. The spectra of compounds **173** and **174** where the position of substitution is unknown have also been reported [23].

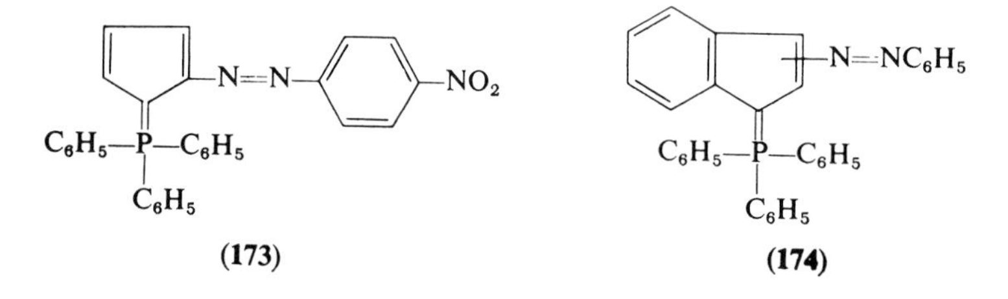

(173) (174)

F. Aryl Derivatives

1. *Arylamines*

The extent of disagreement in the literature over the symmetry assignments of the electronic transitions of the aryl derivatives of Group V highlights the complications which can be introduced into the interpretation of even a "simple" monosubstituted benzene. On the one hand, Klevens and Platt [52] correlate the 251 mμ *N,N*-dimethylaniline transition (Column II of Table XIX) with the 256 ^1L_b benzene transition, as do Jaffe and Orchin [50] who make the comment that the 298 mμ transition has no counterpart in the benzene spectrum. Others [11, 65a, 57], however, have assigned the long wavelength 298 mμ absorption to ^1L_b and the column II maximum (Table XIX) to either an ^1L_a transition shifted strongly toward the red or to an intramolecular charge transfer transition of the same symmetry, that is, long axis polarized. The transitions under column II of Table XX have also been discussed [36, 75] as $n \rightarrow \pi^*$, which is conceptually the same as the intramolecular charge transfer assignment, except that perhaps one expects a spectral blue shift on a solvent change from aprotic to protic solvents.

Since the charge transfer model of the aniline spectrum in its treatment by Murrell [65a] will be of particular interest as we proceed, the approach is outlined below. First, an intramolecular charge transfer from nitrogen to

TABLE XIX

ABSORPTION MAXIMA OF ARYL GROUP V DERIVATIVES

Compound	λ_{max}, mμ ($\epsilon \times 10^{-3}$)		
	I	II	III
$C_6H_5NH_2$[a]	197 (32)	234 (9.1)	288 (2.0)
$C_6H_5N(CH_3)_2$[a]	200 (~25)	251 (15.5)	298 (2.4)
$(C_6H_5)_3N$[b]	228 (7.3)	297 (22.1)	—
$C_6H_5PH_2$[b]	—	234 (3.5)	266 (0.69)
$C_6H_5P(CH_3)_2$[b]	230 (sh) (14)	266 (2.2)	
$(C_6H_5)_3P$	215 (29)	262 (11)	—
$(C_6H_5P(C_4H_9\text{-}n)_2$[b]	—	251 (2.8)	—
$(C_6H_5CH_2)_3P$[b]	—	247 (2.9)	
$C_6H_5As(CH_3)_2$[b]	—	239 (6.8)	
$C_6H_5As(CF_3)_2$	—	224 (6.0)	265 (0.65)
$C_6H_5Sb(CH_3)_2$[b]		250 (3.7)	

[a] Hydrocarbon solvent.
[b] Alcohol solvent.

the two lowest vacant benzene molecular orbitals gives the excited state charge distribution of structures E_{CT^1} and E_{CT^2} below. The energy of each of the charge transfer states is calculated by means of Eq. (8) in Chapter I from the ionization potential of NH_3 (10.52 eV), the electron affinity of benzene (−0.54 eV), and the point charge distribution for E_{CT^1} and E_{CT^2} to give $C = e^2/r$; for E_{CT^1}, $C = 6.03$ eV; and for E_{CT^2}, $C = 4.84$ eV. The perturbation interaction matrices between states of the same symmetry are given below, and include the ground state. Matrix a mixes states of A (C_{2v}) symmetry and Matrix b mixes states of B_2 symmetry. The off-diagonal resonance integral, β_{rs}, reduces to the integral $\int \theta(i)H(i)\phi_r(i)$, where ϕ_r is the atomic orbital function of the phenyl atom bonded to nitrogen and θ_i is the substituent donor molecular orbital. After configuration interaction the calculated (observed) transition energies for $\beta_{rs} = 1.6$ eV are ^1L_b 4.91 (4.31), CT 5.16 (5.27), CT 6.34, and ^1L_a 6.67 (6.30).

$$\begin{pmatrix} E_{L_a} & 0 & 0 & (-\sqrt{\tfrac{1}{6}}\beta_{rs}) \\ 0 & E_{B_a} & 0 & (-\sqrt{\tfrac{1}{6}}\beta_{rs}) \\ 0 & 0 & E_{GS} & (\sqrt{\tfrac{2}{3}}\beta_{rs}) \\ (-\sqrt{\tfrac{1}{6}}\beta_{rs}) & (-\sqrt{\tfrac{1}{6}}\beta_{rs}) & \sqrt{\tfrac{2}{3}}\beta_{rs} & (I - A - 6.03) \end{pmatrix}$$

Matrix a

$$\begin{pmatrix} E_{L_b} & 0 & (-\sqrt{\tfrac{1}{6}}\beta_{rs}) \\ 0 & E_{B_b} & 0 \\ (-\sqrt{\tfrac{1}{6}}\beta_{rs}) & 0 & (I - A - 4.84) \end{pmatrix}$$

Matrix b

Godfrey and Murrell [34a] have modified the original calculations on aniline by introducing a correction for the electrostatic repulsion between the excess electron density at carbon one and remaining electron density on nitrogen. This correction, called the $+I_\pi$ effect, increases the zero order CT_1 state energy and reduces the CT character in the second aniline transition.

More recently Mangini *et al.* [57] also concluded from an analysis of the vapor phase and *n*-hexane solution spectra of aniline, *N,N*-dimethylaniline, and other monosubstituted benzenes that on the basis of vibrational analysis, oscillator strength, etc., the 288 and 298 mμ absorptions of aniline and dimethylaniline should be assigned to ^1L_b. It was further concluded that the band system in column II of Table XIX was dominated by charge transfer states which would break down the analogy with the benzene ^1L_a transition.

If the intense 251 mμ transition of dimethylaniline correlates with the ^1L_b transition of benzene the transition should be polarized perpendicular to the C_{2v} axis as indicated by structure **175**. In the case of either ^1L_a or the intramolecular charge transfer assignment, the transition moment should be parallel to the Z axis. Either Pariser-Parr-Pople SCF molecular orbital calculations or charge transfer models agree that the moderately intense transition found in the region 220–240 mμ in the spectra of *para* X-substituted phenyl boronic acids or anhydrides is long axis polarized ($A_1 \rightarrow A_1$) and probably the ^1L_a transition. An excellent linear correlation of this transition energy with mass spectroscopic XCH_2^+ stabilization energies, where X is H, CH_3, F, Br, and CH_3O, predicts that where X is $N(CH_3)_2$, the boronic acid characteristic transition will exactly overlap the *N,N*-dimethylaniline transition of 40,000 cm^{-1} (λ_{max} 250 mμ). Rather than overlapping absorption bands, the observed spectrum of the *p-N,N*-dimethylaminophenylboronic acid anhydride in hexane instead exhibits strong transitions (ϵ 10^4) at 45,000 and 35,000 cm^{-1}. This result is interpreted as strong configuration interaction between the $(CH_3)_2NC_6H_4$— and $C_6H_4B(OH)_2$ excited states, demonstrating A_1 symmetry (C_{2v} point group) for the excited states of the isolated chromophores.

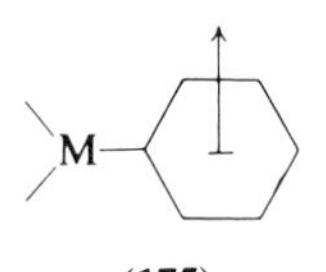

(175)

For the above reasons, in subsequent discussion we will regard the absorption maxima of $C_6H_5NH_2$, $C_6H_5N(CH_3)_2$, $C_6H_5PH_2$, etc. listed under column III of Table XIX as being correlated with the benzene ^1L_b transition. The second (column II, Table XIX) more intense transition will be regarded as polarized along the long axis and as being correlated with either the ^1L_a benzene transition of perhaps intramolecular charge transfer of the same symmetry. Certainly the amount of charge transfer character in this transition may vary a good deal as one proceeds in the series $C_6H_5MR_2$ from M = N through P, As, Sb, and Bi.

For example, very recent molecular orbital calculations on aniline which include σ electrons and incorporate the simplifying assumptions of the Pariser-Parr-Pople method, have been reported by Jaffe [46]. The calculated transition energies (electron volts) and oscillator strengths 4.4 (0.023), 4.7 (0.041), and 6.5 (0.491) eV may be compared with the observed transitions 4.4 (0.028), 5.4 (0.144), and 6.4 (0.510). The second transition (labeled charge transfer [65a] by Murrell) is found by Jaffe, however, to have only about 15% charge transfer character. It may be significant however that with only 15% charge transfer character introduced into the second transition, the calculated transition energy was 0.7 eV too low and the ratio of oscillator strengths too small by a factor of 3.

The controversy surrounding the transitions of aniline lead to similar problems in any discussion of the spectra of $(C_6H_5)_3N$:. The data in Fig. 57

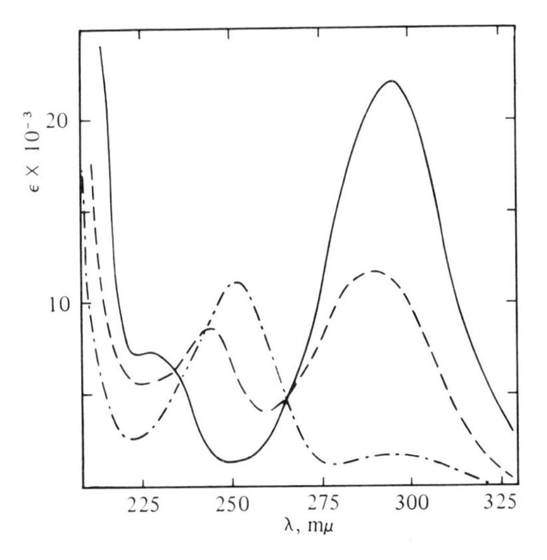

FIG. 57. The spectra (in 95% ethanol) of: (———) $(C_6H_5)_3N$; (----) $(C_6H_5)_2NCH_3$; (-·-·-) $C_6H_5N(CH_3)_2$ [47].

led Jaffe and Orchin [50] to correlate the 297 mμ band of $(C_6H_5)_3N$: and the 298 mμ absorption of $C_6H_5N(CH_3)_2$, with the conclusion that this and similar transitions in other triphenyl derivatives of Group V cannot be clearly identified with either the ^1L_a or ^1L_b band of benzene because of the large perturbation by the nitrogen lone pair electrons. The 228 mμ shoulder of $(C_6H_5)_3N$: is shifted toward the blue from the 251 mμ transition of $C_6H_5N(CH_3)_2$ in this interpretation.

An alternative approach begins with the previously outlined Murrell assignment [65a] of the 234 mμ $C_6H_4NH_2$ transition to $^1A_1$–$^1A_1$ charge transfer. The intense long wavelength transitions (291–298 mμ) in the spectra of $(C_6H_5)_2NCH$ and $(C_6H_5)_3N$ may then be described as the result of configuration interaction between the degenerate, locally excited charge transfer states.

Configuration interaction in the case of C_{2v} planar $(C_6H_5)_2NCH_3$ would give rise to two in-plane allowed charge transfer transitions, and the 245 mμ transition of $(C_6H_5)_2NCH_3$ might represent this second charge transfer transition.

It now remains to be seen what observations and conclusions reached with regard to the spectra of the phenylamines can be carried over into a discussion of the spectra of the Group V organometalloids, and further whether a consideration of the spectra of these will shed any light on the controversies surrounding the aniline and triphenylamine spectra.

2. *Spectra of Aryl*-MR_2

The earliest study of the ultraviolet spectra of arylphosphines, -arsines, and -stibines was made by Bowden and Braude [11] on the series $C_6H_5M(CH_3)_2$. Bowden and Braude assigned the transitions in column II of Table XIX to an excited state structure (**176**) (see Addendum, last page).

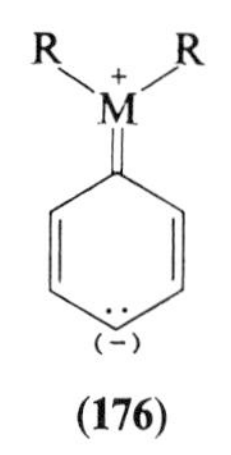

(**176**)

Schindlbauer [75] has somewhat more recently obtained the ultraviolet spectra of a series of interesting phosphines. The vibrational structure found in the 266 mμ transition of $C_6H_5PH_2$ clearly identifies this transition as ^1L_b and strongly supports similar assignment of the corresponding 288 mμ aniline transition. Further comparison of the spectra of series $C_6H_5MH_2$, $(C_6H_5)_2MH$, and $(C_6H_5)_3M$, where M is N and P as in Fig. 58, supports

TABLE XX

ABSORPTION MAXIMA OF TRIARYLMETALLOIDS OF GROUP V

Compound	λ_{max}, mμ ($\epsilon \times 10^{-3}$)		
$(C_6H_5)_3P$	251 (29)		264 (10)
p-$ClC_6H_4P(C_6H_5)_2$	219 (26);	(230) (10);	266 (12)
p-$BrC_6H_4P(C_6H_5)_2$	214 (36)		263 (14)
$(p$-$CH_3OC_6H_4)P(C_6H_5)_2$	216 (25);	230 (15);	261 (12)
p-$(CH_3)_2NC_6H_4P(C_6H_5)_2$	215 (30)		283 (26)
$[p$-$(CH_3)_2NC_6H_4]_2PC_6H_5$			285 (39)
$[p$-$(CH_3)_2NC_6H_4]_3P$			285 (70)

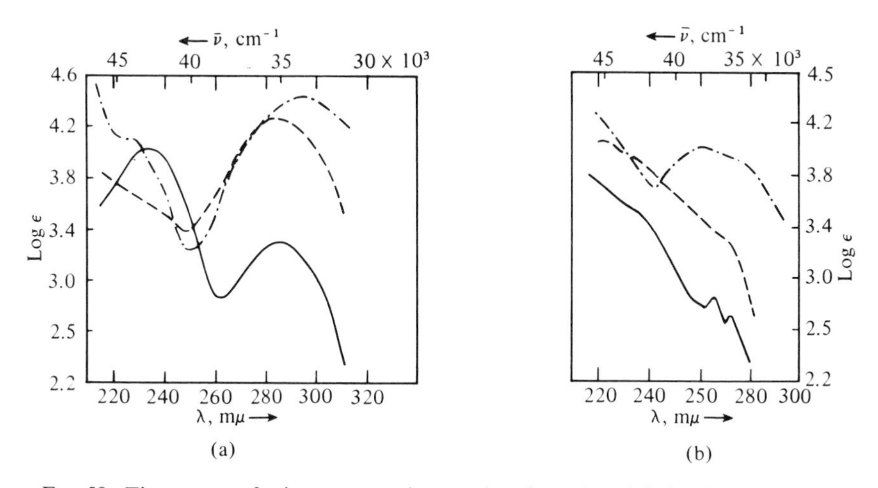

FIG. 58. The spectra of primary, secondary, and tertiary phenyl derivatives of ammonia and phosphine (in C_2H_5OH): (a) (———) $C_6H_5NH_2$; (----) $(C_6H_5)_2NH$; (-·-·-) $(C_6H_5)_3N$. (b) (———) $C_6H_5PH_2$; (----) $(C_6H_5)_2PH$; (-·-·-) $(C_6H_5)_3P$ [75].

correlation of the 262 mμ absorption of $(C_6H_5)_3P$: with the 234 mμ absorption of $C_6H_5PH_2$, and the $(C_6H_5)_3N$: 297 mμ absorption with the $C_6H_5NH_2$ 234 mμ transitions rather than with the less intense ^1L_b long wavelength transitions.

A particularly interesting observation is the 15 mμ *red* shift of the $C_6H_5P(CH_3)_2$ transition relative to $C_6H_5P(n$-$C_4H_9)_2$. Based on an inductive order of electron release the first transition [16] energy of *N*,*N*-diethylaniline (λ_{max} 261 mμ), as expected, is lower than *N*,*N*-dimethylaniline. The dramatically reversed order in the case of phosphines suggests some additional red shift mechanism present in the phosphines but absent in the amines. The possibility which comes to mind most readily is involvement of phosphorus *d* orbitals either in ground state structures such as **177** which increase the electron density on phosphorus and/or excitation to a lowest vacant molecular orbital including phosphorus *d* and antibonding π^* C$\equiv$(H$_3$) orbitals as in structure **178**.

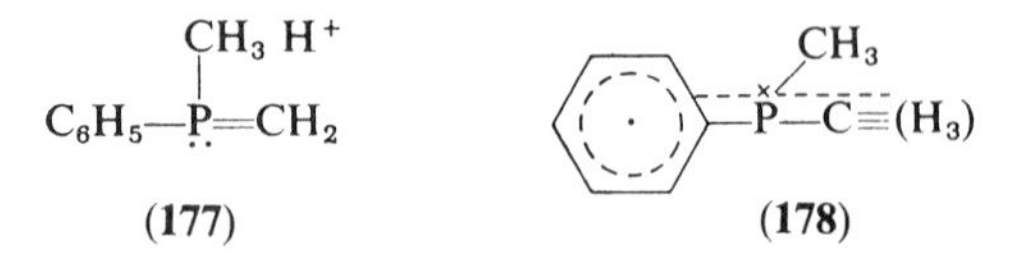

Bissey and Goldwhite [6] have demonstrated phosphorus *d–p* π participation in the electronic transitions of phosphines by the Goodman substituent interference method discussed in detail in Chapter IV. For example, the smoothed extinction coefficients for the ^1L_b transitions are 590 for C_6H_5-PH_2, 1630 for $CH_3OC_6H_5$, and 1400 for $CH_3OC_6H_4PH_2$, which demonstrates that in the excited state where CH_3O is electron donating, —PH_2 is net electron withdrawing Evidence of ground state *p–d* π bonding has been obtained from F NMR studies of $C_6F_5P(CH_3)_2$, $(C_6F_5)_3P$, and related compounds by Hogben and co-workers [41] with the conclusion that even in the ground state the P of C_6F_5P is net π electron withdrawing.

Schindlbauer [75] further finds that $(C_6H_5CH_2)_3P$ possess a moderately strong transition at 247 mμ (5.0 eV) (ϵ 2.9 $\times$ 10^3). Interestingly enough a similar transition is also clearly evident in the spectrum [76] of $(C_6H_5CH_2)_3N$ as evidenced by a shoulder at 248 mμ with strongly increasing absorption in the neighborhood of 240 mμ which disappears on complexing with $AlBr_3$. This result provides an opportunity to test the assignment of the 262 mμ $(C_6H_5)_3P$ and 297 mμ $(C_6H_5)_3N$: transition to charge transfer from N or P to the lowest vacant π^* orbital of phenyl. If the charge transfer model is applicable, we can predict the change in the charge transfer transition energy between $C_6H_5MR_2$ and $C_6H_5CH_2MR_2$ from Eq. (8) in Chapter I. The largest effect of the intervening CH_2 is to decrease the Coulomb term e^2/r by about 1.3–1.4 eV. The appropriate ionization potential of P or N as the electron donor in $C_6H_5CH_2MR_2$ should be a few tenths of an electron volt smaller because of the electron withdrawing inductive effect of C_6H_5 relative to $C_6H_5CH_2$. Therefore, on the basis of this very crude charge transfer model one would predict charge transfer transitions in the spectra of tribenzylphosphine near 210 mμ (4.7 + 1.2 = 5.9 eV). The predicted charge transfer transition would be very close in energy to the ^1L_a benzene transition and again we are faced with problem of configuration interaction between the charge transfer and ^1L_a states. Without very good molecular orbital calculations it is not possible to say whether the 247 mμ transition is predominantly charge transfer or predominantly ^1L_a. In either case the result supports describing the transitions between 230 and 266 mμ in column II of Table XX in the spectra of $C_6H_5MR_2$, where M is P, As, Sb, or Bi, as an intramolecular charge transfer process.

In the series $C_6H_5MR_2$ the energy of the charge transfer transition decreases in the order As > Sb > P and a similar order As > Sb > P > Bi can be constructed in the series $(C_6H_5)_3M$, as will be discussed later. Explanations for the lower energy of charge transfer transitions of phosphines relative to arsines were discussed earlier with regard to an order of charge transfer transition energy in the vinyl derivatives where the order N > As > P is found.

In the series $CH_2{=}CH{-}MR_2$ *the vinylamine (enamine) charge transfer transition is at highest energy, whereas in the* $C_6H_5MR_2$, *except where R is* CH_3, *the nitrogen compounds absorb at the same or longer wavelength than the phosphorus derivatives. This result is unexpected.* A little reflection however offers a ready explanation for this result, while retaining the charge transfer model for the metalloids. The C—N p–p π resonance integral is much greater than that for C—P, etc., which, along with better energy matching between the benzene ^1L_a transition excited state and the charge transfer state, will lead to very strong configuration interaction between these states and a substantial red shift in the transition energy. In fact, considering the relative ionization potentials of benzene (9.24 eV) and ammonia (10.15 eV), we could easily suspect that the charge transfer state might lie higher in energy than the ^1L_a state. This would be the case provided the Coulomb term e^2/r of Eq. (8) became small. If we take as our donor orbital a simple molecular orbital with a nitrogen atomic orbital mixing coefficient of the order of magnitude 0.7, not the localized nitrogen $2p$ atomic orbital, the maximum transfer of charge from nitrogen to ring would be 0.5 electrons. This means then that the Coulomb term used by Murrell [65a] for example in his treatment of the aniline spectra should be reduced by a factor of $(0.5)^2$, i.e., 0.25, giving for Eq. (8) $C = 1.5$ eV. The above line of argument is intended to show the sensitivity of the requirement by the intramolecular charge transfer model that the donating and/or acceptor orbitals be indeed localized orbitals. Thus, the intense long wavelength transition in the spectra of $(C_6H_5)_3N$ and the 234 mμ transition of $C_6H_5NH_2$ represent a mixture of ^1L_a and charge transfer states, which on the basis of Jaffe's calculations is predominantly ^1L_a. This in no way precludes assignment of the similar transitions in the spectra of the phenyl Group V metalloids to intramolecular charge transfer, as we were led to believe for reasons given earlier.

Jaffe [47] obtained the spectra of the triphenyl Group V series $(C_6H_5)_3M$ (shown in Fig. 59) and suggested that the transitions where M is P (261 mμ), As (248 mμ), Sb (256 mμ), or Bi (248 mμ) could not clearly be identified with either the ^1L_b or ^1L_a transition. It was considered that the 280 mμ shoulder

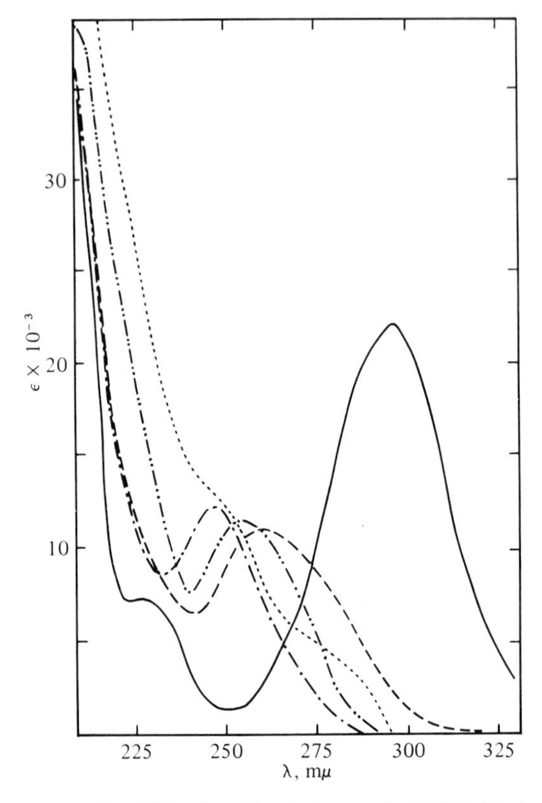

FIG. 59. The spectra (in 95% ethanol) of: (———) $(C_6H_5)_3N$; (----) $(C_6H_5)_3P$; (-·-·-) $(C_6H_5)_2As$; (-··-··-) $(C_6H_5)_3Sb$; and (·····) $(C_6H_5)_3Bi$ [47].

in the spectrum of $(C_6H_5)_3Bi$, and the hint of a similar one in $(C_6H_5)Sb$, might represent $n \rightarrow \pi^*$ transitions.

Rao *et al.* [72] later tentatively assigned these intense bands between 253 and 300 mμ in the spectra of $(C_6H_5)_3N$: to benzene ^1L_b type transitions. This assignment followed from the correlation by Jaffe of the 297 mμ $(C_6H_5)_3N$ transition with the 298 mμ band of $C_6H_5N(CH_3)_2$, and the absence of vibrational structure could be ascribed to p–π interaction of the lone pair electrons with the benzene ring.

Goetz *et al.* [36] prefer to interpret the ultraviolet spectra of p-XC_6H_4P-$(C_6H_5)_2$ where X is H, Cl, Br, OCH_3, or $N(CH_3)_2$ in terms of $n \rightarrow \pi^*$ transitions and the valence bond structures **179** and **180**, although this does not seem to be a particularly appropriate description since there is little or no solvent effect on the 262 mμ transition of $(C_6H_5)_3P$ on changing solvent from ethanol to hexane–cyclohexane [48]. The results of Goetz and co-workers

[36] are combined in Table XX with those of Schiemenz [74] in the series $[p\text{-}(CH_3)_2N{-}C_6H_4]_nP(C_6H_5)_{3-n}$.

Goetz and co-workers [36] consider the major contributing valence bond structure to the excited state of the 261–264 mμ transition to be that of structure **179** and further assign the 230 mμ shoulder in the spectra of $p\text{-}CH_3OC_6H_4P(C_6H_5)_2$ and $p\text{-}ClC_6H_4P(C_6H_5)_2$ to a second $n \rightarrow \pi^*$ transition represented by structure **180**, which implies correlation with the ^1L_a XC_6H_5 transitions at 217 and 210 mμ. The small red shift where X is Cl and blue shift where X is CH_3O in the spectra of $XC_6H_4P(C_6H_5)_2$ is rationalized as the result of inductive stabilization when X is Cl or π electron donating destabilization when X is CH_3O. In molecular orbital charge transfer terminology, Cl inductively lowers the energy of the lowest π^* orbital, increasing its electron affinity whereas *p–p* interaction of CH_3O with the π^* orbital electrons raises the π^* energy and decreases its electron affinity.

The number of possible transitions in a nonplanar 22 π electron system would seem to make it unlikely that any of the higher excited states of $p\text{-}XC_6H_4P(C_6H_5)_2$ may be simply related to localized benzene transitions in anything more than an approximate fashion. Nonetheless, if we accept a charge transfer assignment for the 261–264 mμ transitions, and assign the 230 mμ shoulders to red shifted ^1L_a transitions of the XC_6H_5 group (or the reverse assignments), it is probable that the 214–219 mμ transitions can be related by configuration interaction to a substantial contribution from the ^1B transitions found, for example, in the spectra of $C_6H_5N(CH_3)_2$ at 200 mμ.

The long wavelength 285 mμ transition of the $[p\text{-}(CH_3)_2NC_6H_4]_nP(C_6H_5)_{3-n}$ series appears qualitatively different from the less intense shorter wavelength transitions of $p\text{-}XC_6H_4P(C_6H_5)_2$. This transition is considered by Schiemenz [74] as a $\pi \rightarrow \pi^*$ transition, and in each case shows a small red shift on changing solvent from cyclohexane to methanol. The best simple description is probably to regard the 283 mμ absorption excited state as a mixture after configuration interaction of a $(CH_3)_2NC_6H_4$— transition near 250 mμ and the $(C_6H_5)_3P$: transition near 264 mμ. In this way not only is the large increase in intensity and lower energy for the first transition of $p\text{-}(CH_3)_2NC_6H_4P(C_6H_5)_3$ accounted for, but also the absence of the 259 mμ region of transition which can be assigned to the chromophore $(CH_3)_2\text{-}NC_6H_4$— is accounted for. The valence bond structure **180** assigned by Goetz and co-workers to this transition, or perhaps structure **181**, would actually not be a bad representation of such a transition.

Cullen and Hochstrasser [25] examined the ultraviolet spectra of a series of phenyl arsines C_6H_5AsRR', where R and R′ were C_6H_5, CH_3, or CF_3. The results clearly support assignment of the intense transitions between 220 and 250 mμ to a charge transfer transition from the arsenic lone pair to the phenyl ring vacant π^* orbitals, where R is not CF_3.

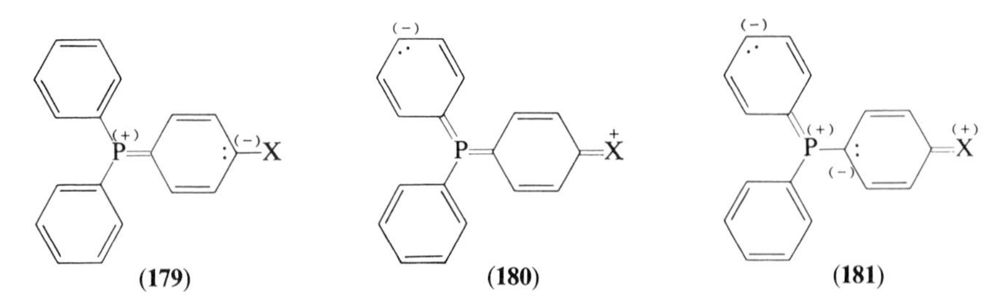

First, the ^1L_b transition, as shown in Figs. 60 and 61, in the spectra of $C_6H_5As(CF_3)_2$ and $(C_6H_5)_2AsCF_3$ is found near 265–269 mμ with an extinction coefficient per phenyl ring of 450–650, which leads to the conclusion that the perturbation of the benzene π system by —$As(CF_3)_n$ is less than that by OCH_3. The absence of significant *p–p* π bonding of the arsenic lone pair is further supported by the reported C—As—C bond angles of 96° in triphenylarsine. Therefore, we can conclude the arsenic lone pair electrons are indeed essentially localized in the arsenic 4*p* atomic orbital, which satisfies the first major requirement for the success of the charge transfer model, i.e., a localized donor orbital. Yet it is clear that the arsenic lone pair is the source of those intense transitions ranging from 224 mμ in $C_6H_5As(CF_3)_2$ to 247 mμ in $(C_6H_5)_5As$ since the spectra exhibit a small 10 Å blue shift with a solvent change hexane to methanol, and in trifluoroacetic acid the triphenylarsine 248 mμ transition disappears altogether, presumably because of the arsenic lone pair protonation. Further evidence of the charge transfer character of this

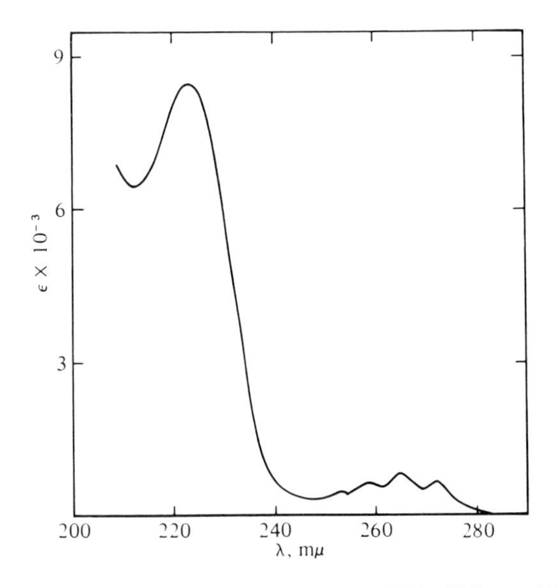

FIG. 60. The absorption spectrum of $C_6H_5As(CF_3)_2$ ([25, pp. 118–132]).

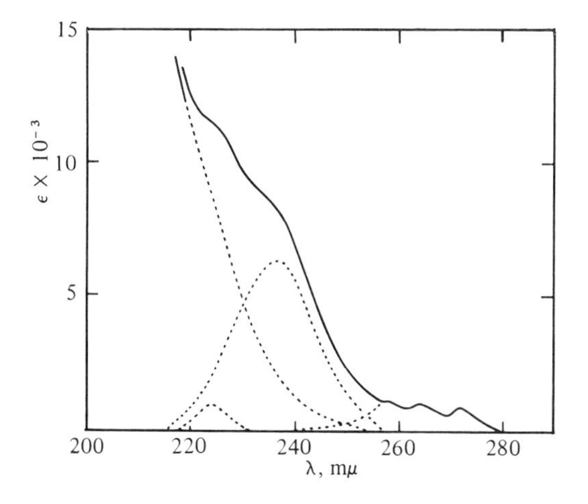

FIG. 61. The absorption spectrum of $(C_6H_5)_2AsCF_3$: (———) the observed spectrum; (----) the derived structure.

transition is the dramatic inductive effect of a substituent bonded to arsenic in the transition energy, for example, λ_{max} $C_6H_5As(CH_3)_2$, 239 mμ and λ_{max} $C_6H_5As(CF_3)_2$, 224 mμ. Inductive effects on aromatic $\pi \rightarrow \pi^*$ transitions are usually small.

Additional proof of the localized properties of the transitions and absence of phenyl conjugation is obtained in the relative constancy of the charge transfer transition energy over the series $C_6H_5As(CH_3)_2$ (239 mμ), $(C_5H_5)_2$-$As(CH_3)_2$ (241 mμ), and $(C_6H_5)_3As$ (248 mμ) as compared with the analogous nitrogen series $C_6H_5N(CH_3)_2$ (251 mμ) and $(C_6H_5)_3N$: (297 mμ).

The observation of two transitions in the charge transfer region of the spectrum of $(C_6H_5)_2AsCF_3$ requires a special explanation since this is the only case for which two such transitions are found. Examination of steric models have led Cullen and Hochstrasser [25] to suggest that the two transitions are derived from nonequivalence of the phenyl rings since the models require one ring with its *ortho* hydrogen coming into close proximity with a fluorine of the CF_3.

In a later paper Cullen and co-workers [24] report the spectra of a series of *o*-, *m*-, and *p*-CH_3- and CF_3-substituted triphenylarsines. There is a red shift of the charge transfer transition in all cases (for example, λ_{max} [2,4,6-$(CH_3)_3C_6H_2]_3As$, 274 mμ) which is explained as the result of better mixing between the localized benzene and charge transfer transitions. However, in keeping with the expected inductive effect on the phenyl π electron affinity $(p\text{-}CF_3C_6H_4)_3As$ (λ_{max} 252.5 mμ) does absorb at a longer wavelength than $(p\text{-}CH_3C_6H_4)_3As$ (λ_{max} 248.0 mμ). An attempt to apply the Murrell [65a] charge transfer model of aniline to the arsines is marred by a mistake in the application of group theory to the problem. First, the local symmetry of one benzene ring after *para* substitution is taken as D_{2h}, which is a very peculiar choice since it excludes the very atomic orbital and lone pair electrons responsible for the transition and renders it impossible to give a symmetry assignment to the charge transfer transitions. In one place we are told that the lowest energy charge transfer state CT_1, $B_{1u}(^1L_a)$ and E_{1u} states mix, which seems logical except, as can be seen from the matrices by Murrell given earlier and as is evident for C_{2v} local symmetry, that this interaction matrix should also contain the ground state. Instead we are told that in spite of A_g symmetry under the D_{2h} point group the ground state mixes with the ^1L_b (B_{2u} for D_{2h} symmetry) and $E_{1u}(B_{2u})$ states [58, p. 642]. At this point we would conclude that only the ground state has been misplaced, but somehow in the final correlation diagram, the L_b transition, for example, ends up with 28% CT_1 character in spite of the fact they were not even in the original matrices diagonalized. Using ionization potentials of R_2AsH and e^2/r terms (Eq. 8) [24], a benzene electron affinity of -0.94 eV [3a] and benzene excited states ($^1E_{1u}$, 6.75; $^1B_{1u}$, 5.95; $^1B_{2u}$, 4.71 eV), a correct Murrell calculation ($\beta = -1.40$ eV) gives the following results: for $C_6H_5As(CF_3)_2$, the energy calculated in electron volts (% CT) is as follows: ^1L_b, 4.6 (5.0 CT_2); ^1L_a, 5.6 (20 CT_1); ^1B_a 6.5 (10 CT_1); ^1B_b, 6.5 (16 CT_2); CT_1, 8.1 (64%); CT_2, 8.4 (70%) compared with the observed 4.7 and 5.5 eV maxima, and for $C_6H_5As(CH_3)_2$: ^1L_b, 4.4 (25 CT_2); CT_1, 4.9 (60%); CT_2, 5.8 (50%); ^1L_a, 6.2 (12 CT_1), ^1B_a ^1B_b, 7.1, with an observed maximum of 5.7 eV. The observed maximum in the spectrum of $C_6H_5As(CH_3)_2$ is close to the calculated average of charge transfer states; CT_2 should be a weak transition. A correction applied to the zero-order CT_1 state for the lone pair electron $+I_\pi$ (π inductive) effect would increase the energy and decrease the CT_1 character in the second transition of $C_6H_5AsR_2$.

The charge transfer transition energies of compound **182** where M is P (λ_{max} 256–257 mμ) and M is As (λ_{max} 246–248 mμ), differ little from the transition energies of $(C_6H_5)_3M$ [59], whereas in the corresponding carbanion where M is C^-:, maxima of 642 mμ and for $(C_6H_5)_3C^{(-)}$: 422 and 488 mμ are found.

The 262 mμ charge transfer transition of triphenyl phosphine is blue shifted to 253 mμ by the fluorine inductive effect in the spectrum [80] of $(C_6F_5)_3P$.

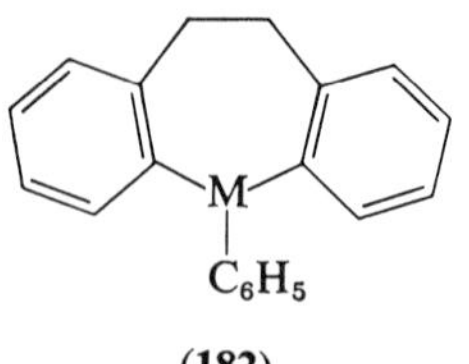

(182)

G. Aryl Tetracoordinate Metalloids

The study of phosphines, arsines, etc. has in general been paralleled by studies of quaternary halides, oxides, sulfides, etc. (see Table XXI). Thus, the ultraviolet spectra of substituted phenylphosphonic, diphenylphosphinic, and phenylalkylphosphinic acids were studied by Jaffe and Freedman [49] in 1952. The spectra of Fig. 62 where aryl is phenyl are taken from this

TABLE XXI

Ultraviolet Maxima of Tetracoordinate Group V Aryl Compounds

Compound	λ_{max}, mμ ($\epsilon \times 10^{-3}$)		
	I	II	III
$C_6H_5PO_3H_2$			264 (0.524)
p-$ClC_6H_4PO_3H_2$	224 (14.4)		264 (0.292)
$(C_6H_5)_2PO_2H$			265.0 (1.20)
p-$Cl(C_6H_4)_2PO_2H$	234 (22.2)		265 (1.00)
$(C_6H_5)_3P$—O	224 (21.3)		266 (1.96)
$(C_6H_5)_3P$—S	221 (28.8)		262 (5.5)
$(p$-$CH_3OC_6H_5)P(O)(C_6H_5)_2$	224 (20.9)	243 (16)	265 (2.9)
$(p$-$CH_3OC_6H_4)P(S)(C_6H_5)$	218 (26.3)	239 (19)	268 (sh) (5)
$[p$-$(CH_3)_2NC_6H_4]P(O)(C_6H_5)_2$	215 (28)	282 (25.5)	
$(C_6H_5)_4PBr$			263 (4.4)
$(C_6H_5)_4SbBr$			258
$(C_6H_5CH_2)_3PO$	219 (18.2)		261 (0.603)

reference. Jaffe [47] later reported the spectra of $(C_6H_5)_3AsO \cdot H_2O$ and $(C_6H_5)_3SbCl_2$, which are compared with the spectra of $(C_6H_5)_3PO$ in Fig. 63. Rao and co-workers [72] have reported the spectra of the series $(C_6H_5)_4M^+X^-$, where M is P, As, and Sb when X is Br or Cl, and related these to the corresponding Group IV metalloid spectra. Goetz *et al.* [36] studied the dipole moments and spectra of a series of *para*-monosubstituted triphenylphosphine oxides and sulfides for the substituents Cl, Br, OCH_3, and $N(CH_3)_2$. The spectroscopic effect of increasing $p(CH_3)_2N$ substitution of the phenyl rings in triphenylphosphine oxide, triphenylphosphine sulfide, the tetraphenylphosphonium cation, and the phenylalkylphosphonium cation has been discussed by Schiemenz [74].

As may be seen from Figs. 62 and 63, removal of the lone pair electrons of arylphosphines, arsines, and stibines removes from the spectra of the compounds the intense charge transfer transition, leaving the ^1L_b transition with its characteristic vibrational structure and the more intense ^1L_a transition at a shorter wavelength. The spectra of $(C_6H_5)_4M^+$ Group V metalloids

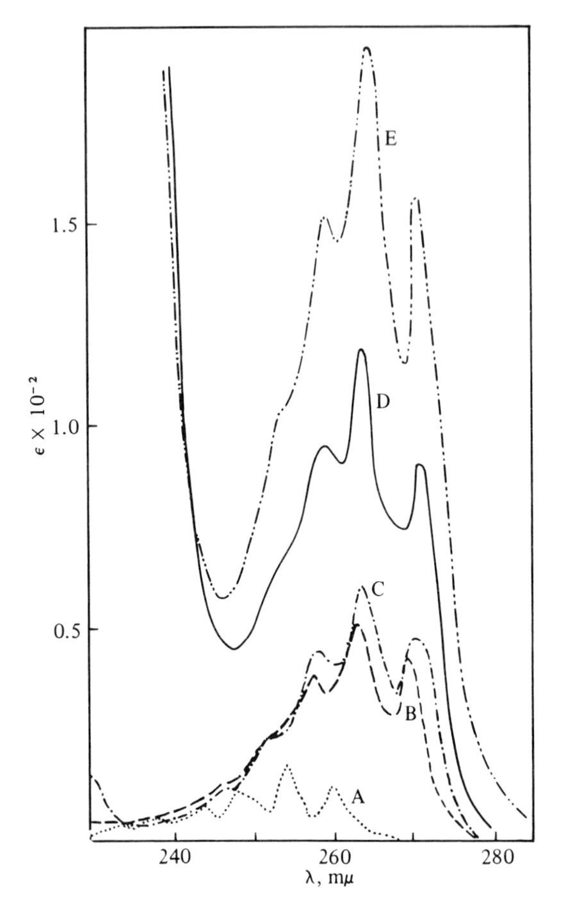

FIG. 62. Absorption spectra: (A) benzene; (B) benzenephosphonic acid; (C) benzenephosphonous acid; (D) diphenylphosphinic acid; (E) triphenylphosphine oxide [49].

are very similar to the spectra of the corresponding C_6H_5M Group IV compounds.

Even though there is good evidence of ring–metal *p–d* π bonding in the aryl tetracoordinate Group V metalloids, the central metal atom acts essentially as an insulator between the aryl rings, in a fashion about as effective as C or N. For example, Jaffe and Freedman [48] found that the absorption spectra of (aryl)(aryl)′PO_2H were very closely approximated by the averaged sum of the spectra of $(aryl)_2PO_2H$ + $(aryl)_2'PO_2H$, and that further the ^1L_b extinction coefficient ratios of $(C_6H_5)_3PO$, $(C_6H_5)_2PO_2H$, and (C_6H_5)-$(C_2H_5)PO_2H$ are close to 3:2:1. A similar conclusion was reached by Schiemenz [74] after observing that for $[p\text{-}(CH_3)_2NC_6H_4]_nP^+(C_6H_5O)_{4-n}$ the

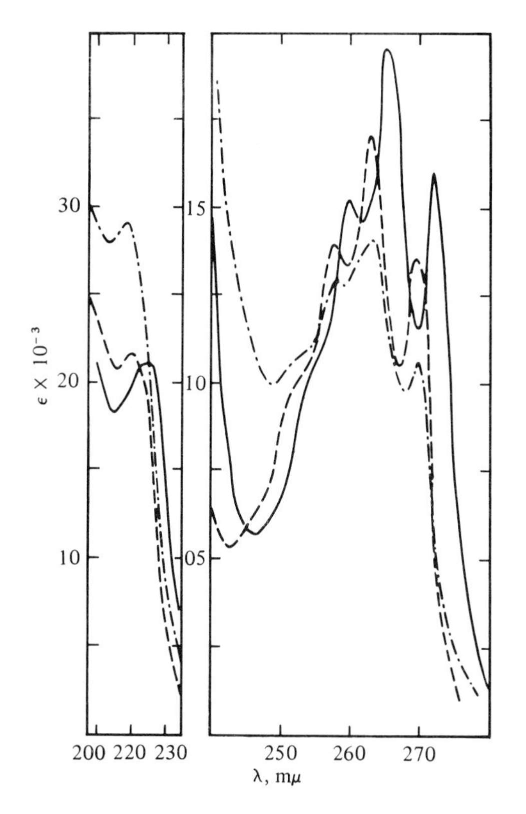

FIG. 63. The spectra in 95% EtOH of: (———) $(C_6H_5)_3PO$; (----) $(C_6H_5)_3AsO$; and (-·-·-) $(C_6H_5)_3SbCl_2$.

transition energy and ϵ/n of the 300 mμ absorption was independent of n [the number of $(CH_3)_2N$ groups].

Solvent effects on the spectra of tetraphenylphosphonium bromide [72], triphenylphosphine oxide, or tri-*p*-tolylphosphine oxide are small or nonexistent [2]. With stronger π electron donor substituents such as CH_3O or $(CH_3)_2N$ in the *para* position of $(C_6H_5)_4P$ or $(C_6H_5)_3P{=}X$, small red shifts of the absorption maxima are found with a change from nonpolar to polar solvents [2, 74].

Evidence for excited state *d–p* π bonding in these compounds is largely limited to the phosphorus derivatives. Only small or ambiguous effects are found on the ^1L_b transition and in the absence of a Goodman substituent interference study of the type which demonstrated *d* orbital participation in phosphines, attention should be concentrated on the ^1L_a transition. The arguments for *d* orbital participation run along the following lines. Since the ^1L_a transition contains a large contribution from structure **183**, *d* orbital participation as indicated by structure **184** should lead to an increase in the intensity and lower the transition energy. However, a strong inductive electron withdrawing group may also be expected to lead to a small ^1L_a red shift so we must be careful in interpreting small red shifts of absorption maxima. With strongly electron donating *para* substituents, contributions from structure **185** become important in the excited state and the *d* orbital effect will be larger. These expectations seem satisfied by the available data.

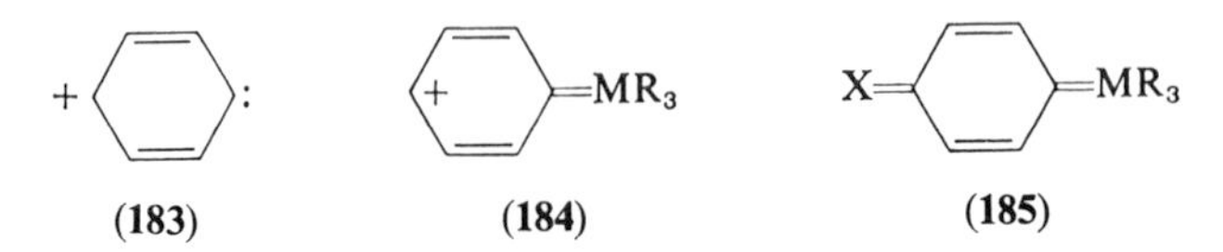

(183) **(184)** **(185)**

Consider the 3000 cm^{-1} difference between the ^1L_a transition energy of $(C_6H_5)_4C$ 47,500 cm^{-1} (211 mμ) and that of $(C_6H_5)_3PO$ 44,500 cm^{-1} (225 mμ) or 44,300 cm^{-1} (226 mμ) for $(C_6H_5)_3P$(cyclopentyl) $Br^{(-)}$, which may be compared with the $(CH_3)_2NC_6H_5$ 39,800 cm^{-1} (251 mμ) and *p*-$(CH_3)_2NC_6H_4P^+$-$(CH_3)_3I^{(-)}$ 35,400 cm^{-1} (282 mμ) transition with a red shift of about 4400 cm^{-1}. Qualitatively, tetracoordinate phosphorus —P^+R_3 appears to be about as effective at π electron withdrawal as the —C≡N group which shifts the benzene ^1L_a transition to the red by some 4000 cm^{-1}. The *para* inductive effect of the $N^+(CH_3)_3$ substituent shifts the ^1L_a $(CH_3)_2NC_6H_5$ transition by almost 2000 cm^{-1}. It is tempting to ascribe the difference between the ^1L_a $(CH_3)_2NC_6H_5$ red shift caused by $P^+(CH_3)_3$ and $N^+(CH_3)_3$ to the phosphorus *d* orbitals, but the greater polarizability of phosphorus over nitrogen must contribute at least something to this spectral red shift difference.

A much more impressive result is found in the spectra of the tri-2-furyl-, tri-2-pyryl-, and tri-2-thienylphosphine oxides (**186–188**) [39]. The 183 and 211 mμ absorptions of pyrole are shifted to 218 and 243 mμ in the spectrum of compound **187**, where R is CH_3, and either the 191 or 205 mμ transition of furan is found shifted toward the red to 238 mμ in the spectrum of compound **186**. The strong π electron donating ability of the pyryl and furyl rings, it is argued, greatly increase the importance of *d* orbital participation. The first absorption maximum of compound **188** is only 7 mμ to red of its position in thiophene (231 mμ) and it is suggested that this reflects the weaker electron donating ability of thiophene in addition to possible *d*

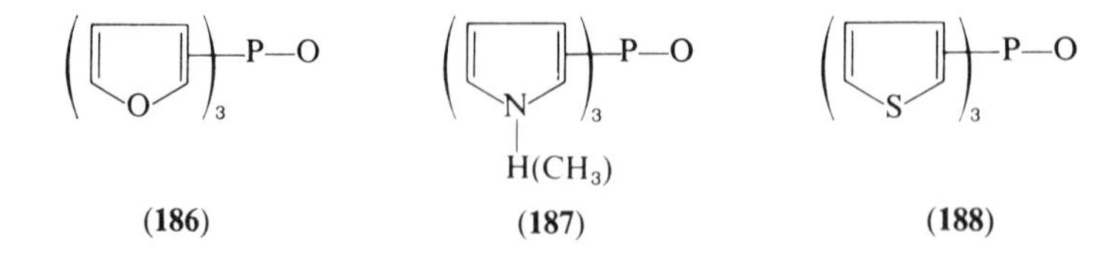

(186) **(187)** **(188)**

orbital effects of sulfur as expressed in structure **189**, which would decrease the effect of structure **190**.

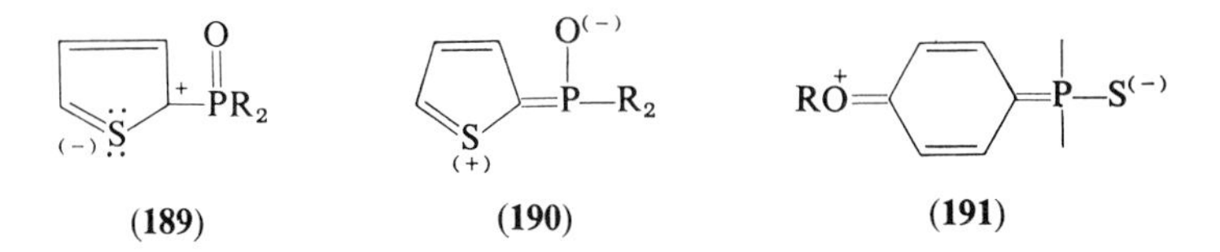

(**189**) (**190**) (**191**)

Contrary to what appears to be misquote [45] of work by Goetz and co-workers, the ^1L_a transition of p-$CH_3OC_6H_4P(S)(C_6H_5)_2$ does not occur at lower energy than that of p-$CH_3OC_6H_4P(O)(C_6H_5)_2$ and therefore does *not* indicate an $n \rightarrow \pi_d^*$ transition involving structure **191**. Goetz and co-workers did assign maxima in the spectra of p-$CH_3OC_6H_4P(X)(C_6H_5)_2$ (X is S, 239 mμ; X is O, 242 mμ) to $n \rightarrow \pi^*$ transitions with the excited states represented by structure **191**.

Some time ago Doub and Vandenbelt [30] found that if hydrogen was treated as a substituent of p-XC_6H_4Y, where X was electron donating and Y was electron withdrawing, the difference $\Delta\lambda_0$ between the ^1L_a absorption wavelength maximum of p-XC_6H_4Y and a hypothetical unsubstituted benzene ^1L_a transition near 180 mμ called λ^0, could be expressed as in Eq. (18) in terms of $\delta\lambda_X$ and $\delta\lambda_Y$, where $\delta\lambda$ represents the difference between the absorption maxima of the monosubstituted benzenes and λ^0. Equation (18) may be rearranged to give Eq. (19).

$$k\,\Delta\lambda_0 = \delta\lambda_X\,\delta\lambda_Y \tag{18}$$

$$\lambda_{max}(p\text{-}XC_6H_4Y) = k^{-1}\,\delta\lambda_X{}^0\,\delta\lambda_Y{}^0 + \lambda_0 \tag{19}$$

If Y remains constant, then a plot of the absorption maxima of *para*-X-substituted XC_6H_4Y against $\delta\lambda_X{}^0$ should be a straight line. According to Doub and Vandenbelt the success of such a relationship as Eq. (19) depends on X and Y not both being π donors or acceptors, and therefore the excellent correlation of λ_{max} for p-$XC_6H_4P(O)(C_6H_5)_2$ with $\delta\lambda_X{}^0$ (see Fig. 64) observed by Monagle and co-workers [65], along with similar results by Schiemenz [73] for p-$XC_6H_4P^+(CH_3)(C_6H_5)_2$, have been offered as evidence of p–d π electron withdrawal by phosphorus. However, a word of caution is necessary here since if the analysis is made in terms of linear energy units (frequency), no segregation into *ortho*- and *para*-directing groups is found [37] unless *both* substituents interact strongly with the ring.

The difference $\Delta\lambda$ between the ^1L_a absorption maximum of *para*-X-substituted triphenylphosphine oxides and the ^1L_a C_6H_5X transition gives an excellent correlation with σ_X [38]. However, this is not a correlation analogous to the well-known Hammet $\sigma\rho$ relationship, and should not be confused

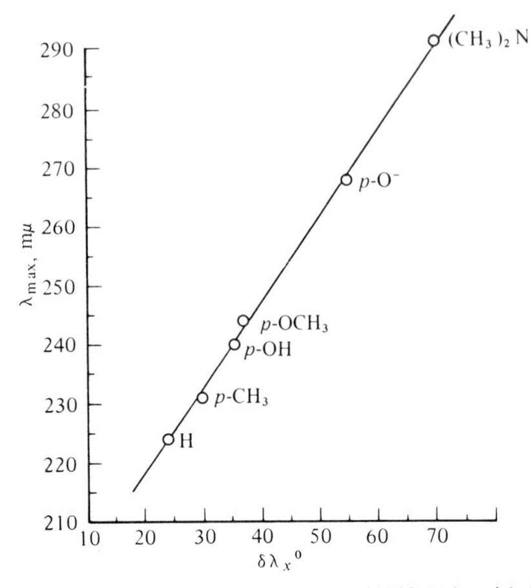

FIG. 64. Correlation of λ_{max} for $XC_6H_4P(O)(C_6H_5)_2$ with $\delta\lambda_x{}^0$.

with it. If the absorption of light by p-$XC_6H_4P^+(O)R_2$ in the ^1L_a transition is regarded as a reaction [see reaction (20)], the usual Hammet relationship would be given by Eq. (21). Consider a similar Hammet relationship to

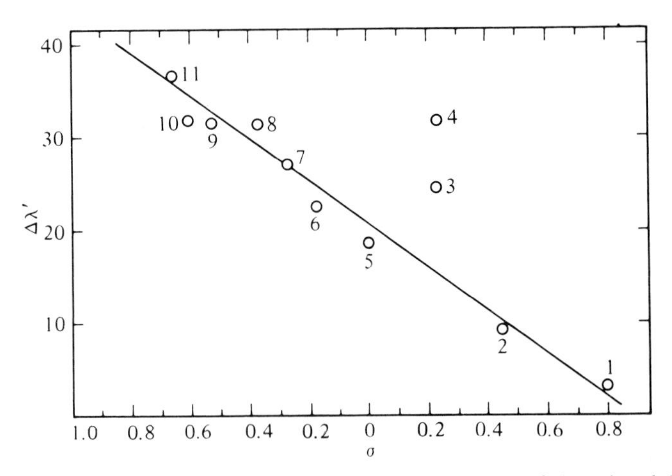

FIG. 65. The relationship between the change in wavelength of the p-band ($\Delta\lambda'$) and the Hammet substituent parameter, σ, for substituted phosphine oxides, where $\Delta\lambda'$ is λ_{max} p-$XC_6H_4PO(C_6H_5)_2$ $-\lambda_{max}$ C_6H_5X. X is: (1) NO_2; (2) CO_2H; (3) Cl; (4) Br; (5) H; (6) CH_3; (7) CH_3O; (8) OH; (9) $O^{(-)}$; (10) $(CH_3)_2N$; (11) NH_2.

express the transition represented by reaction (22). Subtracting Eq. (23) from Eq. (21) gives Eq. (24) which expresses the relationship between $E_{sX} - E_s$ and σ_s, which is the plot of Fig. 65. The ^1L_a transition energies of p-$XC_6H_4PO(C_6H_5)_2$ and XC_6H_5 do not however correlate well with the Hammet σ_X. It is obvious however from Eqs. (21) and (23) that if the corrections to σ_X required to fit the substituent X to the line in Figs. 66 and 67 are the same for both reactions (20) and (22), these corrections will cancel out in a plot such as that in Fig. 65 and we should obtain correlation of $E_{sX} - E_s$ with the incorrect σ. As may be seen from Figs. 66 and 67, this is precisely the case. The conclusion to be reached is that the plot of Fig. 65 says nothing about the importance of a structure such as **184** in the excited ^1L_a transition of phosphine oxides, but merely indicates that except for chlorine and bromine, in general whatever causes a substituent X to deviate from a Hammet plot of the ^1L_a transition of benzene also causes it to deviate from a similar plot of the p-$XC_6H_4P(O)(C_6H_5)_2$ ^1L_a transition, a not very surprising conclusion. Since the use of the ^1L_a maximum of 225 mμ reported by Goetz *et al.*

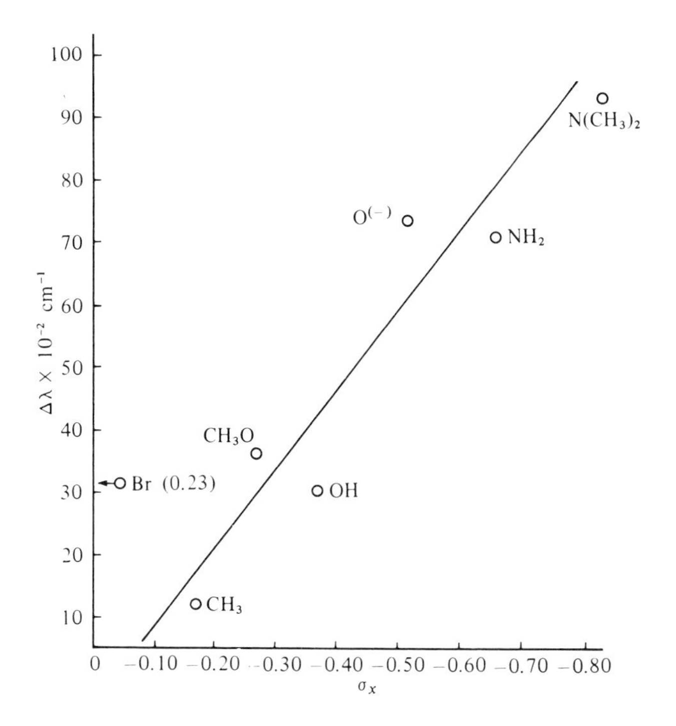

FIG. 66. $E_t[p\text{-}XC_6H_4P(O)(C_6H_5)_2] - E_t[(C_6H_5)_3PO]$ against σ_X.

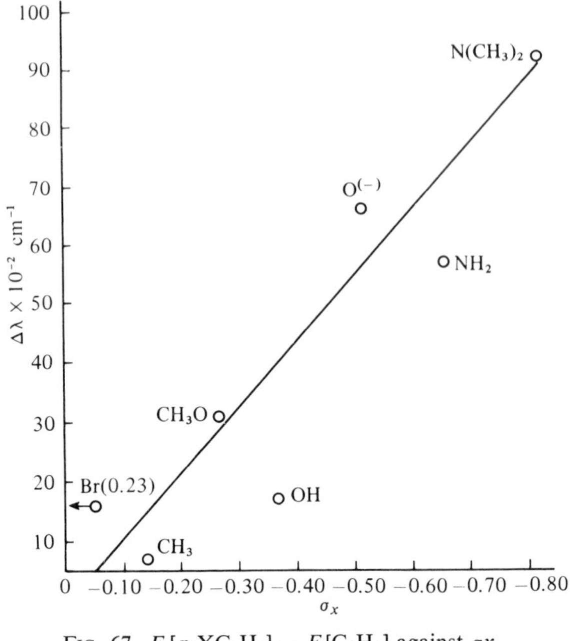

FIG. 67. $E_t[p\text{-}XC_6H_5] - E_t[C_6H_6]$ against σx.

[36] for p-ClC$_6$H$_4$P(O)(C$_6$H$_5$) brings the substituent Cl on to the line in Fig. 60 with a $\Delta\lambda'$ of 15, we suspect the ^1L_a transition, where X is Br, has also been mistakenly assigned and no special role for the d orbitals of Cl or Br need be considered.

$$\text{S—C}_6\text{H}_4\text{—X} \xrightarrow{E=h\nu} \text{S—}\overset{+}{\text{C}_6\text{H}_4}\text{=X}^{(-)} \tag{20}$$

$$E_{sX} - E_X = \sigma_s\rho \tag{21}$$

$$\text{S—C}_6\text{H}_4\text{—H} \xrightarrow{E=h\nu} \text{S—}\overset{+}{\text{C}_6\text{H}_4}\text{:—H} \tag{22}$$

$$E_s - E_{\phi H} = \sigma_s\rho' \tag{23}$$

$$E_{sX} - E_s = \sigma_s(\rho - \rho') + E_X - E_H \tag{24}$$

In Fig. 63, the ^1L_a transition energies of $(C_6H_5)_3PO(C_6H_5)_3AsO \cdot H_2O$, and $(C_6H_5)_3SbCl_3$ increase in the order P < As < Sb, which is the expected order of decreasing p–d π bonding between the phenyl rings and metal atom.

Beg and Samiuzzaman [4] have suggested that the 233 mμ (log $\epsilon \sim 4.6$) transition common to $(4\text{-}CH_3C_6H_4)_3P^+RI^-$, where R is alkyl (CH$_3$ through

C_6H_{13}), may either be the result of steric interactions between rings or a charge transfer transition from the iodide ion to the π_d^* orbitals of the phosphonium ion. Experimentally the question should be easily answered since if the transition is intermolecular charge transfer, a change of the anion to Cl^-, $BF_4^{(-)}$, etc. should cause the transition to disappear, and the intensity of 233 the mμ transition itself in $(4\text{-}CH_3C_6H_4)_3P^+RI^-$ should not obey Beer's law. Since cyclopentyltriphenylphosphonium bromide [69] absorbs at 226 mμ (log ε 4.40) and 268 mμ (log ε 3.68), and the ^1L_b transition of $(p\text{-}CH_3C_6H_4)_3P^+RI^-$ is found at 276 mμ, we are inclined to correlate the 226 mμ and 233 mμ transitions and assign them both to the ^1L_a transition rather than intermolecular charge transfer. Tetraphenylphosphonium bromide absorbs very strongly in the region of 230 mμ. An additional argument against an intramolecular charge transfer assignment is the absence of a corresponding transition in compound **192** for which maxima of 220 mμ (log ε 4.5) and 275 mμ (log ε 3.55) are found.

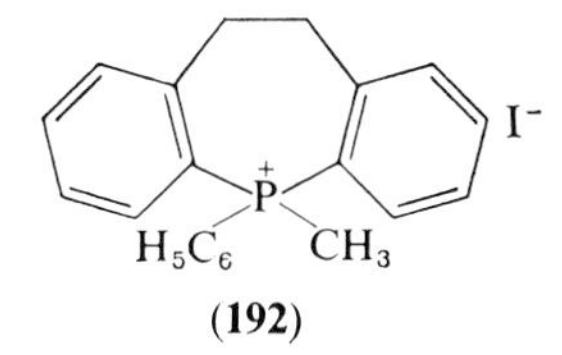

(192)

H. Miscellaneous Metalloids

The ultraviolet spectra [5] of the acylphosphonates $RC(O)P(O)(OR')_2$, where R′ is alkyl and R varies from CH_3 to $p\text{-}XC_6H_5$, have been compared with spectra of the corresponding aldehydes. [For example, compare $CH_3C(O)P(O)(OC_2H_5)_2$, λ_{max} 334 mμ (log ε 1.72), or $C_6H_5C(O)P(O)(OC_2H_5)$, λ_{max} 258 mμ (log ε 4.05), with CH_3CHO, λ_{max} 290 mμ (log ε 1.23), and C_6H_5CHO, λ_{max} 241 mμ (log ε 4.15).] These results clearly indicate a lowering of the π^* orbital of RC=O by the phosphorus —$P(O)(OR_2)$ *d* orbitals through *p–d* π bonding. As expected, the red shift of the $n \rightarrow \pi^*$ transitions where R is alkyl is greater than the $\pi \rightarrow \pi^*$ red shift in the spectra of RCHO if hydrogen is replaced by —$P(O)(OR)_2$. Note that if an electron withdrawing inductive effect of $P(O)(OR)_2$ had been important, the $n \rightarrow \pi^*$ transition would have shown a blue shift.

The effect of α-phosphine substitution on the $n \rightarrow \pi^*$ transition of a carbonyl contrasts sharply with the relative $n \rightarrow \pi^*$ transition energies of an amide (λ_{max} $CH_3C(O)NH_2$, 215 mμ) and a ketone (λ_{max} $(CH_3)_2CO$, 279 mμ). The $n \rightarrow \pi^*$ transition of $(C_6H_5)_2PC(O)C_6H_5$, found at 390 mμ, shows a very large *red* shift similar to that observed for silyl and other Group IV metal ketones. The low $n \rightarrow \pi^*$ transition energy of the α-silyl ketones, etc.

is thought to be the result of a combination of inductive release which raises the oxygen nonbonding orbital energy and a metal *d–p* π interaction which lowers π^* (see Chapter IV). Although the electronegativity of P is now thought to be about the same as C, presumably a similar explanation in terms of *P–d* orbitals and greater polarizability, applies to the low $n \rightarrow \pi^*$ transition energy of $(C_6H_5)_2PC(O)(C_6H_5)$ (see Addendum).

The spectra of the phosphacyanines **193** and **194** possess [29] an intense absorption maximum log ϵ (4.4–4.1) between 472 mμ, where R^1 is CH_3, R^2 is H, in structure **193** and 587 mμ where R^1 is C_2H_5 and R^2 is CH_3, for Structure **194**. The spectra are similar to analogous azo- and methinecyanines.

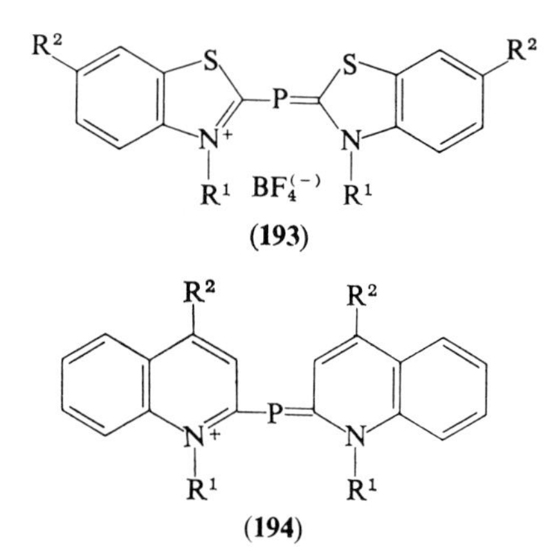

(193)

(194)

The ultraviolet spectra of the biphenyl derivatives of the phosphorus and arsenic compounds **195** and **196** provide evidence of steric hindrance to non-coplanarity of the rings in the *ortho* derivatives where the intensity of the biphenyl absorption is greatly reduced [34].

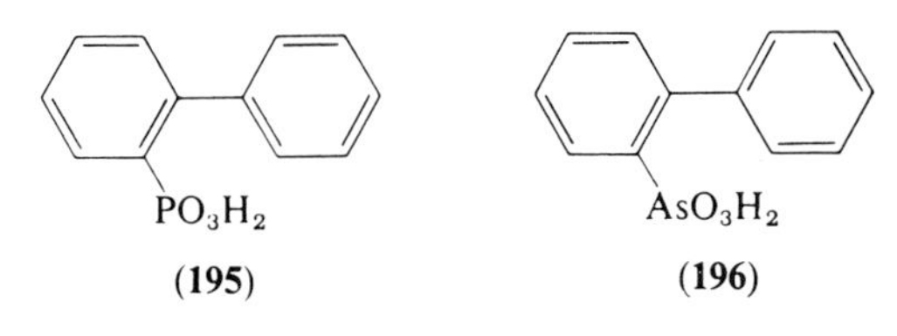

(195) **(196)**

ORGANOMETALLIC DERIVATIVES OF TELLURIUM

The ultraviolet spectra of ethylphenyltelluride [11] and the tellurophene [13] **197** as would be expected resemble closely the spectra of the corresponding sulfur and selenium compounds (see Fig. 68). The spectrum of methylphenylselenide taken from the work of Mangini and co-workers [57]

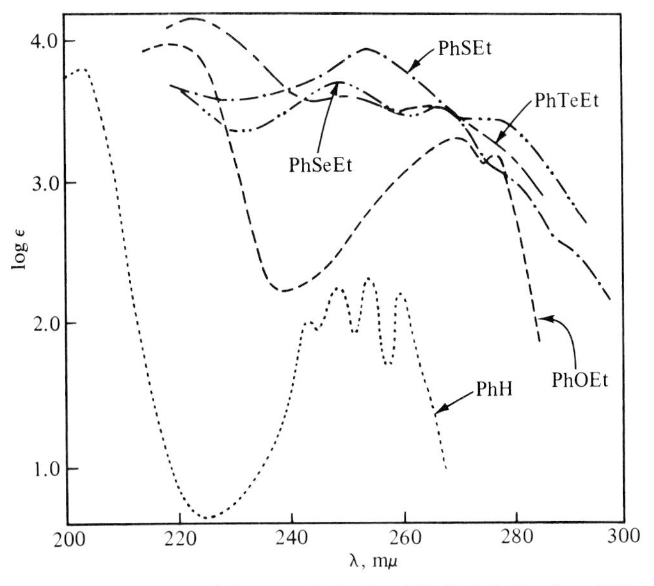

FIG. 68. The spectra of Group VI C_6H_5—M-alkyl derivatives [11].

is shown in Fig. 69. The 37,200 cm^{-1} band of $C_6H_5SeCH_3$ was assigned to the ^1L_b transition and the more intense 48,000 cm^{-1} was assigned to charge transfer which broke down to correlation with the ^1L_a transition. It may be that the 44,000 cm^{-1} band, which is conspicuous in the vapor phase spectrum

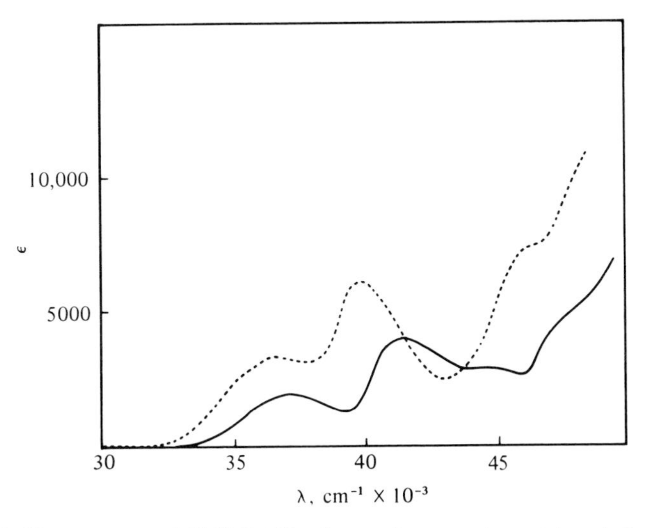

FIG. 69. The spectrum of $C_6H_5SeCH_3$: (———) vapor; (----) in solution of cyclohexane [57].

of $C_6H_5SeCH_3$, can be attributed to partially dissociative states involving the X—CH_3 bond, since a similar assignment has been made for a continuum absorption in the spectra of aniline near 40,000 cm^{-1} (250 mμ) and in the 37,000 cm^{-1} region of the dimethylaniline spectrum. Presumably a similar analysis of the C_6H_5TeR spectrum can be made.

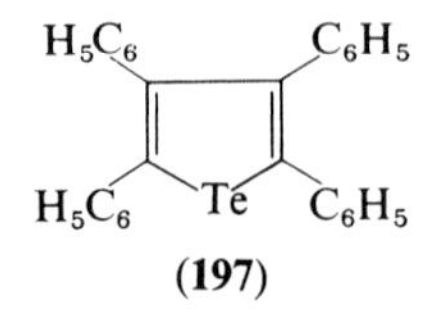

(197)

The ultraviolet spectra of $(C_6H_5)_2Se$, $(C_6H_5)SeCl_2$, and $(C_6H_5)_3SeCl$ have been discussed by Arshad and co-workers [1b] but the spectra of corresponding Te compounds do not seem to be available.

REFERENCES

1a. A. M. Aguiar, K. C. Hansen, and G. S. Reddy, *J. Am. Chem. Soc.* **89**, 3067 (1967).
1b. M. Arshad, A. Beg, and A. R. Shaikh, *Tetrahedron* **22**, 653 (1966).
2. V. B. Baliah and P. S. Subbarayan, *J. Org. Chem.* **25**, 1833 (1960).
3. S. H. Bauer and C. F. Aten, *J. Chem. Phys.* **39**, 1254 (1963).
3a. R. S. Becker and E. Chen, *J. Chem. Phys.* **45**, 2403 (1966).
4. M. A. A. Beg and Samiuzzaman, *Tetrahedron* **24**, 194 (1968).
5. K. D. Berlin and D. Burpo, *J. Org. Chem.* **31**, 1304 (1966).
6. J. E. Bissey and H. Goldwhite, *Tetrahedron Letters* **1966**, 3247.
7. H. Bock and E. Baltin, *Chem. Ber.* **98**, 2844 (1965).
8. H. Bock and G. Rudolph, *Chem. Ber.* **98**, 2273 (1965).
9. H. Bock, G. Rudolph, E. Baltin and J. Kroner, *Angew. Chem.* **77**, 469 (1965).
10. G. M. Boglyubov and A. A. Petrov, *Chem. Rev.* (*Russian*) **36**, 258 (1967).
11. K. Bowden and E. A. Braude, *J. Chem. Soc.* **1952**, 1068.
12. E. H. Braye and W. Hubel, *Chem. Ind.* (*London*) **1959**, 1250.
13. E. H. Braye, W. Hubel, and I. Caplier, *J. Am. Chem. Soc.* **83**, 4406 (1961).
14. A. G. Brook, K. Kivisk, and G. LeGrow, *Can. J. Chem.* **43**, 1175 (1965).
14a. D. A. Brown, *J. Chem. Soc.* **1962**, 929.
15. E. P. Carr and H. Stucken, quoted in G. O. Burr and E. S. Miller, *Chem. Rev.* **29**, 422 (1941).
16. E. Y. C. Chang and C. Price, *J. Am. Chem. Soc.* **83**, 4650 (1961).
17. C. Charrier, W. Chodkiewicz, and P. Cadiot, *Bull. Chim. Soc. France*, **1966**, 1002.
18. C. Charrier, M. P. Simonnin, W. Chodkiewicz, and P. Cadiot, *Compt. Rend.* **258**, 1537 (1964).
19. A. H. Cowley, A. Burg, and W. R. Cullen, *J. Am. Chem. Soc.* **88**, 3178 (1966).
20. A. H. Cowley and W. D. White, Meeting Am. Chem. Soc., 153rd, Miami Beach (1967).
21. D. P. Craig and N. L. Paddock, *J. Chem. Soc.*, **1962**, 4118.
22. D. P. Craig, *J. Chem. Soc.*, **1959**, 997.

23. P. E. Crofts and M. P. Williamson, *J. Chem. Soc. C* **1967**, 1093.
24. W. R. Cullen, B. R. Green, and R. M. Hochstrasser, *J. Inorg. Nucl. Chem.* **27**, 641 (1965).
25. W. R. Cullen and R. M. Hochstrasser, *J. Mol. Spectry.* **5**, 118 (1960).
26. P. de Koe and F. Bickelhaupt, *Angew. Chem. Intern. Ed. Engl.* **6**, 567 (1967).
27. W. B. De Moore, H. O. Pritchard, and N. Davidson, *J. Am. Chem. Soc.* **81**, 8575 (1959).
28. M. J. S. Dewar, E. A. C. Lucken, and M. A. Whitehead, *J. Chem. Soc.* **1960**, 2423.
28a. M. J. S. Dewar and V. P. Kubba, *J. Am. Chem. Soc.* **82**, 5685 (1960).
29. K. Dimroth and P. Hoffman, *Angew. Chem.* **76**, 433 (1964).
30. L. Doub and J. M. Vandenbelt, *J. Am. Chem. Soc.* **69**, 2714 (1947).
31. J. S. Driscoll, D. Grisley, J. Pustinger, J. Harris, and C. N. Malthews, *J. Org. Chem.* **29**, 2427 (1964).
32. M. I. Ermakova, L. N. Vorontsova, and N. Latosh, *Zh. Obshch. Khim.* **37**, 649 (1967).
33. H. Fischer and H. Fischer, *Chem. Ber.* **99**, 658 (1966).
34. L. D. Freedman, *J. Am. Chem. Soc.* **77**, 6223 (1955).
34a. M. Godfrey and J. N. Murrell, *Proc. Roy. Soc. (London) Ser. A* **278,** 57, 64 (1964).
35. H. Goetz and H. Juds, *Ann. Chem.* **698**, 1 (1966).
36. H. Goetz, F. Nerdel, and K. H. Wiechel, *Ann. Chem.* **665**, 1 (1963).
37. B. G. Gowenlock and K. J. Morgan, *Spectrochim. Acta* **17**, 310 (1961).
38. C. E. Griffin and H. H. Hsieh, quoted in Hudson [45].
39. C. E. Griffin, R. P. Peller, K. Martin, and J. Peters, *J. Org. Chem.* **30** 97 (1965).
40. A. A. Grigorenko, M. I. Shevchuk, and A. V. Dombrovski, *Zh. Obsch. Khim.* **36**, 1121 (1966).
41. M. G. Hogben, R. S. Gay and W. Graham, *J. Am. Chem. Soc.* **88**, 3458 (1966).
42. M. Halmann, *J. Chem. Soc.*, 2853 (1963).
43. W. Henderson Jr., M. Epstein, and F. Seichter, *J. Am. Chem. Soc.* **85**, 2463 (1965).
44. L. Horner and H. Oediger, *Ann. Chem.* **627**, 142 (1959).
44a. L. Horner and H. Oediger, *Chem. Ber.* **91**, 437 (1958).
45. R. F. Hudson, "Structure and Mechanism in Organo-Phosphorous Chemistry." Academic Press, New York, 1965.
46. H. H. Jaffe, Meeting Am. Chem. Soc., 155th, San Francisco (1968).
47. H. H. Jaffe, *J. Chem. Phys.* **22**, 1430 (1954).
48. H. H. Jaffe and L. D. Freedman, *J. Am. Chem. Soc.* **74**, 2930 (1952).
49. H. H. Jaffe and L. D. Freedman, *J. Am. Chem. Soc.* **74**, 1069 (1952).
50. H. H. Jaffe and M. Orchin, "Theory and Applications of Ultraviolet Spectroscopy" Wiley, 1962.
51. M. I. Kabachnik, *Tetrahedron* **20**, 655 (1964).
52. H. B. Klevens and J. R. Platt, *J. Am. Chem. Soc.* **71**, 1714 (1949).
52a. E. M. Kosower and B. G. Ramsey, *J. Am. Chem. Soc.* **81**, 856 (1959).
53. V. Lambert and D. Mueller, *J. Am. Chem. Soc.* **88**, 3669 (1966).
54. D. Lloyd and M. I. C. Singer, *Chem. Ind. (London)* **1967**, 787.
55. D. Lloyd and M. I. C. Singer, *Chem. Ind. (London)* **1967**, 510.
56. W. Mahler and A. Burg, *J. Am. Chem. Soc.* **80**, 6161 (1958).
57. A. Mangini, A. Trombetti, and C. Zauli, *J. Chem. Soc. B*, **1967**, 153.
58. F. G. Mann, *Prog. Org. Chem.* **4**, 239 (1958).
59. F. G. Mann, I. T. Millar, and B. P. Smith, *J. Chem. Soc.* **1953**, 1130.
60. G. Märkl, *Angew. Chem. Intern. Ed. Engl.* **5**, 846 (1967).
61. G. Märkl, *Angew. Chem. Intern. Ed. Engl.* **2**, 153, 479 (1963).
62. G. Märkl, F. Lieb, and A. Mertz, *Angew. Chem. Intern. Ed. Engl.* **6**, 458 (1967).

63. G. Märkl, F. Lieb, and A. Mertz, *Angew. Chem. Intern. Ed. Engl.* **6**, 88 (1967).
63a. S. F. Mason and G. W. Vane, *Chem. Commun.* **1965**, 540.
64. H. Mohler and J. Sorge, *Helv. Chim. Acta* **22**, 235 (1939).
65. J. J. Monagle, J. V. Mengenhauser, and D. Jones, Jr., *J. Org. Chem.* **32**, 2477 (1967).
65a. J. N. Murrell, *Tetrahedron Suppl. 2* **19**, 277 (1963).
66. N. A. Nesmayanov and O. A. Reutov, *Dokl. Akad. Nauk. SSSR* **171**, 111 (1966).
67. G. Opitz, H. Hellman, and H. W. Schubert, *Ann. Chem.*, **623**, 112 (1959).
68. F. Ramirez, N. B. Desai, B. Hansen, and N. McKelvie, *J. Am. Chem. Soc.* **83**, 3539 (1961).
69. F. Ramirez and S. Levy, *J. Am. Chem. Soc.* **79**, 67 (1957).
70. W. Reid and H. Appel, *Ann. Chem.* **679**, 56 (1964).
71. M. B. Robin, R. R. Hart, and N. A. Kuebler, *J. Am. Chem. Soc.* **89**, 1564 (1967).
72. C. N. R. Rao, J. Ramachandran, and A. Balasubramanian, *Can. J. Chem.* **39**, 171 (1961).
73. G. P. Schiemenz, *Angew. Chem. Intern. Ed. Engl.* **6**, 564 (1967).
74. G. P. Schiemenz, *Tetrahedron Letters*, **38**, 2729 (1964).
75. H. Schindlbauer, *Monatsh. Chem.* **94**, 99 (1963).
76. R. Shula and S. T. Zenchelsky, *J. Am. Chem. Soc.* **82**, 4138 (1960).
77. M. I. Shevchuk and A. V. Dombrovski, *Zh. Obshch. Khim.* **34**, 2717 (1964).
78. E. Tannenbaum, E. M. Coffin, and A. J. Harrison, *J. Chem. Phys.* **21**, 311 (1953).
79. H. W. Thompson and J. J. Frewing, *Nature* **135**, 507 (1935).
80. L. A. Wall, R. E. Donadio, and W. J. Pummer, *J. Am. Chem. Soc.* **82**, 4847 (1960).
81. M. A. Weiner and G. Pasternack, *J. Org. Chem.* **32**, 3707 (1967).
82. N. Zhmurova and A. V. Kirsanov, *Zh. Obshch. Khim.* **36**, 1248 (1966).
83. R. West, private communication; J. Gaidis, Ph.d. Thesis, Univ. of Wisconsin, 1967.

APPENDIX A

ELECTRONIC TRANSITIONS OF GROUP I ORGANOMETALLIC COMPOUNDS[a]

Organic radical	Metal	Solvent[b]	λ_{max}, mμ	(log ϵ)	Ref.[c]
Vinyl	Li	THF	280 (sh)	(2.71)	47
Allyl	Li	THF	315	(3.66)	47, 49
	Li	THF	317		10
γ- and α-Methylallyl	Li	THF–10% ether	291	(3.79)	47
α-Methylallyl	Li	THF	291		49
γ-Methylallyl	Li	THF	291		49
β-Methylallyl	Li	THF	330		49
	Li	THF	332		10
3-Butenyl	Li	Isopentane	208, 212, 218, 225, 251		36
n-Butyl	Li	Hexane	221, 225		36
Phenyl	Li	THF	292	(2.88)	47
			268	(2.99)	
			261	(3.02)	
Benzyl	Li	THF	330	(3.98)	47, 48, 49
		Ether	328		48
	Na	THF	355	(4.08)	1
p-Tolyl	Li	THF	268, 273, 292		49
α-Methylbenzyl	Li	THF	333		49
		THF–10% ether	333	(4.08)	47
		THF	335		48
		Ether	335		48
		Benzene	344		48
p-Methylbenzyl	Li	THF	315		49
1-Phenyl-2-propenyl	Li	THF	395	(4.38)	47
	K	NH_3	420	(4.18)	2

[a] Compounds are listed in order of increasing C, H in their empirical formula.

[b] THF is tetrahydrofuran; CHA is cyclohexylamine; HMP is hexamethylphosphortriamide; MeTHF is 2-methyltetrahydrofuran; Py is pyridine; DMSO is dimethylsulfoxide; DME is ethyleneglycol dimethyl ether.

[c] Reference numbers refer to those listed at the end of Chapter II.

[d] All maxima of spectra are not reported, or spectrum presented for limited range.

[e] Assigned by reference to contact ion pair.

[f] Assigned by reference to solvent separated ion pair.

[g] Does not obey Beer's Law.

[h] Parentheses and italic numbers refer to dipole strength *not* log ϵ.

[i] Time dependent.

[j] Concentration dependent.

Organic radical	Metal	Solvent[b]	λ_{max}, mμ	(log ϵ)	Ref.[c]
Cumyl	K	THF	338	(4.25)	1
Naphthalene[d]	Li	THF	322[e]		25
			326[f]		
	Na	THF	800	(3.4)	37
			435	(3.5)	
			370	(3.7)	
Naphthalene	Na	THF	323		25
			364		
at −60°C	Na	THF	318 (sh)		25
			326[f]		
			369[f]		
Biphenyl	Li	THF	639	(4.1)	9
			405	(4.6)	
	Na	THF	613	(4.1)	9
			400	(4.6)	
9-Fluorenyl	Li[d]	THF	349[e]		25
			373[f]	(3.98)	
			488		
			523		
		MeTHF	347[e]		25
			373[f]		
	Li	CHA	452	(3.03)	44, 43
			477	(3.11)	
			510	(2.92)	
		Dioxane	346[e]		25
			437		
		DME	373[f]	(3.9)	25
			455		
			486	(3.1)	
			521		
		Py	373[f]		25
		Ether	415	(3.03)	43
			435	(3.08)	
			457	(2.94)	
		DMSO	373[f]		25
		Toluene	348[e]		25
	Na	THF	356[e]	(4.03)	25
			370 (sh)[f]		
			430	(2.6)	
			452	(2.7)	
			486	(2.6)	
			521 (sh)		
9-Fluorenyl at −50°C	Na	THF	357 (sh)		25
			373[f]	(4.20)	
			430		
			450		
			486	(3.0)	

Organic radical	Metal	Solvent[b]	λ_{max}, mμ	(log ϵ)	Ref.[c]
9-Fluorenyl at $-50°C$			521[f]	(2.7)	
		MeTHF	355[e]		25
at $-80°C$		MeTHF	373[f]		25
at 50°C		DME	358[e]		25
		DME	373[f]		25
		DMSO	373[f]		25
	Na	Dioxane	353[e]		25
		Py	373[f]		25
	$(n\text{-}C_4H_9)_4N^+$	THF	368[e]		25
	Cs	THF	364[f]	(4.08)	25
		DME	364[e]		25
	K	THF	362[f]		25
		DME	362[e]		25
	Cs	CHA	447	(2.99)	44
			472	(3.08)	
			504	(2.94)	
Xanthenyl	Li	CHA	470	(3.89)	43
Diphenylmethyl	Li	CHA	443	( )[g]	44
	Cs	CHA	443	(4.57)	44
	K	NH_3	440	(4.6)	2
Anthracene	Li	MeTHF	730	*(0.92)*[h]	9
			369	*(0.60)*[h]	
			327	*(1.10)*[h]	
			258	*(2.18)*[h]	
	2Li	MeTHF	556	*(1.33)*[h]	9
			313	*(3.44)*[h]	
	Na	MeTHF	704	*(0.79)*[h]	9
			366	*(0.45)*[h]	
			325	*(1.08)*[h]	
			256	*(1.84)*[h]	
	2Na	MeTHF	592	*(1.45)*[h]	9
			326	*(3.70)*[h]	
	K	MeTHF	714	*(0.83)*[h]	9
			366	*(0.54)*[h]	
			326	*(1.10)*[h]	
			256	*(2.08)*[h]	
	2K	MeTHF	617	*(1.45)*[h]	9
			330	*(3.78)*[h]	
	Na	HMP	764		4
			740		
			695		
			656		
			641 (sh)		
			597		
			586 (sh)		
			548		
	Na	THF	720	(4.02)	28, 37

Organic radical	Metal	Solvent[b]	λ_{max}, mμ	(log ϵ)	Ref.[c]
Anthracene			595	(3.65)	
			550	(3.48)	
			360	(4.39)	
			325	(4.51)	
	2Na	THF	606	(4.26)	28
			325	(4.77)	
	K	THF	724		13
			695		
			654		
			638		
			595		
			546		
Phenanthrene	Na	THF	1100	(3.5)	37
			625	(3.2)	
			435	(3.9)	
			400	(3.8)	
9-Methylfluorenyl	Li	CHA	480	(2.96)	43
			512	(3.05)	
			551	(2.89)	
1-Methoxyfluorenyl	Li	MeTHF	344[e]		25
at −50 to −70°C			368[f]		25
			498		
4,5-Methylenephenanthryl	Li	Ether	455	(3.83)	43
	Li	CHA	505	(3.87)	44
4,5-Methylenephenanthryl	Cs	CHA	505	(3.84)	44
1,1-Diphenyl-3-butenyl	Li	THF	495		47
1,2-Benzofluorenyl	Li	Ether	392	(3.97)	43
			405 (sh)	(3.91)	
		CHA	425	(3.91)	44
			450 (sh)	(3.89)	
2,3-Benzofluorenyl	Li	Ether	387	(4.21)	43
			535	(3.78)	
	Li	CHA	420	(4.34)	44
			605	(3.23)	
			657	(3.11)	
	Cs	CHA	418	(4.28)	44
			599	(3.19)	
			648	(3.12)	
2,3-Benzofluorenyl	Na	THF	407[e]		25
			431		
			567[e]		
			611[e]		
			675 (sh)		
at −50°C	Na	THF	408 (sh)		25
			431[f]		
			567		
			611[f]		

Organic radical	Metal	Solvent[b]	λ_{max}, mμ	(log ϵ)	Ref.[c]
2,3-Benzofluorenyl at −50°C			676		
3,4-Benzofluorenyl	Li	CHA	397	(3.68)	44
			465	(3.31)	
			487	(3.41)	
			519	(3.31)	
		DME	406	(3.68)	43
			470	(3.25)	
			496	(3.34)	
			533	(3.21)	
	Li	Ether	420	(3.36)	43
			438	(3.46)	
			460	(3.35)	
	Cs (?)	CHA	455	(3.33)	44
			477	(3.39)	
			508	(3.29)	
	Na	THF	324[e]		25
			368		
			388		
at −50 to −70°C	Na	THF	340[f]		25
			388		
			410		
Benzanthrenyl	Li	Ether	428	(4.29)	43
			540	(3.81)	
		CHA	444	(4.54)	43
			518	(3.72)	
			550	(3.81)	
			592	(3.84)	
			640	(3.83)	
$C_6H_5(CH{=}CH)_2\ddot{C}HC_6H_5$	Na	Ether	568		24
Triphenylene	Na	THF	714	(3.5)	37
			526	(3.4)	
			416	(3.9)	
			345	(3.9)	
Naphthacene	Na	THF	625	(4.2)	37
			351	(4.7)	
Tetracene	Li	MeTHF	806	*(1.41)*[h]	9
			402	*(0.90)*[h]	
			357	*(1.45)*[h]	
			284	*(2.74)*[h]	
	2Li	MeTHF	585	*(1.98)*[h]	9
			340	*(4.34)*[h]	
at −195°	2Li	MeTHF	671	*(2.61)*[h]	9
			364	*(4.30)*[h]	
			416		
	2Na	MeTHF	606	*(1.99)*[h]	9
			352	*(4.03)*[h]	
	K	MeTHF	794	*(1.00)*[h]	9

Organic radical	Metal	Solvent[b]	λ_{max}, mμ	(log ϵ)	Ref.[c]
Tetracene at −195°C			398	*(0.67)*[h]	
			356	*(1.33)*[h]	
			283	*(2.62)*[h]	
	2K	MeTHF	621	*(2.14)*[h]	9
			356	*(4.47)*[h]	
	Na	MeTHF	794	*(0.91)*[h]	9
			398	*(0.60)*[h]	
			356	*(1.31)*[h]	
			283	*(2.62)*[h]	
Terphenyl	Li	THF	885	(4.5)	9
			485	(4.6)	
			441 (sh)		
	Na	THF	806	(4.4)	9
			461	(4.5)	
			424 (sh)		
1,6-Diphenylhexatriene	Na	THF	893	(3.7)	12, 14
			787	(3.7)	
			629	(4.65)	
	2Na	THF	637	(4.73)	
1,1-Diphenyl-*n*-hexyl	Li	Hexane	410		46
		Benzene	415		46
	Li·2THF	Benzene	450		46
	Li·4THF	THF	496		46
	Li	2,3-diMe-THF	443		46
		2-MeTHF	451		46
		Tetrahydro-pyran	436		46
		n-Butyl-amine	445		46
		N,N-Di-methyl-aniline	447		46
		Pyridine	484		46
		HMP	485		46
		n-Propyl ether	436		46
	Li	THF	496		48
		DME	495		48
		Ether	438		48
		Isopropyl ether	428		48
		n-Butyl ether	435		48
		Benzene–butyl-Li	428	(4.20)	21, 48
		Hexane–butyl-Li	415		46, 48

Organic radical	Metal	Solvent[b]	λ_{max}, mμ	(log ϵ)	Ref.[c]
1,1-Diphenyl-*n*-hexyl		THF	496	(4.26)	47, 49
			315	(3.5)	
Fluoradenyl	Li	CHA	522	(3.60)	43
			562	(3.67)	
Tris(*p*-nitrophenyl)-methyl	Na	DME	685	(4.53)	27
		THF	668	(4.50)	
		CH_3CN	800	(4.77)	
		Py	820	(4.77)	
	K	DME	770	(4.68)	27
		THF	780	(4.62)	
		CH_3CN	800	(4.78)	
		Py	825	(4.77)	
Triphenylmethyl	Li	CHA	488	( )[g]	44
	Cs	CHA	488	(4.45)	44
	Li	THF	425	(3.97)	47
			500	(4.27)	
	K	NH_3	495	(4.40)	
	Na	THF	485		4
			412		
		HMP	505		4
			432		
$C_6H_5(CH{=}CH)_3\ddot{C}HC_6H_5$	Na	Ether	600		24
1,1-di-*o*-tolyl-*n*-hexyl	Li	THF	501		49
			328		
1,1-di-*p*-tolyl-*n*-hexyl	Li	THF	490		49
			318		
1,1-di-*m*-tolyl-*n*-hexyl	Li	THF	498		49
			310 (sh)		
$C_6H_5{-}(CH{=}CH)_4{-}\ddot{C}HC_6H_5$	Na	Ether	635		24
1,3,3-Triphenylpropenyl	Li	CHA	470	(4.33)	44
			556	(4.66)	
Pentacene	Li	MeTHF	870	(*1.66*)[h]	9
			435	(*0.88*)[h]	
			397	(*1.7*)[h]	
			300	(*3.52*)[h]	
	K	MeTHF	847	(*1.40*)[h]	9
			435	(*0.71*)[h]	
			394	(*1.7*)[h]	
			301	(*3.50*)[h]	
	Na	MeTHF	847	(*1.42*)[h]	9
			431	(*0.59*)[h]	
			392	(*1.7*)[h]	
			300	(*3.58*)[h]	
	2Li	MeTHF	680	(*3.31*)[h]	9
			450		

Organic radical	Metal	Solvent[b]	λ_{max}, mμ	(log ϵ)	Ref.[c]
Pentacene			376	*(6.26)*[h]	
	2Na	MeTHF	685	*(3.27)*[h]	9
			457		
			379	*(6.54)*[h]	
	2K	MeTHF	694	*(3.28)*[h]	9
			461		
			382	*(6.30)*[h]	
at − 195°C	2Li	MeTHF	735	*(3.2)*[h]	9
			485		
			394	*(6.1)*[h]	
Dibenz[*a*,*h*]anthracene	Na	THF	1100	(4.0)	37
			714	(4.4)	
			526	(4.3)	
			435	(4.7)	
			392	(4.7)	
9,10-Dihydro-9,9-dimethyl-10-phenyl-anthracenyl	Li	CHA	454	(4.50)	44
	Cs	CHA	445	(4.53)	44
Quarterphenyl	Li	THF	1220	(4.7)	9
			1090 (sh)	(4.6)	
			524	(4.5)	
			488	(4.5)	
	Na	THF	961	(4.4)	9
			502	(4.4)	
			469	(4.4)	
p-Biphenylyldiphenyl-methyl	Li	CHA	587	( )[g]	44
Bis-*p*-biphenylylmethyl	Cs	CHA	573	(4.65)	44
1,1-Bis(*p*-biphenyl)ethyl	K	Ether	545	(5.19)	23
	K	Ether	590	(5.16)	23
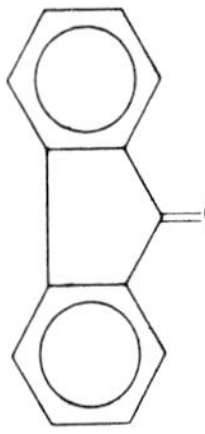	Na ?	THF	557		29
$=CH—\ddot{C}(C_6H_5)_2$	Na ?	THF	518		29
			546		
1,1,4,4-Tetraphenylbutyl	2Na	THF	472[e]		25
at 50°C	2Li	THF	460[e]		25
1,1,4,4-Tetraphenylbutyl	2Cs	THF	481[e]		25
at − 50 to − 70°C	2Li	THF	500[f]		25
	2Na	THF	501[f]		25
9-Phenylfluorenyl	Li	CHA	452	(3.29)	43, 44
			487	(3.36)	
			520	(3.23)	

Organic radical	Metal	Solvent[b]	λ_{max}, mμ	(log ϵ)	Ref.[c]
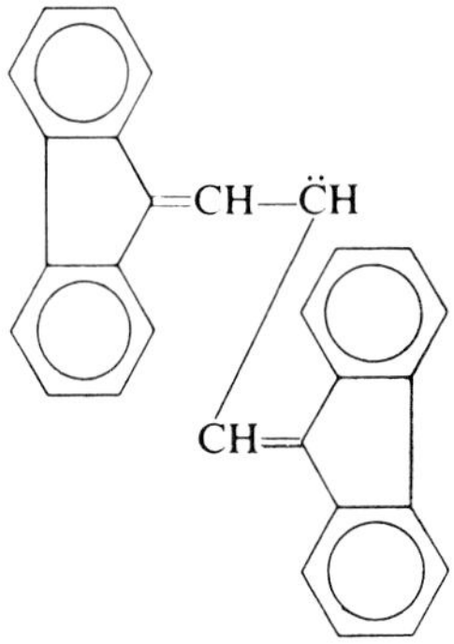 $=CH-\ddot{C}H$ / $CH=$	Na?	DMF	(310) 355 633	(4.5) (4.3) (5.3)	29
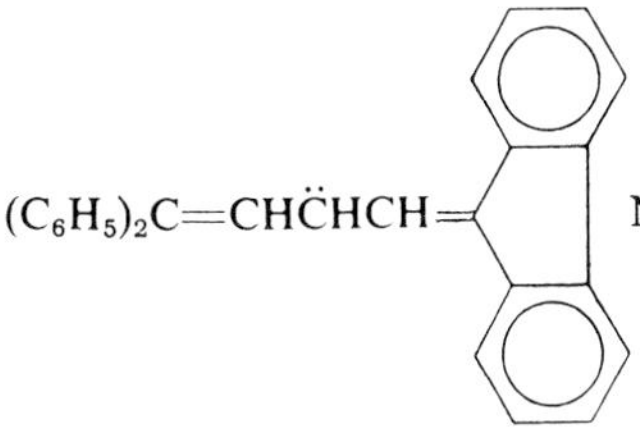 $(C_6H_5)_2C{=}CH\ddot{C}HCH{=}$	Na?	THF	575		29
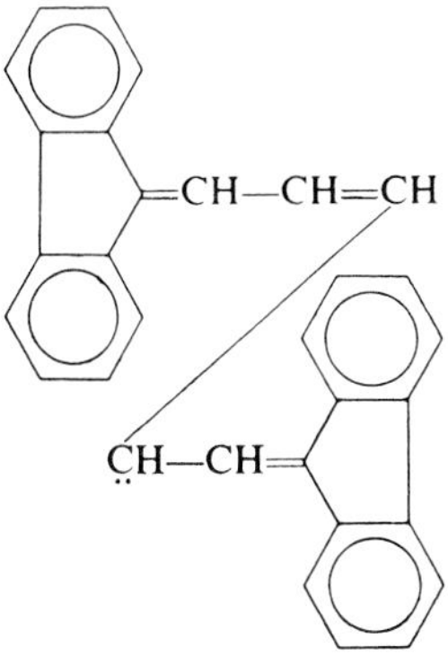 $=CH-CH=CH$ / $\ddot{C}H-CH=$	Na?	THF	740		29
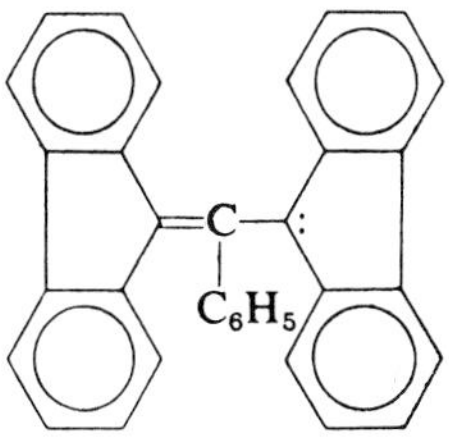 $=\underset{C_6H_5}{C}-\colon$	Na?	THF	600		29
Tris-*p*-biphenylylmethyl	K	Ether	580	(4.86)	23
Tris-*p*-biphenylylmethyl	Li	CHA	605		43

Organic radical	Metal	Solvent[b]	λ_{max}, mμ	(log ϵ)	Ref.[c]
Tris-*m*-biphenylylmethyl	K	Ether	485	(4.66)	23
Polybutadienyl	Li	THF	288		10
	K	THF[i]	312		10
	Li	CH	275		10
Polyisoprenyl	Li	THF[i]	287		10
	Na	THF[i]	(300)		10
	K	THF[i]	318		10
	Li	CH	270		10
	Na	CH	297		10
	K	CH	(363)		10
Polystyryl	Li	THF	335	(4.00)	47, 49
		THF	337–338		10, 25
		Benzene	334		10
		CH	328		10
		Toluene	(318–333)[j]		38
	Na	THF	343		10
			340	(4.1)	1, 25
	K	THF	346		10
	Cs	THF	343–345		25
Poly-α-methylstyryl	Li	THF	340		47, 48, 49
		Ether	337		48
		Benzene	349		48
at 10^{-3} *M*	Li	Benzene	338		38
			349 (sh)		
at 10^{-4} *M*	Li	Benzene	324		38
Poly-*p*-methylstyryl	Li	THF	327		49
Poly-*o*-methylstyryl	Li	THF	325		49
Poly-*m*-methylstyryl	Li	THF	338		49

APPENDIX B

ELECTRONIC TRANSITIONS OF ORGANOBORON COMPOUNDS

Compound	Solvent[a]	λ_{max}, mμ[b] (log ϵ)	Ref.[c]
$CH_2{=}CHBCl_2$	Gas	207 (3.8)	21
$CH_2{=}CHBCl(CH_3)$		206 (4.03)	43
$(CH_3)_3B$	Gas	(175) (3), 180–195 (sh) (2.3), 206 (1.8), 207–216 (sh) (1.2)	83
$(CD_3)_3B$	Gas	– 179 (2.7), 178–200 (sh) (1.9), 201–207 (1.6), 209–220 (sh) (1.4)	83
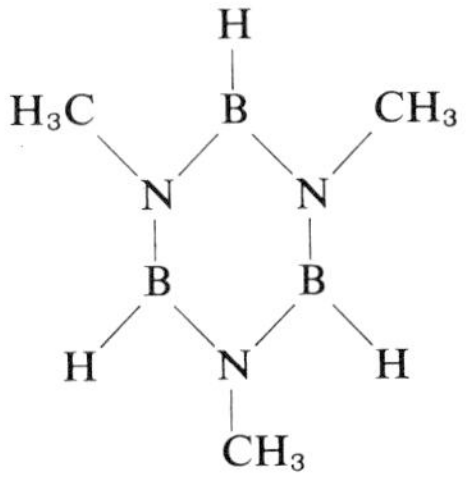 H H_3C B CH_3 N N B B H N H CH_3	Heptane	186 (4.4), 190 (4.4), 194 (4.4), (217), 220 (3.08), 224 (3.08), 227 (3.08)	88
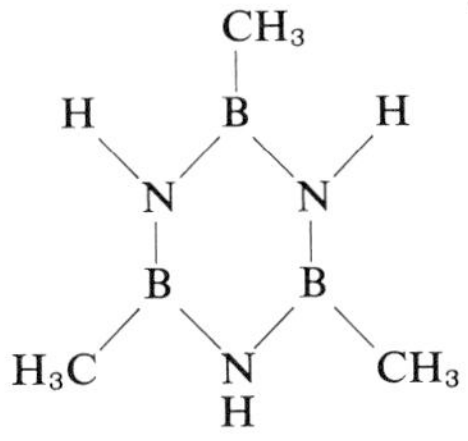 CH_3 H B H N N B B H_3C N CH_3 H	Heptane	176 (3.8), (185), 190 (3.3)	88
$(CH_2{=}CH)_2BCl$	Gas	222 (4.22)	43
$(CH_2{=}CH)_2BCl$	Gas	210 (3.9)	21
$CH_2{=}CHB(CH_3)_2$	Gas	196 (4.0)	43
$CH_2{=}CHB(Br)N(CH_3)_2$		223	76
$(CH_3)_2C{=}CHB(OH)_2$	CH	223 (3.0)	64
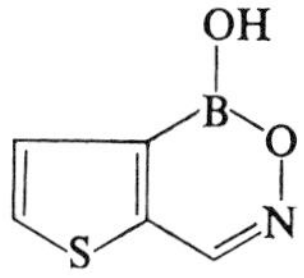 OH B O N S	Ethanol	209 (3.91), 278 (4.01)	46

[a] CH is cyclohexane; MeCH is methylcyclohexane.

[b] Maxima enclosed in parentheses ( ) were read from small figures and are only approximate.

[c] Reference numbers refer to those listed at the end of Chapter III.

Compound	Solvent[a]	λ_{max}, mμ[b] (log ϵ)	Ref.[c]
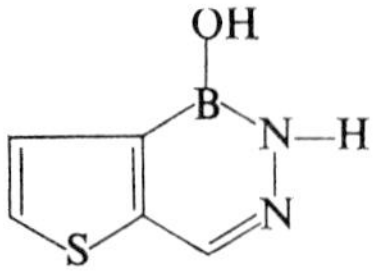 OH, B, N—H, N, S	Ethanol	213 (4.09), 221 (sh) (3.95), 262 (sh) (3.72), 274 (sh) (3.97), 285 (4.14), 293 (4.05)	46
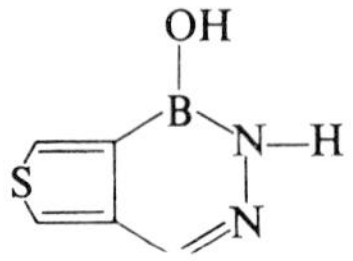 OH, B, N—H, N, S	Ethanol	222 (4.37), 295 (3.85)	46
$(CH_2{=}CH)_2BCH_3$	Gas	220 (4.12)	43
$C_6H_5BCl_2$		229 (4.37), 233 (4.37), 238 (4.37), 264 (3.04), 270 (3.04), 274 (3.04), 280 (3.04), 282 (3.04)	5
$C_6H_5BF_2$		229 (4.02), 233 (4.02), 238 (4.02), 265 (2.99), 268 (2.99), 271 (2.99), 274 (2.99), 278 (2.99), 281 (2.99)	5
o-$NO_2C_6H_4B(OH)_2$	Water	273	91
$C_6H_5B(OH)_2$	Heptane	222 (3.95), 235 (3.80), 268 (2.87), 272 (2.89), 280 (2.77)	78
$(CH_2{=}CH)_3B$	Gas	234 (4.28)	43
$CH_2{=}CHB[N(CH_3)_2]_2$		218	76
$(C_2H_5S)_3B$	Pentane	195 (3.69), 235 (4.00)	77
$(C_3H_7)_2BNH_2$	Pentane	181 (3.2), 190 (sh) (3.0)	77
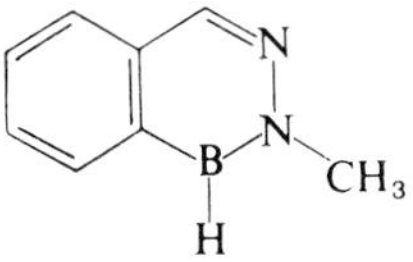 N, N, B, CH_3, H	Ethanol	212 (4.7), 268 (3.7), 272 (3.7), 279 (3.6), 284 (3.6), 292 (3.7), 295 (3.6), 305 (4.0)	31
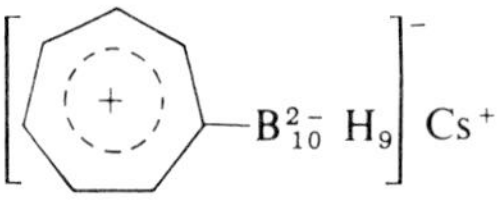 [(+)—$B_{10}^{2-}H_9$]$^-$ Cs^+	Water	217 (4.60), 262 (4.04), 439 (4.13)	48
	CH_3CN	222, 265, 513	
	12 N HCl	217, 262, 427	
	$CH_3CN\cdot H_2O$ 50%	221, 263, 485	
	96% H_2SO_4	217, 262, 373	
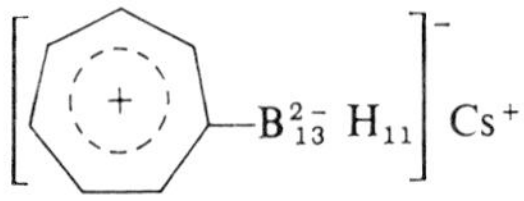 [(+)—$B_{13}^{2-}H_{11}$]$^-$ Cs^+	Water	215 (4.52), 262 (4.11), 349 (4.01)	49

Compound	Solvent[a]	λ_{max}, mμ[b] (log ϵ)	Ref.[c]
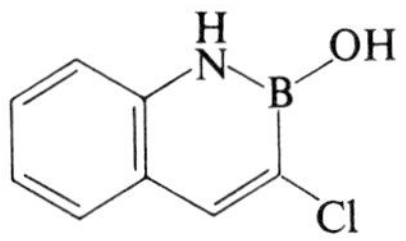 H N B OH Cl	$C_2H_5OH \cdot H_2O$	223, 264, 273, 313, 324	29
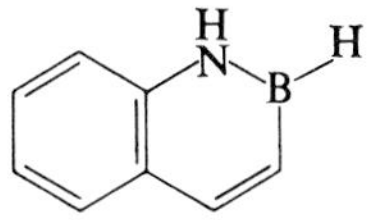 H N B H	CH	220 (4.6), (260) (4.0), (310) (3.8)	26
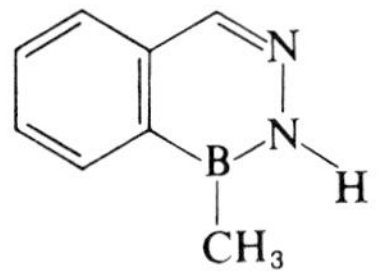 N N H B CH_3	Ethanol	210 (4.55), 265 (3.88), 290 (3.46), 301 (3.53)	30
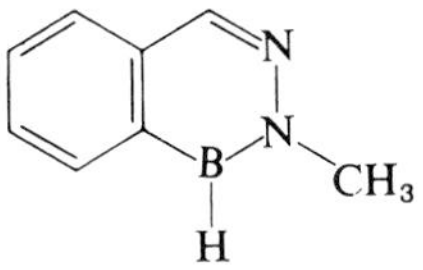 N N CH_3 B H	Ethanol	210 (4.6), 275 (brd) (3.7), 290 (3.8), 295 (sh), 304 (3.9)	31
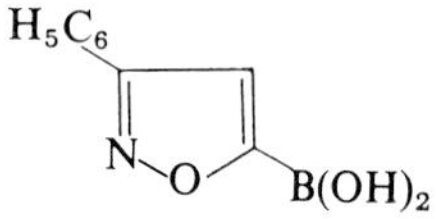 H_5C_6 N O $B(OH)_2$	Ethanol	242 (4.14)	13
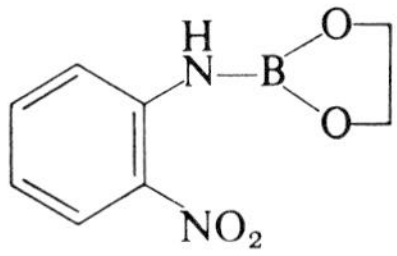 H N B O O NO_2		535	65
$(C_2H_5)_2BSC_4H_9$	Pentane	198 (3.3), 225 (2.9), 245 (2.6)	77
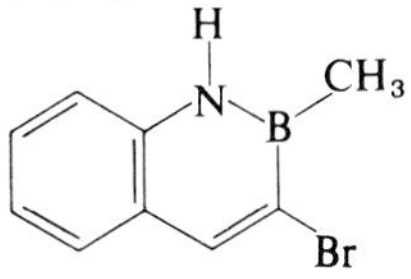 H N B CH_3 Br		243 (sh) (4.0), 256 (sh) (4.0), 267 (4.2), 278 (4.2), 297 (sh) (3.8), 313 (4.0), 327 (4.0)	28
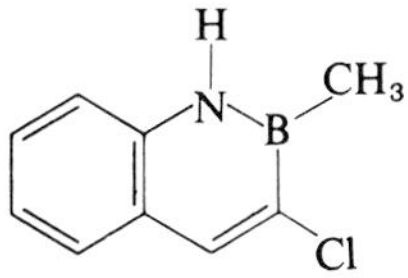 H N B CH_3 Cl		240 (sh) (4.0), 256 (sh) (4.0), 265 (4.2), 275 (4.2), 297 (sh) (3.8), 311 (4.0), 325 (4.0)	28
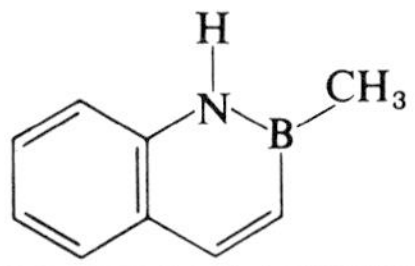 H N B CH_3	CH	(260) (3.9), (310) (3.8)	26

Compound	Solvent[a]	λ_{max}, mμ[b] (log ϵ)	Ref.[c]
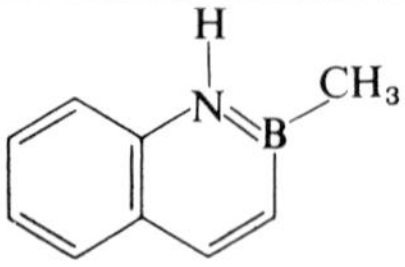 H, N, B, CH_3		250 (sh) (3.8), 255 (3.9), 264 (3.9), 295 (sh) (3.6), 305 (3.75), 320 (sh) (3.7)	28
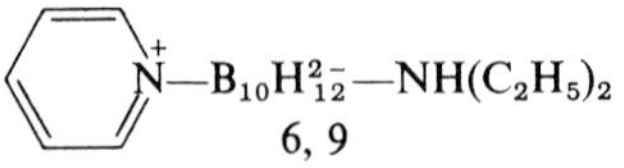 $\overset{+}{N}$—$B_{10}H_{12}^{2-}$—$NH(C_2H_5)_2$ 6, 9	CH_3CN	245 (3.92), 381 (3.48)	45
$C_6H_5BN(C_2H_5)_2$	CH	240 (4.17)	22
$(C_2H_5)_2NB(C{\equiv}CCH_3)_2$	CH	220 (4.30), 245 (4.23)	92
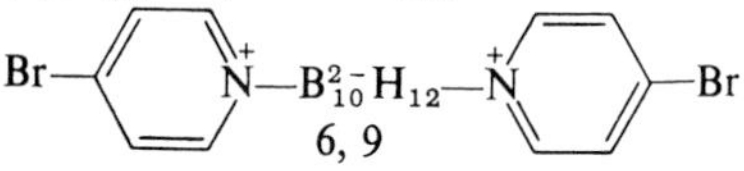 Br—$\overset{+}{N}$—$B_{10}^{2-}H_{12}$—$\overset{+}{N}$—Br 6, 9	CH_3CN	239 (3.90), 398 (3.31)	45
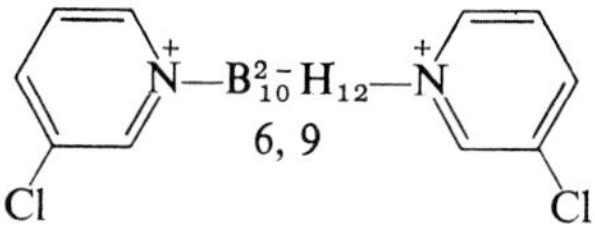 $\overset{+}{N}$—$B_{10}^{2-}H_{12}$—$\overset{+}{N}$ 6, 9; Cl, Cl	CH_3CN	236 (sh), 264 (sh), 405 (3.83)	45
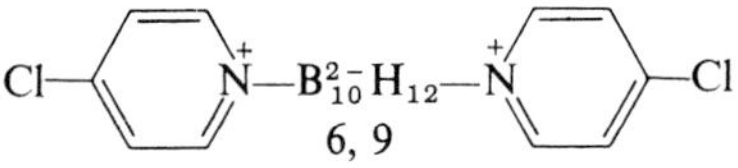 Cl—$\overset{+}{N}$—$B_{10}^{2-}H_{12}$—$\overset{+}{N}$—Cl 6, 9	CH_3CN	224 (4.08), 400 (3.55)	45
$CH_2{=}CHB(OC_4H_9)_2$	Pentane	185 (4.08)	77
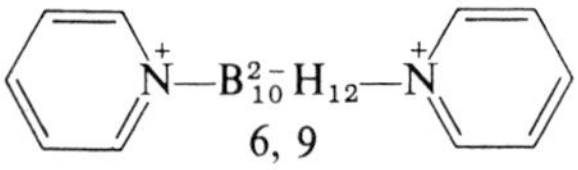 $\overset{+}{N}$—$B_{10}^{2-}H_{12}$—$\overset{+}{N}$ 6, 9	CH_3CN	241 (3.91), 380 (3.59)	45
$(C_3H_7)_2B(OC_4H_9)$	Pentane	None above 175	77
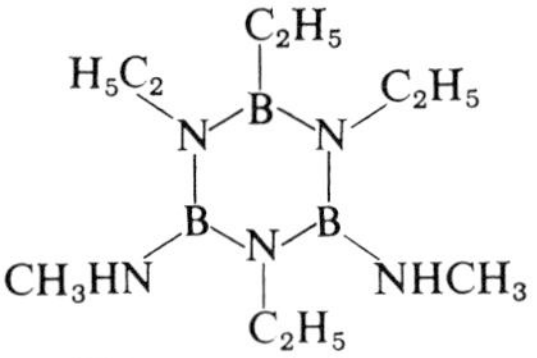 C_2H_5, H_5C_2, C_2H_5, N, B, N, B, N, B, CH_3HN, $NHCH_3$, C_2H_5	Pentane	190 (sh) (2.6)	77
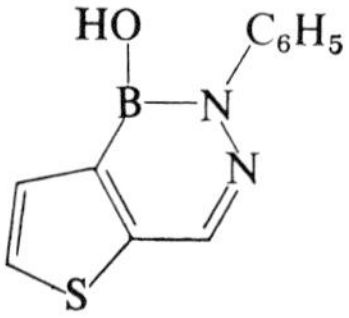 HO, C_6H_5, B—N, N, S	Ethanol	208 (4.14), 306 (4.12)	53
$(CH_3)_2NB(C_6H_5)(C{\equiv}CCH_3)$	CH	225 (4.23)	92
$(CH_3C{\equiv}C)_2BN(CH_3)C(CH_3)_2$	CH	220 (4.34), 230 (4.30)	92
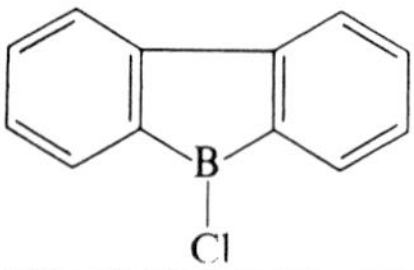 B, Cl	Hexane	397	60

Compound	Solvent[a]	λ_{max}, mμ[b] (log ϵ)	Ref.[c]
O, O; B—C_6H_4Cl-*p*	CH	272 (4.19), 278 (4.18)	52
O, O; B—C_6H_4Cl-*o*	CH	273 (4.02), 281 (3.95)	52
O, O; B—C_6H_4Cl-*m*	CH	272 (4.06), 278 (4.03)	52
H—N, N—H; B—$C_6H_4NO_2$-*o*	CH	(250) (4.0), (280) (4.0), (350) (sh) (3.4)	52
H—N, N—H; B—$C_6H_4NO_2$-*p*	CH	(220) (sh) (4.3), (250) (4.0), (350) (4.1)	52
NH, NH; B—$C_6H_4NO_2$-*m*	CH	(250) (4.1), (298) (4.1), (345) (3.3)	52
O, O; B—$C_6H_4NO_2$-*p*	CH	(250) (3.8), (290) (4.2)	52
O, O; B—$C_4H_4NO_2$-*o*	CH	(270) (4.0), (300) (sh) (3.2)	52
O, O; B—$C_6H_4NO_2$-*m*	CH	(245) (4.1), (270) (4.1), (300) (sh) (3.1)	52
Cl—; —NHB(Cl)(C_6H_5); NO_2	Ether	523 (3.62)	65
	Benzene	530 (3.46)	
O_2N—; —NHB(Cl)(C_6H_5); NO_2	Ether	506 (3.59)	65
	Benzene	510 (3.56)	

Compound	Solvent[a]	λ_{max}, mμ[b] (log ϵ)	Ref.[c]
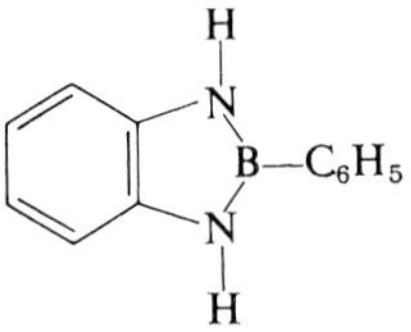 H, N, B—C_6H_5, N, H	CH	(245) (3.9), 290 (4.18), 297 (4.22), 309 (3.87)	52
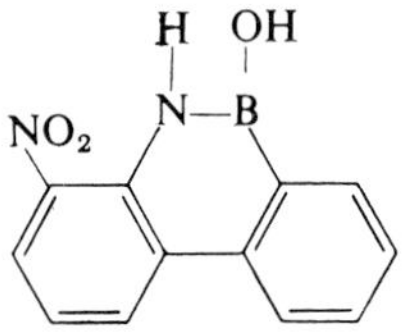 H, OH, N—B, NO_2		232 (sh) (4.3), 260 (4.3), 276 (sh) (3.8), 304 (3.5), 322 (3.5), 393 (3.7)	35
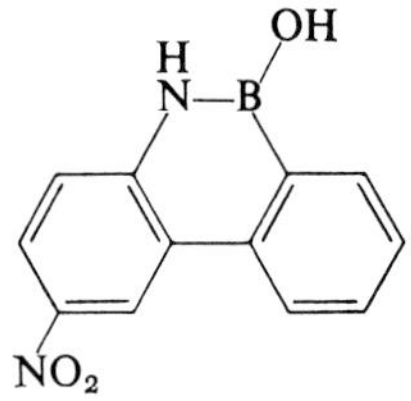 H, OH, N—B, NO_2		240 (sh) (4.2), 252 (4.2), 260 (4.2), 278 (4.1), 340 (4.0)	35
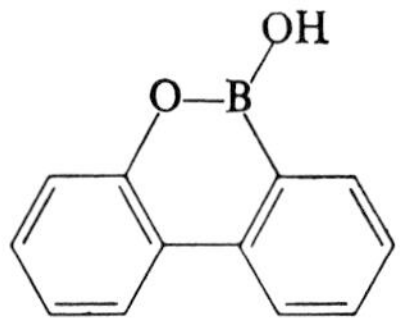 OH, O—B	Ethanol	258, 268, 283, 294	29
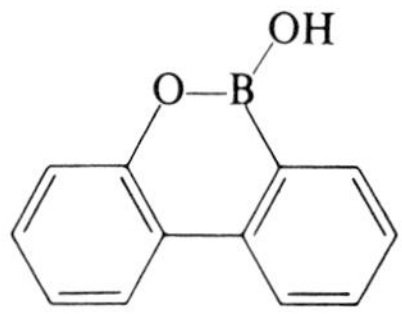 OH, O—B	Ethanol	(245) (sh) (2.9), (250) (sh) (4.0), 260 (4.2), 268 (4.2), 285 (3.6), 296 (3.8), 306 (3.7)	27
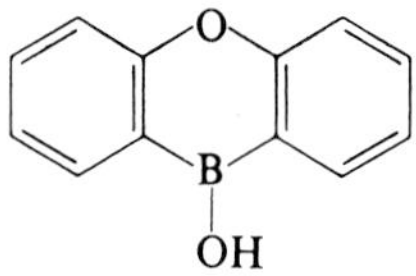 O, B, OH	CH	231 (4.89), 250 (sh) (4.1), 267 (3.10), 277 (3.11), 321 (3.90)	22
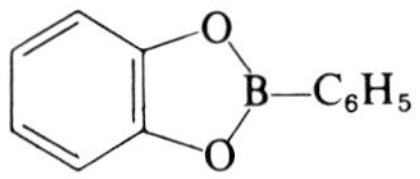 O, B—C_6H_5, O	CH	(228) (4.2), (271) (4.1), 279 (4.0)	52, 36
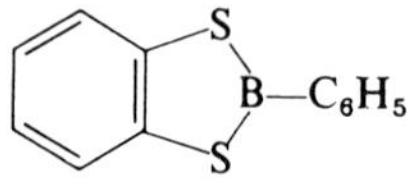 S, B—C_6H_5, S		260 (4.2), 295 (4.1)	36

Compound	Solvent[a]	λ_{max}, mμ[b] (log ϵ)	Ref.[c]
H, Br, N—B, C_6H_5, NO_2		528	65
$(C_6H_5)_2BCl$	CH_3CN	230 (4.11)	42
H, N, B—C_6H_4Cl-*m*, N, H	CH	295 (4.22), 303 (4.19), 317 (3.75)	52
H, N, B—C_6H_4Cl-*o*, N, H	CH	297 (4.16), 308 (4.11), 322 (3.71)	52
H, N, B—C_6H_4Cl-*p*, N, H	CH	295 (4.27), 303 (4.26), 316 (3.82)	52
H, Cl, N—B, C_6H_5, NO_2	CH	504	65
	CCl_4	511–512	
	Ether	514 (3.62)	
	Benzene	516 (3.43)	
	ClC_6H_4	518 (3.43)	
H, OH, N—B	Ethanol	230 (4.3), 238 (4.3), 252 (4.3), 258 (sh) (4.3), 268 (4.1), 300 (sh) (3.7), 315 (3.9), 325 (3.9)	29, 34, 35
H, H, N—B	CH	(250) (4.3), (320) (3.6)	36
N, B	Ethanol	235 (sh) (4.2), 258 (4.8) 266 (sh) (4.6), 290 (sh) (3.6), 298 (3.7), 310 (3.7), 325 (3.5)	33

Compound	Solvent[a]	λ_{max}, mμ[b] (log ϵ)	Ref.[c]
N, B—C_6H_5, O		238 (3.9), 245 (3.9), 258 (3.8), 285 (4.1)	36
$C_6H_5NHB(Cl)C_6H_5$		516	65
H, N, B—C_6H_5, N, H		250 (3.9), 300 (4.2)	36
H OH, N—B, NH_2		240 (sh) (4.8), (260) (sh) (4.1), 271 (4.0), 318 (3.8), 330 (3.8)	35
H OH, NH_2 N—B		237 (4.5), 274 (4.1), 326 (3.7)	35
$(C_6H_5)_2BOH$	CH_3CN	237 (4.3), 265 (3.2), 275 (3.0)	42
o-$C_6H_5OC_6H_4B(OH)_2$	$CHCl_3$	281 (3.45)	22
$(C_6H_5)_2BNH$	Heptane	240 (4.25), 272 (3.04)	78
H, N, N, H, B, HO OH	Ethanol	243 (4.0), 250 (3.9), 282 (sh) (3.9), 292 (4.1), 315 (4.2), 326 (sh) (4.1)	62
$C_6H_5S(C_3H_7)_2$	Pentane	188 (4.33), 204 (4.34), 234 (3.90), 242 (3.85), 270 (3.00)	77
N⋮C—$\overset{+}{N}$—$B_{10}^{2-}H_{12}$—$\overset{+}{N}$—C⋮N 6, 9	CH_3CN	273 (4.04), 467 (3.91)	45
$C_6H_5NHB(C_3H_7)_2$	Pentane	198 (4.51), 235 (3.90), 285 (3.18)	77
CH_3—$\overset{+}{N}$—$B_{10}^{2-}H_{12}$—$\overset{+}{N}$—CH_3 6, 9	CH_3CN	237 (4.01), 372 (3.67)	45

Compound	Solvent[a]	λ_{max}, mμ[b] (log ε)	Ref.[c]
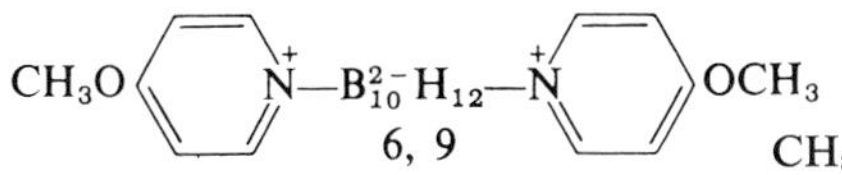 $CH_3O-C_5H_4\overset{+}{N}-B_{10}^{2-}H_{12}-\overset{+}{N}C_5H_4-OCH_3$ 6, 9	CH_3CN	243 (4.42), 351 (3.82)	45
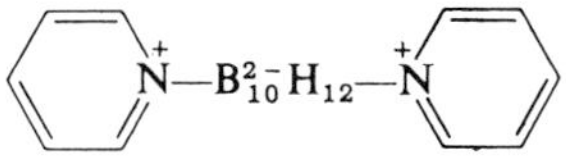 $C_5H_5\overset{+}{N}-B_{10}^{2-}H_{12}-\overset{+}{N}C_5H_5$	CH_3CN	268 (4.02), 380 (3.74)	45
$(C_4H_9)_3B$	Pentane	185 (sh) (3.0)	45
$(sec\text{-}C_4H_9)_3B$		214	95
$(n\text{-}C_4H_9)_3B$		214	24, 95
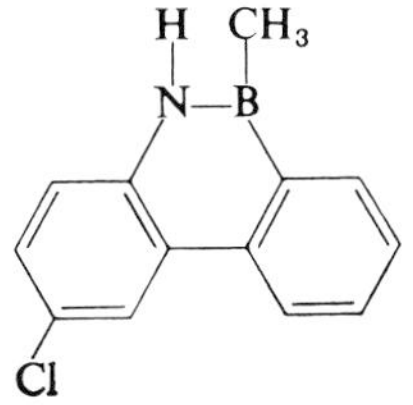 H, CH_3, N—B (Cl-substituted)		236 (4.6), 244 (4.5), 252 (4.4), (260) (sh) (4.1), (290) (sh) (3.4), (300) (sh) (3.5), (310) (sh) (3.7), 322 (3.8), 334 (3.9)	35
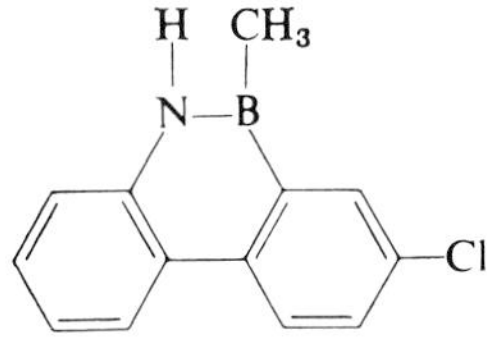 H, CH_3, N—B, —Cl		232 (4.7), 240 (4.7), 252 (4.5), 276 (4.1), 294 (3.7), 302 (3.7), 312 (4.0), 326 (4.2)	35
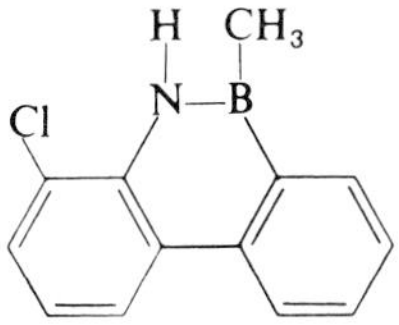 H, CH_3, N—B, Cl		236 (4.6), 251 (4.4), (274) (sh) (3.8), 296 (3.5), 300 (3.5), 312 (3.7), 326 (3.8), 334 (3.7)	35
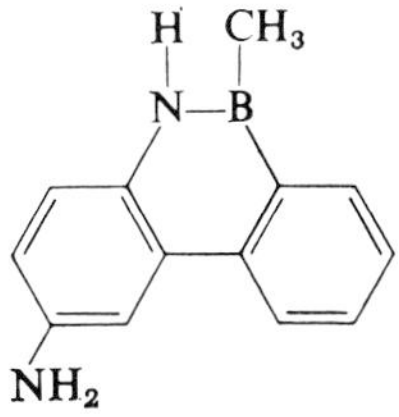 H, CH_3, N—B, NH_2		222 (4.5), 234 (4.5), 248 (4.4), 274 (sh) (4.1), 316 (sh) (3.6), 348 (3.8)	35
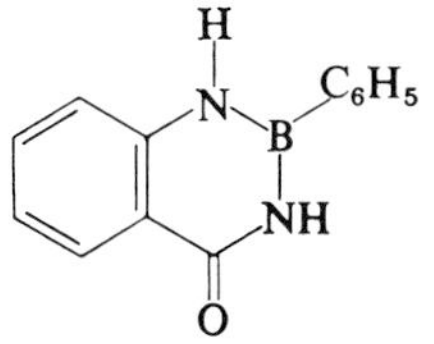 H, N, B, C_6H_5, NH, O	Ethanol	260 (4.1), 312 (3.5)	16

Compound	Solvent[a]	λ_{max}, mμ[b] (log ϵ)	Ref.[c]
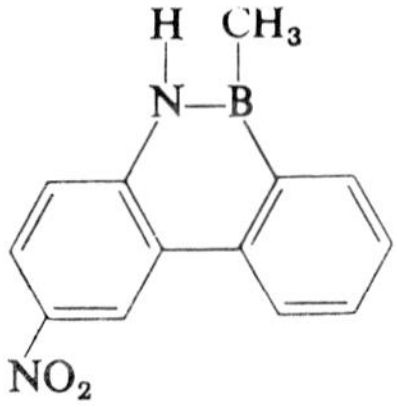 H CH_3 N—B; NO_2		224 (4.5), 244 (4.4), 264 (4.4), 280 (4.3), 322 (4.0), 350 (4.1)	35
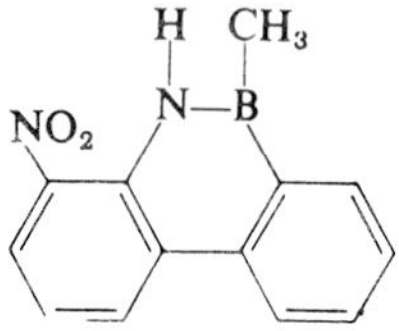 H CH_3 N—B; NO_2		224 (4.5), 237 (4.5), 260 (4.2), 276 (3.9), 306 (3.6), 322 (3.6), 380 (3.8)	35
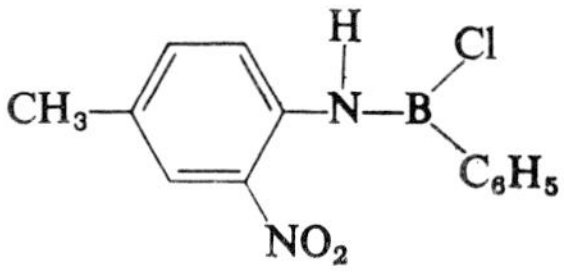 CH_3—; H N—B Cl C_6H_5; NO_2	Ether	519 (3.59)	65
	Benzene	524 (3.45)	65
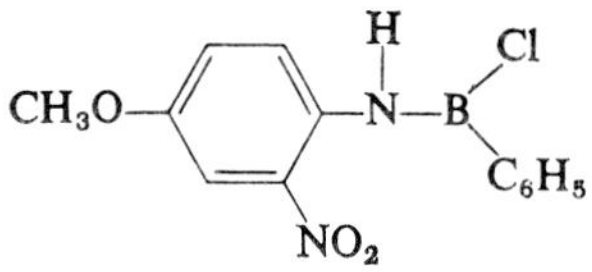 CH_3O—; H N—B Cl C_6H_5; NO_2	Ether	532 (3.62)	65
	Benzene	542 (3.51)	65
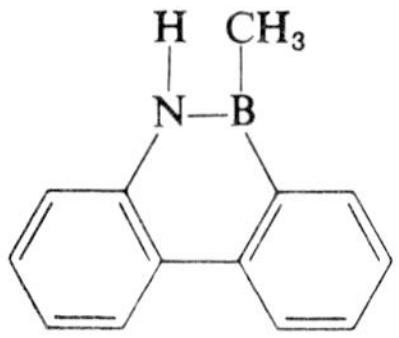 H CH_3 N—B	CH	220 (4.20), 224 (4.19), 230 (4.19), 236 (4.19), 240 (4.20), 246 (4.21), 250 (4.20), 298 (3.72), 312 (3.94), 324 (3.99)	35, 36, 37, 38, 40
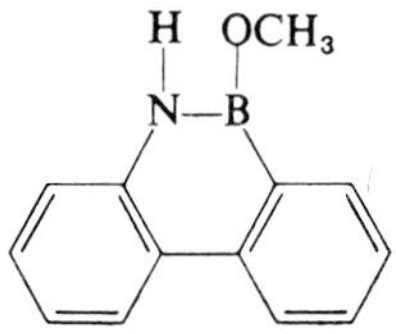 H OCH_3 N—B	$CHCl_3$	262 (4.23), 272 (4.23), 300 (sh) (3.79), 312 (3.99), 325 (4.03)	40
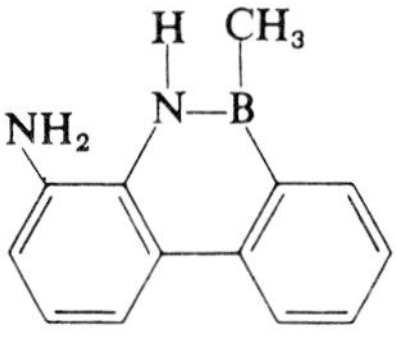 H CH_3 N—B; NH_2		228 (4.9), 238 (4.5), 270 (4.2), 322 (3.6), 330 (3.6)	35
$(CH_3C{\equiv}C)_2BN(CH_3)C_6H_5$	CH	225 (4.50), 250 (4.38)	92

Compound	Solvent[a]	λ_{max}, mμ[b] (log ϵ)	Ref.[c]
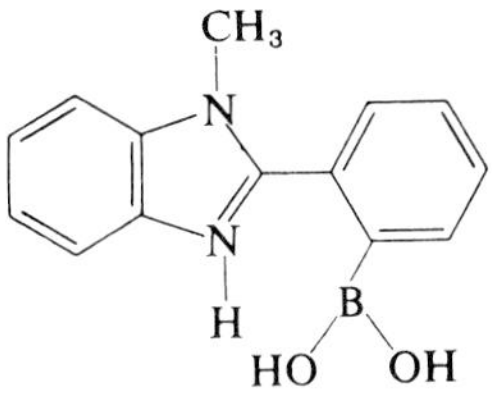 CH_3, N, N, H, B, HO, OH	Ethanol	246 (4.1), 252 (4.0), 285 (sh) (4.0), 294 (4.2), 318 (4.3), 330 (4.1)	62
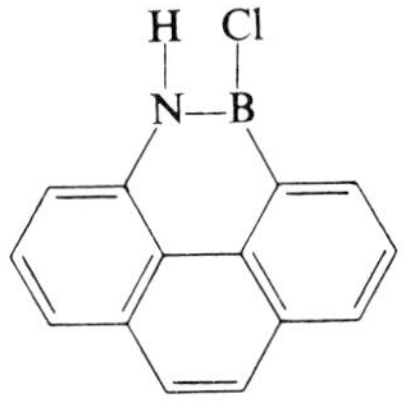 H, Cl, N—B	$CHCl_3$	252 (4.60), 270 (4.41), 278 (4.39), 295 (3.91), 307 (4.05), 331 (sh) (3.54), 340 (sh) (3.38), 348 (sh) (3.76), 359 (sh) (3.43), 365 (3.92)	40
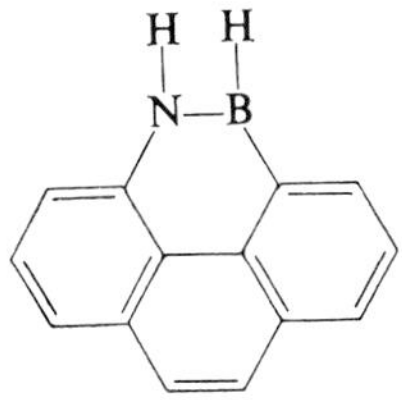 H, H, N—B	MeCH	238 (4.53), 250 (4.57), 267 (4.36), 295 (3.88), 307 (4.09), 317 (4.09), 320 (4.18), 331 (3.54), 340 (3.40), 347 (3.92), 351 (sh) (3.38), 358 (3.54), 365 (4.14)	40
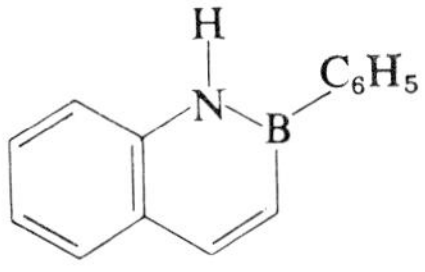 H, N, B, C_6H_5	CH	(260) (4.5), (321) (4.0)	26
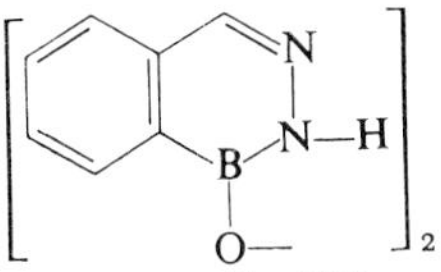 [N, N—H, B, O—]$_2$	Ethanol	269 (4.26), 280 (3.85), 300 (3.64)	30
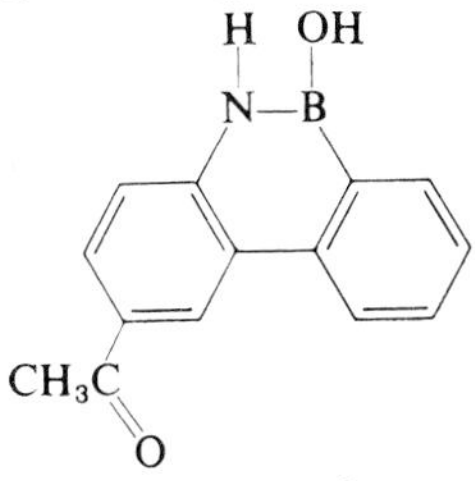 H, OH, N—B, CH_3C, O		210 (4.3), 252 (sh) (4.3), 261 (4.4), 284 (4.2), 300 (4.2), 320 (4.0), 334 (4.0)	34
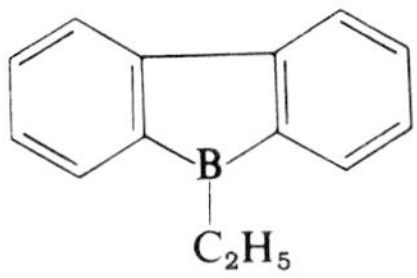 B, C_2H_5	Hexane	387	59

Compound	Solvent[a]	λ_{max}, mμ[b] (log ϵ)	Ref.[c]
$(C_6H_5O)_2BCH{=}CH_2$	Pentane	190 (4.98), 208 (4.14), 264 (3.43), 270 (3.57), 277 (3.49)	77
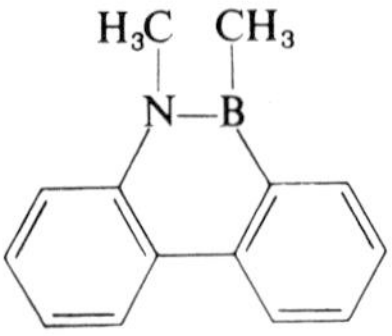 H_3C CH_3 N—B		(240) (4.5), (320) (3.9)	37
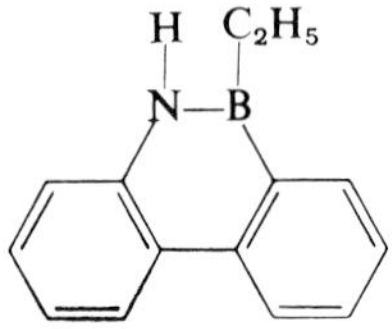 H C_2H_5 N—B	CH	220 (4.33), 230 (4.29), 238 (4.31), 250 (4.32), 298 (3.85), 313 (3.92), 325 (4.05)	36
$C_6H_5NHB(C_6H_5)OCH_2CH_2Cl$		520	65
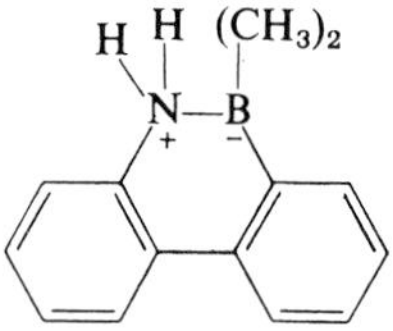 H H $(CH_3)_2$ $\overset{+}{N}$—$\overset{-}{B}$		(220) (4.3), (270) (4.2)	37
$(C_6H_5)_2BNHC_2H_5$	Heptane	240 (4.23), 260–270 (2.9)	78
$(C_6H_5)_2BOCH_2CH_2NH_2$	CH	233 (3.98)	22
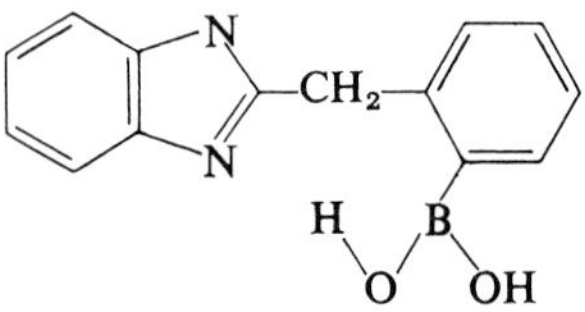 N, CH_2, N, H, B, O, OH	Ethanol	270 (3.8), 278 (3.9)	62
(p-$CH_3C_6H_4)_2BOH$	Heptane	250 (4.38)	78
(o-$CH_2C_6H_4)_2BNH_2$	Pentane	192 (4.78), 204 (4.65), 236 (4.20), 278 (3.26)	77
$[CH_2{=}C(CH_3)C{\equiv}C]_2BN(C_2H_5)_2$	CH	225 (4.15), 260 (4.25)	92
$C_6H_5B(n\text{-}C_4H_9)OC_4H_9\text{-}n$	Pentane	199 (4.64), 225 (4.04), 235 (3.72), 263 (2.83), 270 (2.67), 278 (2.36)	77
$C_6H_5B(OC_4H_9)_2$	Heptane	220 (3.93), 235 (3.40), 255 (2.69), 260 (2.72), 268 (2.79), 275 (2.68)	77

Compound	Solvent[a]	λ_{max}, mμ[b] (log ε)	Ref.[c]
$C_6H_5B(OC_4H_9)_2$	Pentane	194 (4.67), 220 (4.02), 235 (3.47), 260 (2.72), 268 (2.79), 275 (2.68)	78
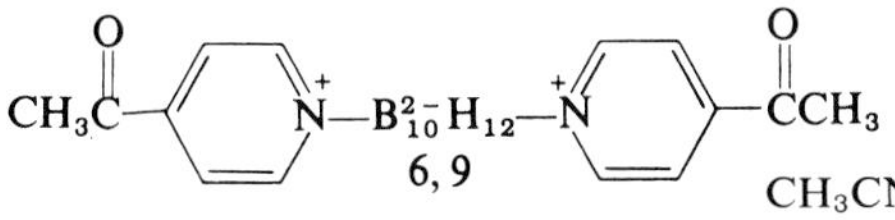 $CH_3C(=O)$–$C_5H_4\overset{+}{N}$–$B_{10}^{2-}H_{12}$–$\overset{+}{N}C_5H_4$–$C(=O)CH_3$ 6, 9	CH_3CN	236 (sh), 272 (sh), 460 (3.81)	45
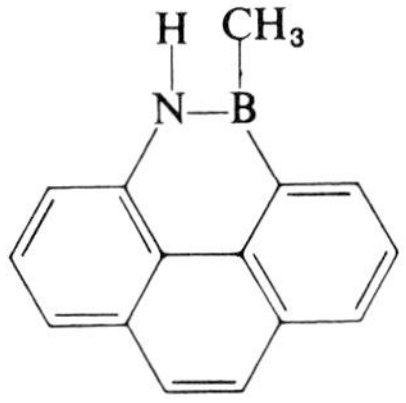 H, CH_3 N–B (structure)	$CHCl_3$	252 (4.61), 270 (4.42), 279 (sh) (4.28), 296 (3.90), 308 (4.10), 332 (sh) (3.60), 341 (sh) (3.38), 348 (3.83), 359 (sh) (3.48), 366 (4.02)	40
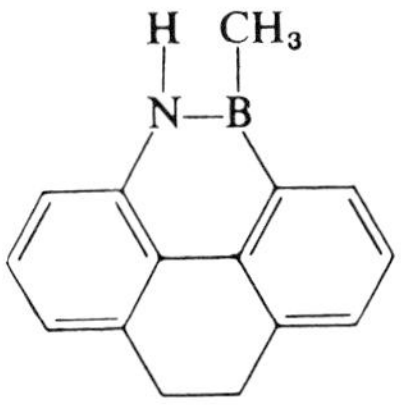 H, CH_3 N–B (structure)	$CHCl_3$	257 (sh) (4.32), 268 (sh) (4.05), 285 (sh) (3.91), 304 (3.60), 316 (3.89), 330 (4.00)	40
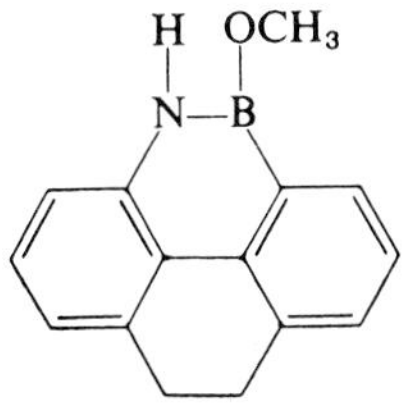 H, OCH_3 N–B (structure)	$CHCl_3$	275 (3.97), 284 (3.97), 300 (3.69), 314 (3.87), 327 (3.97)	40
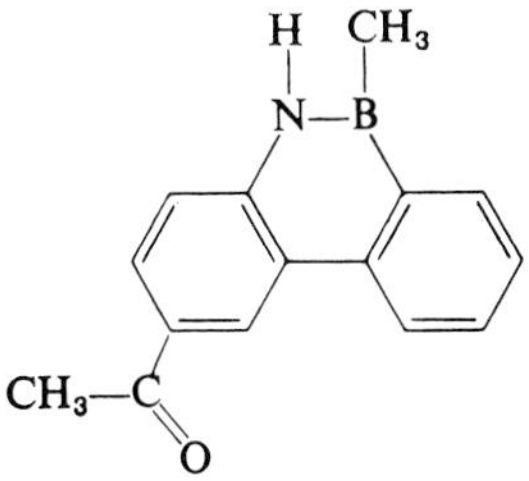 H, CH_3 N–B, CH_3–C=O (structure)		250 (4.6), 263 (4.6), 272 (4.3), 300 (4.2), 322 (4.0), 335 (4.0)	34
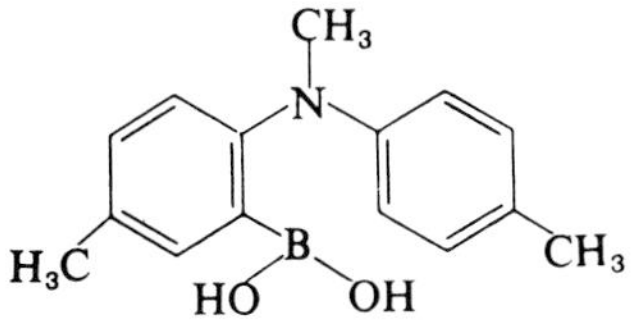 CH_3 N, H_3C, B, HO, OH, CH_3 (structure)		(235) (sh) (4.0), (325) (sh) (3.6)	67

Compound	Solvent[a]	λ_{max}, $m\mu$[b] (log ϵ)	Ref.[c]
1,3,5-Trimethyl-2-butyl-4,6-bis(butylthio)borazine (C_4H_9 on B; H_3C, CH_3 on N; C_4H_9S, SC_4H_9 on B; CH_3 on N)	Pentane	195 (sh) (2.8)	77
Naphtho[2,3-d]-1,3,2-dioxaborole, B—C_6H_4Cl-*o*	CH	306 (3.73), 312 (3.69), 320 (3.96)	52
Naphtho[2,3-d]-1,3,2-dioxaborole, B—C_6H_4Cl-*m*	CH	305 (3.74), 311 (3.71), 319 (4.00)	52
Naphtho[2,3-d]-1,3,2-dioxaborole, B—C_6H_4Cl-*p*	CH	305 (3.86), 312 (3.81), 319 (4.12)	52
Naphtho[2,3-d]-1,3,2-dioxaborole, B—$C_6H_4NO_2$-*o*	CH	303 (3.57), 309 (3.53), 317 (3.57), 323 (3.42)	52
Naphtho[2,3-d]-1,3,2-dioxaborole, $BC_6H_4NO_2$-*m*	CH	306 (3.92), 312 (3.86), 320 (4.12)	52
Naphtho[2,3-d]-1,3,2-dioxaborole, $BC_6H_4NO_2$-*p*	CH	310 (3.94), 315 (3.93), 322 (4.04)	52
Naphtho[1,2-d]-1,3,2-dioxaborole, B—C_6H_5	CH	(225) (4.7), (260) (4.2), (320) (3.5)	51
Naphtho[2,3-d]-1,3,2-dioxaborole, B—C_6H_5	CH	304 (3.76), 311 (3.73), 318 (4.04)	52
Phenyl(1-naphthyl)chloroborane, C_6H_5—B(Cl)—$C_{10}H_7$	Heptane	240 (4.27), 282 (3.91)	78

Compound	Solvent[a]	λ_{max}, mμ[b] (log ϵ)	Ref.[c]
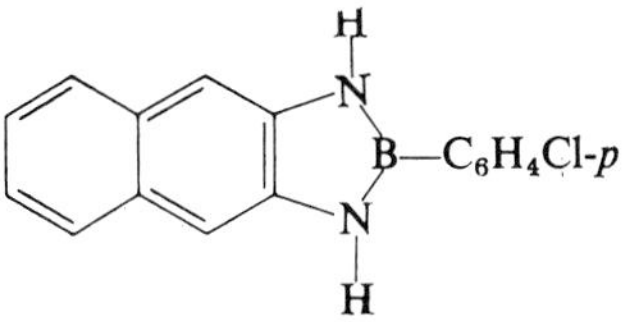 B—C_6H_4Cl-*p*	CH	322 (4.28), 329 (4.22), 337 (4.52)	52
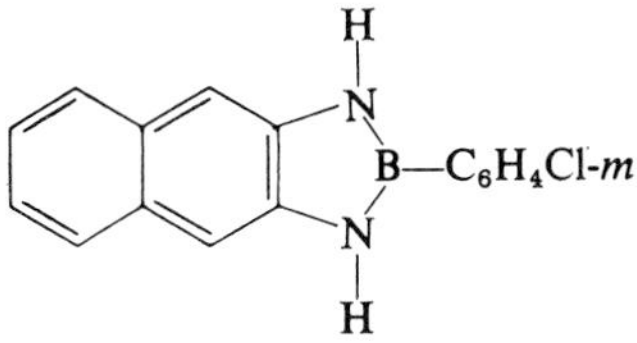 B—C_6H_4Cl-*m*	CH	322 (4.22), 329 (4.17), 337 (4.43)	52
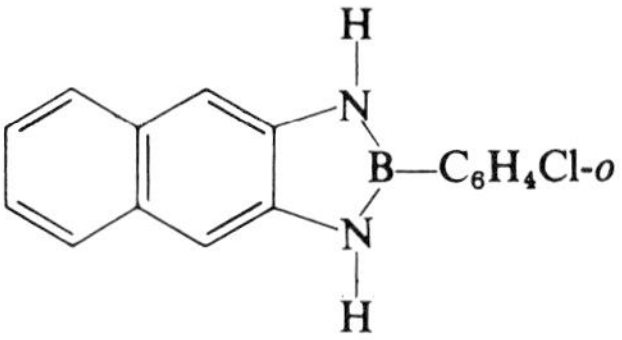 B—C_6H_4Cl-*o*	CH	325 (4.10), 332 (4.09), 340 (4.26)	52
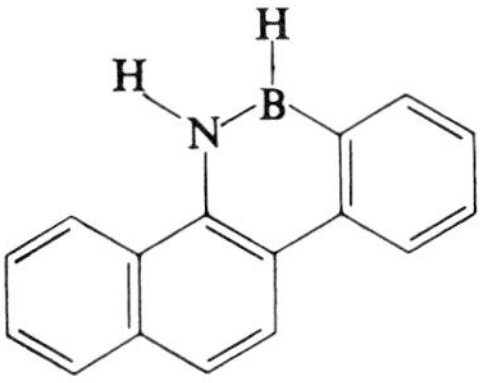	$CHCl_3$	268 (4.64), 276 (4.67), 309 (3.98), 322 (3.95), 337 (3.98), 353 (4.04)	40
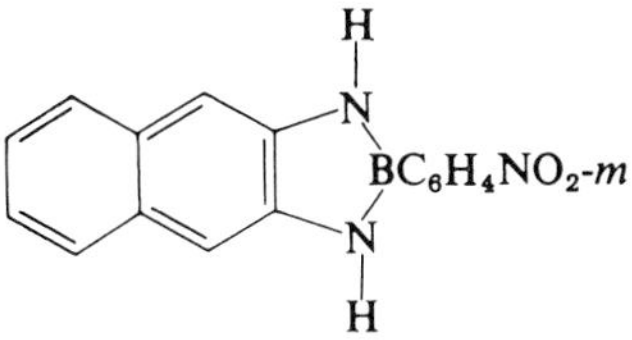 $BC_6H_4NO_2$-*m*	CH	325 (4.09), 336 (4.16)	52
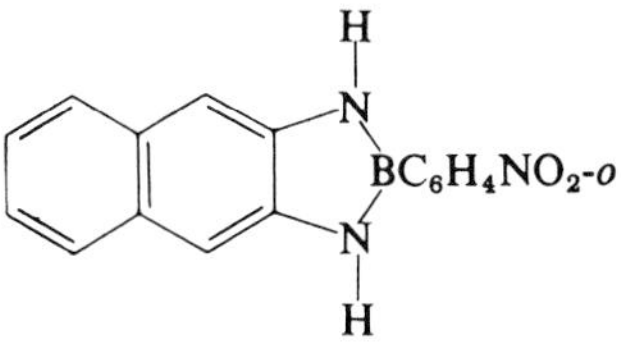 $BC_6H_4NO_2$-*o*	CH	315 (3.78), 322 (3.79), 329 (3.96)	52
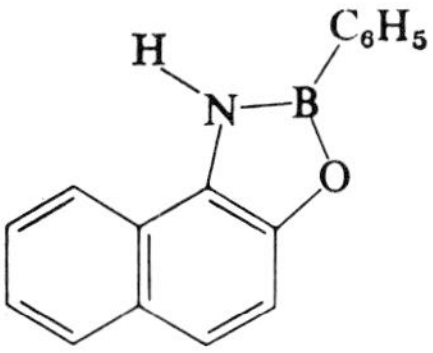 C_6H_5	CH	(240) (4.5), (340) (3.5), (280) (4.0)	51

Compound	Solvent[a]	λ_{max}, mμ[b] (log ϵ)	Ref.[c]
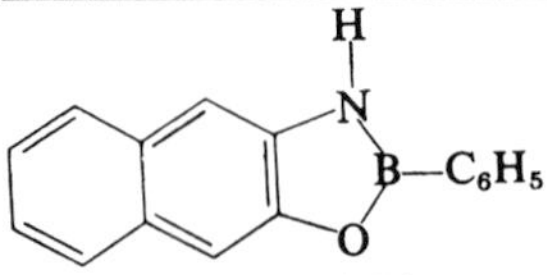 [structure: naphtho-fused oxazaborole, N–H, B–C_6H_5, O]	CH	(220) (4.6), (315) (3.7)	51
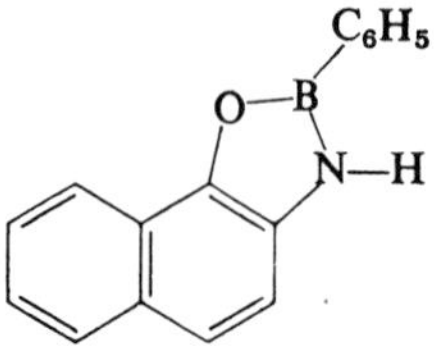 [structure: naphtho-fused oxazaborole, C_6H_5–B, O, N–H]	CH	(230) (4.5), (270) (4.1), (330) (3.7)	51
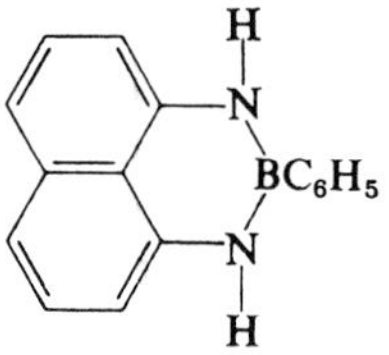 [structure: naphthalene-1,8-diamine boron derivative, BC_6H_5, two N–H]	Ethanol HCl	325 (4.0), 350 (sh) (3.9)	16
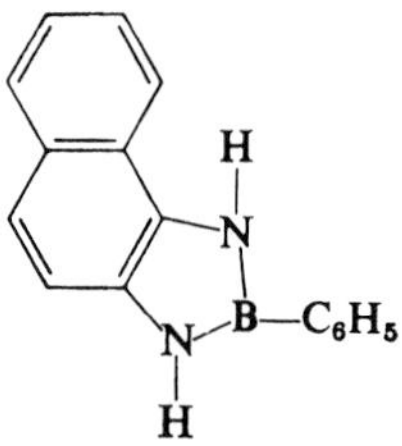 [structure: naphtho-fused diazaborole, two N–H, B–C_6H_5]	CH	(230) (4.7), (270) (4.3), (325) (4.0)	51
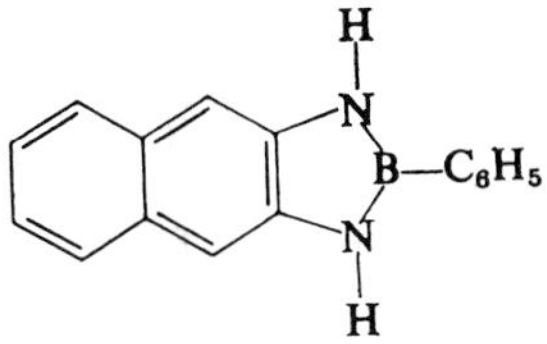 [structure: naphtho-fused diazaborole, two N–H, B–C_6H_5]	CH	(240) (4.5), 321 (4.13), 327 (4.07), 336 (4.40)	52
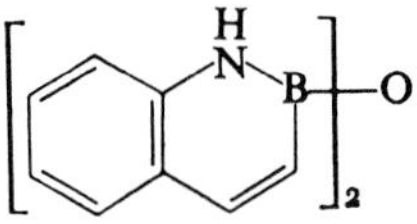 [structure: [benzazaborine N–H, B]$_2$O]	CH	225 (4.7), 257 (4.3), 267 (4.3), 312 (4.1), 323 (4.1)	29
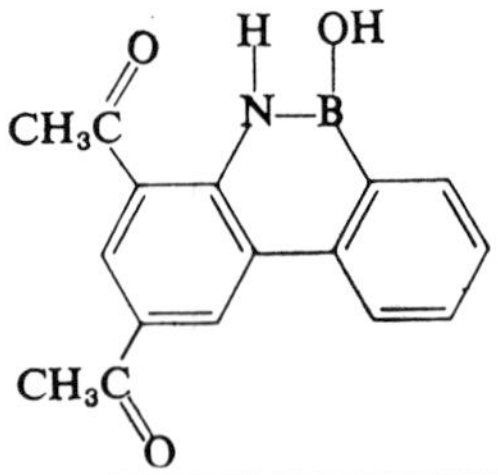 [structure: diacetyl dibenzazaborine, N–H, B–OH, two CH_3C=O]		222 (4.0), 280 (4.5), 298 (sh) (4.2), 347 (4.0)	34

Compound	Solvent[a]	λ_{max}, mμ[b] (log ϵ)	Ref.[c]
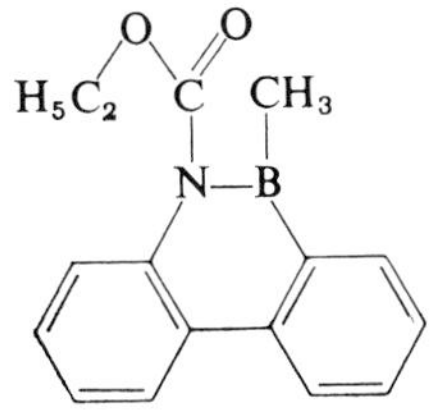 O O H_5C_2 C CH_3 N—B	Ethanol	235 (sh) (4.3), 280 (sh) (3.8), 315 (3.21), 325 (3.2)	38
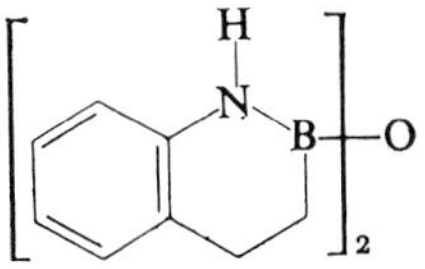 [H N B —O]$_2$	CH	240 (4.2), 277 (3.7)	29
$(p\text{-ClC}_6H_4)_2BOC_4H_9$	Heptane	250 (4.51), 280–285 (3.36–3.39)	78
$(C_6H_5)_2BOC_4H_9\text{-}i$	Pentane	183 (4.48), 202 (4.41), 238 (4.24), 265 (3.07), 270 (3.06), 280 (2.87)	77, 78
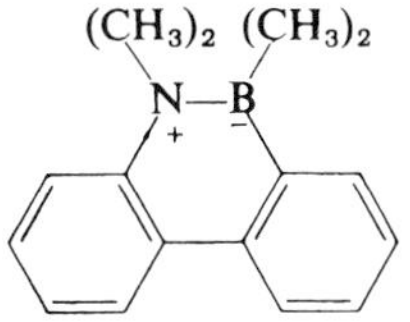 $(CH_3)_2$ $(CH_3)_2$ N—B + −		(270) (brd) (4.0)	37
$(C_6H_5)_2BNHC_4H_9\text{-}i$	Heptane	240 (4.25), 260–270 (3.02)	78
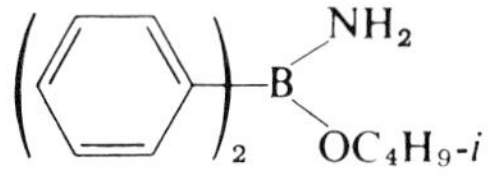 ( )$_2$ B NH_2 $OC_4H_9\text{-}i$	Pentane	184 (4.78), 200 (4.61), 238 (4.32), 273 (3.30)	77, 78
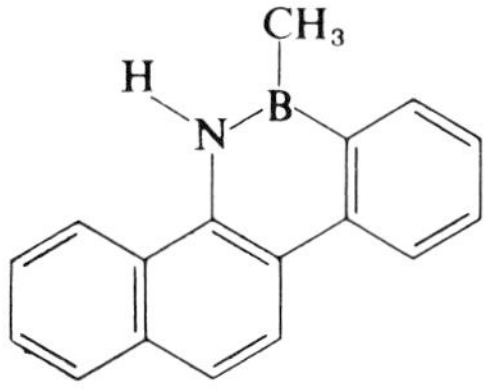 CH_3 H N B	$CHCl_3$	258 (sh) (4.51), 267 (4.71), 277 (4.77), 309 (4.01), 322 (4.01), 338 (4.04), 354 (4.11)	40
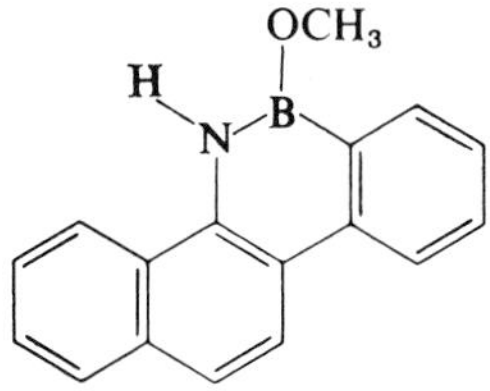 OCH_3 H N B	$CHCl_3$	251 (4.31), 259 (sh) (4.41), 268 (4.62), 278 (4.83), 311 (3.97), 324 (3.96), 339 (4.00), 356 (4.05)	40

Compound	Solvent[a]	λ_{max}, mμ[b] (log ϵ)	Ref.[c]
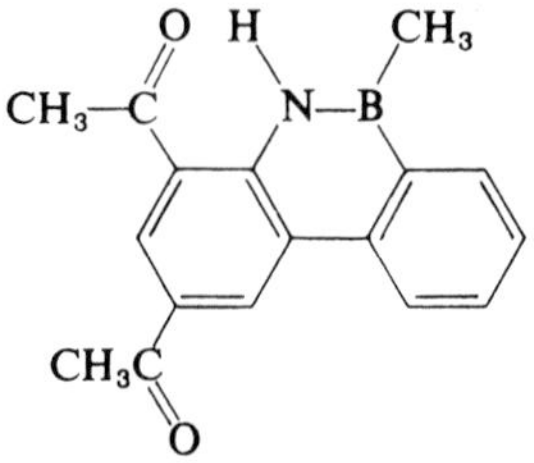 [structure: 1,3-diacetyl-substituted B-methyl-10,9-borazarophenanthrene; labels O, H, CH_3, CH_3—C, N—B, CH_3C, O]	$CHCl_3$	235 (4.2), 243 (4.0), 278 (4.6), 298 (sh) (4.3), 340 (sh) (4.0), 350 (4.0)	40
$[CH_2{=}CH(CH_3)C{\equiv}C]_2BN(CH_3)C_6H_5$	CH	230 (4.34), 270 (4.23)	92
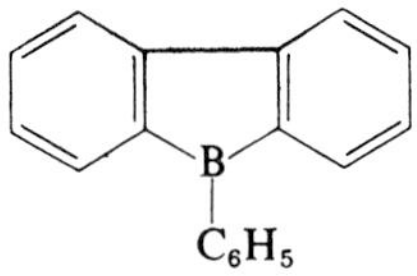 [structure: 9-phenyl-9-borafluorene; labels B, C_6H_5]	Hexane	405	60
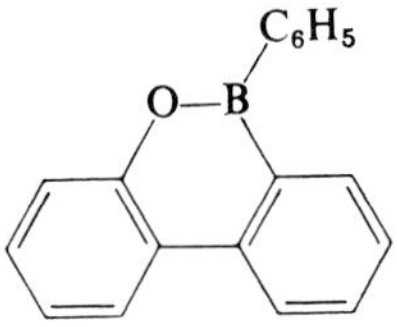 [structure: 10-phenyl-10,9-boroxarophenanthrene; labels C_6H_5, O—B]	CH	244 (4.4), 258 (4.4), 270 (sh) (4.2), 282 (3.7), 295 (3.9), 312 (3.8), 325 (3.8)	27
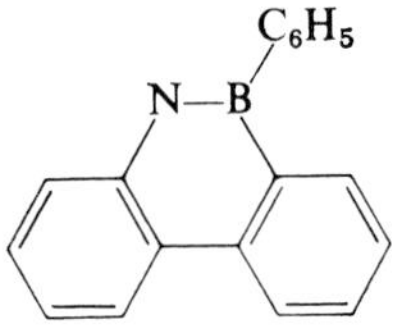 [structure: 10-phenyl-10,9-borazarophenanthrene; labels C_6H_5, N—B]	CH	224 (4.32), 238 (4.35), 256 (4.39), 300 (3.79), 314 (3.96), 328 (4.05)	36
$(C_6H_5)_3B$	MeCH	238 (4.28), 276 (4.54), 287 (4.59)	82
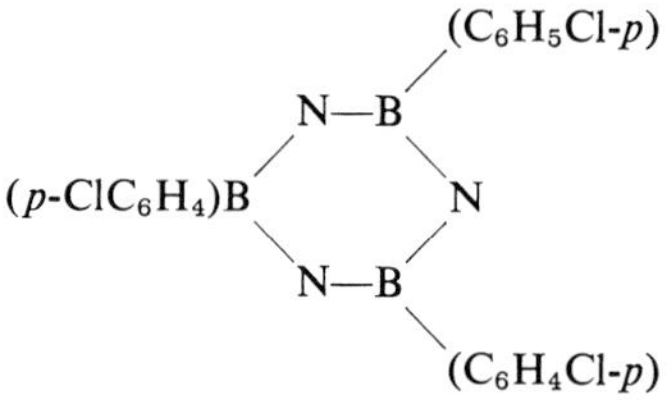 [structure: borazine ring with $(p\text{-}ClC_6H_4)B$, $(C_6H_5Cl\text{-}p)$, $(C_6H_4Cl\text{-}p)$; labels N—B, N, N—B]	Heptane	240 (4.73), 267.5 (3.07)	78
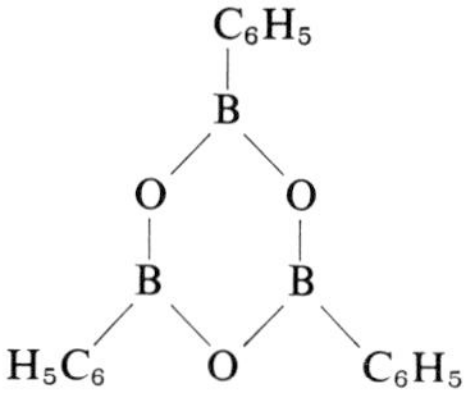 [structure: triphenylboroxine; labels C_6H_5, B, O, O, B, B, H_5C_6, O, C_6H_5]	MeCH	228 (4.9), 232 (4.9), 268 (3.3), 272 (3.4), 279 (3.2)	87

Compound	Solvent[a]	λ_{max}, mμ[b] (log ϵ)	Ref.[c]
$(C_6H_5)_2BOC_6H_5$	CH_3CN	237 (4.3), 265 (3.2), 275 (3.0)	42
H H_5C_6 B C_6H_5 N N HB BH N C_6H_5	Ether	263 (3.3), 271 (3.2), 280 (sh) (2.3)	97
H C_6H_5 N—B H_5C_6—B N—H N—B H C_6H_5	Heptane	230 (4.65), 262–265, (2.97)	78
$(C_6H_5)_3BNH_3$	Ethanol	200 (4.86), 255 (3.04), 262 (3.09), 268 (2.93)	77, 78
$[2,4,6\text{-}(CH_3)_3C_6H_2]_2BOH$	MeCH	204 (5.0), 249 (4.4), 280 (3.3)	87
$(C_6H_5NH)_2BC_6H_{13}$	Pentane	200 (4.70), 235 (4.24), 285 (3.54)	77
$(p\text{-}CH_3C_6H_4)_2BOC_4H_9\text{-}i$	Heptane	250 (4.36)	77
$(o\text{-}CH_3C_6H_4)_2BOC_4H_9\text{-}i$	Pentane	192 (4.66), 202 (4.59), 236 (4.14), 272 (3.12), 278 (3.12), 287 (3.02)	77
$(p\text{-}CH_3C_6H_4)_2BOC_4H_9\text{-}i$	Pentane	250 (3.39), 278 (3.15)	77
SC_4H_9 H_5C_2 B C_2H_5 N N B B C_4H_9S N SC_4H_9 C_2H_5	Pentane	195 (3.2), 245 (2.8)	77
CH_3 C_6H_5 N—B (biphenyl-bridged ring)		(240) (4.5), (320) (3.9)	37
$(1\text{-}C_{10}H_7)_2BCl$	Heptane	222 (4.79), 300 (3.94)	78
$(1\text{-}C_{10}H_7)_2BOH$	Heptane	222 (5.0), 300 (4.15)	78

Compound	Solvent[a]	λ_{max}, mμ[b] (log ϵ)	Ref.[c]
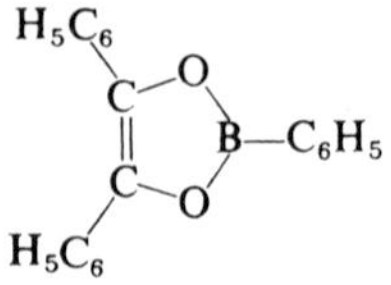	CH	220 (4.40), 242 (4.11), 285 (4.22)	63
$(1\text{-}C_{10}H_7)_2BNH_2$	Heptane	222 (4.96), 295–300 (4.13)	78
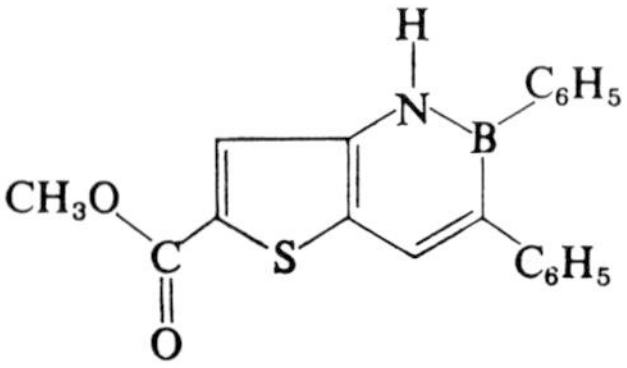	Ethanol	222 (4.30), 240 (4.23), 286 (4.17), 310 (4.01), 355 (4.40)	39
$(C_6H_5)_3BNCCH_3$	CH_3CN	225 (4.15), 267 (2.60), 275 (2.70)	42, 85
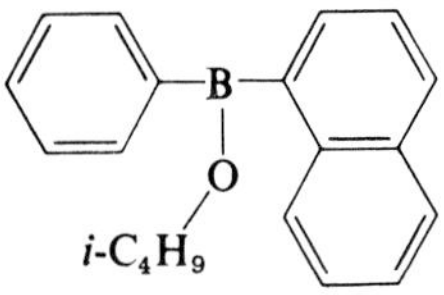	Heptane	240 (4.28), 282 (3.92)	78
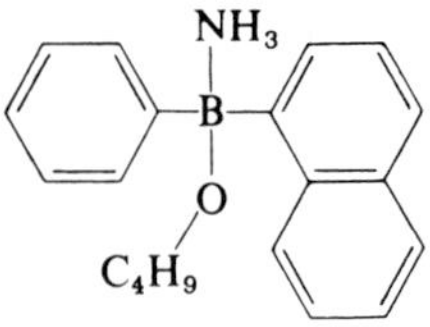	Heptane	240 (4.25), 282 (3.85)	78
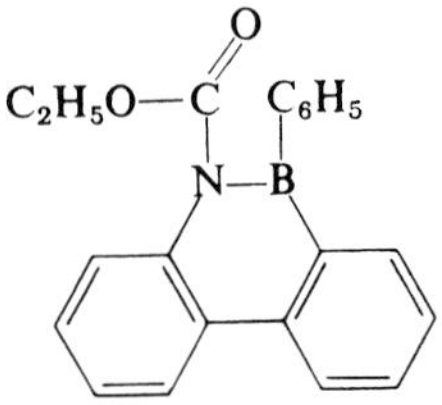		230 (sh) (4.4), 250 (sh) (4.2), 270 (sh) (3.9), 295 (3.7), 310 (3.8), 322 (3.8)	38
$(1\text{-}C_{10}H_7)_2BOCH_3$	Heptane	220 (4.92), 300 (4.05)	78
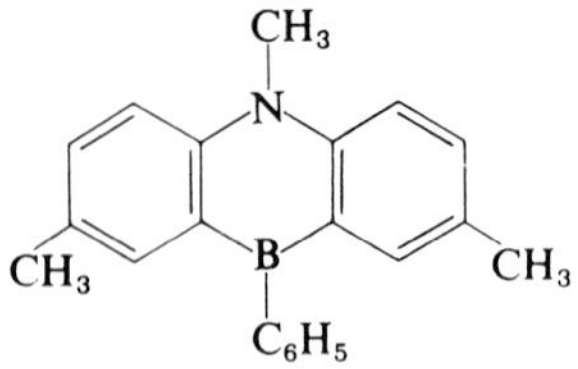	Ethanol	260 (4.9), 298 (3.4), 392 (sh) (3.9), 410 (4.0)	67
$(p\text{-}CH_3C_6H_4)_3B$	MeCH	235 (4.2), 249 (4.3), 282 (sh) (4.3), 297 (4.5)	82

Compound	Solvent[a]	λ_{max}, mμ[b] (log ε)	Ref.[c]
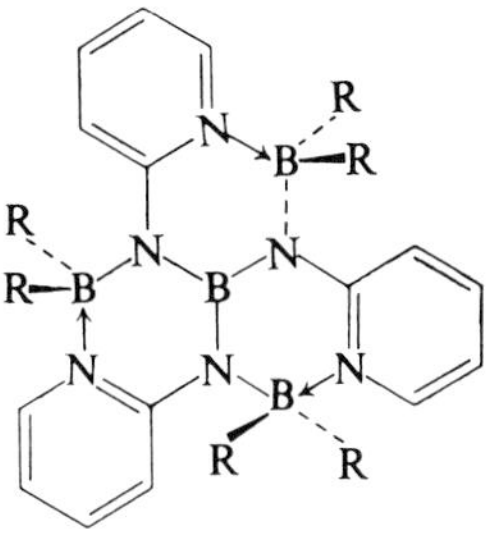 [structure: tris(2-pyridylamino)boron ring with BR_2 groups] $R = CH_3, C_2H_5, C_4H_9$	Ethanol	266 (4.6), 352 (4.22)	96
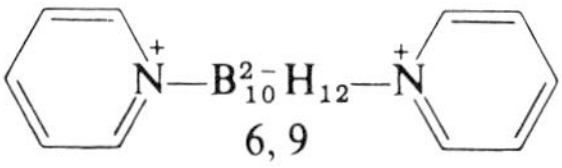 $C_5H_5N^+—B_{10}^{2-}H_{12}—{}^+NC_5H_5$ (6, 9)	CH_3CN	255 (sh), 410 (3.46)	45
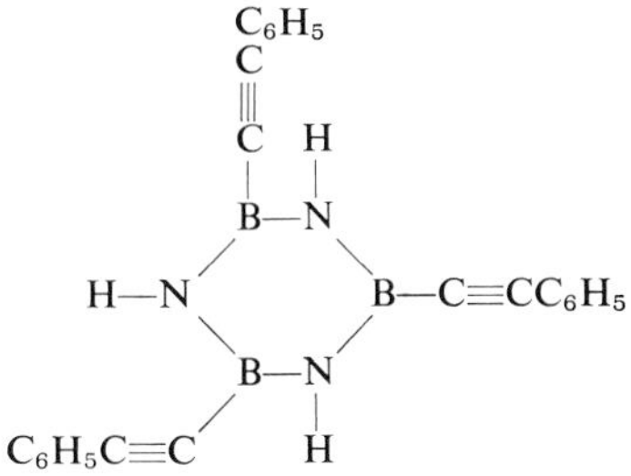 [structure: $(HNB—C{\equiv}CC_6H_5)_3$ borazine]	Ether	240 (sh), (4.6), 252 (4.7), 262 (sh) (4.8), 267 (4.8), 273 (4.7), 280 (4.8)	97
$[(C_6H_5)_2B]_2O$	CH	240 (4.34), 270 (3.15)	22, 42
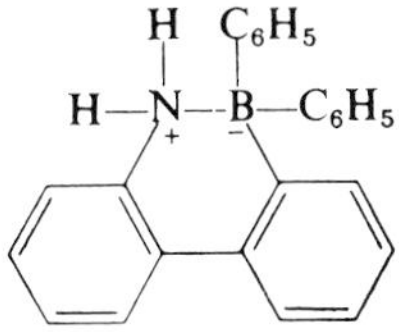 [structure: $H_2N^+—B^-(C_6H_5)_2$ bridged biphenyl]		(220) (4.4), (270) (4.2)	37
$(C_6H_5)_2BN(C_6H_5)_2$	CH	282 (4.26)	19
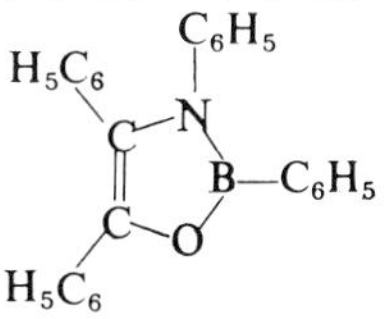 [structure: 3,4,5-triphenyl-2-phenyl-1,3,2-oxazaborole ring, H_5C_6, C_6H_5]	CH	284 (4.38)	63
$(C_6H_5)_4B^-Na^+$	CH_3CN	(240) (sh) (4), 266 (3.5), 274 (3.2)	42
$(1\text{-}C_{10}H_7)_2BOC_4H_9\text{-}i$	Heptane	222 (4.93), 295–300 (4.14)	78
$(1\text{-}C_{10}H_7)_2BSC_4H_9$	Heptane	222 (4.88), 295–300 (4.07)	78
$(1\text{-}C_{10}H_7)_2BNHC_4H_9\text{-}i$	Heptane	222 (4.98), 285–288 (4.14)	78

Compound	Solvent[a]	λ_{max}, mμ[b] (log ϵ)	Ref.[c]
$(1\text{-}C_{10}H_7)_2BN(C_2H_5)_2$	Heptane	222 (5.00), 288 (4.20)	78
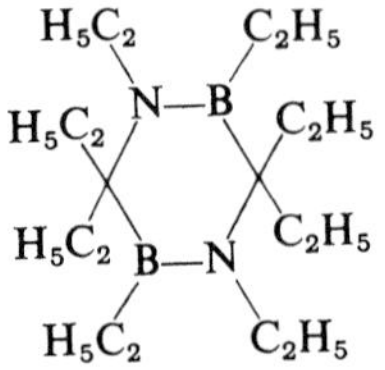 $[(H_5C_2)N-B(C_2H_5)-C(C_2H_5)_2-]_2$ (ring: N—B, B—N; H_5C_2, C_2H_5 substituents)	Ethanol	225 (3.76)	14
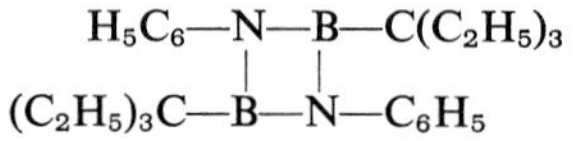 $H_5C_6-N-B-C(C_2H_5)_3$ / $(C_2H_5)_3C-B-N-C_6H_5$	Ethanol	253 (2.29), 259 (2.27), 265 (2.14)	14
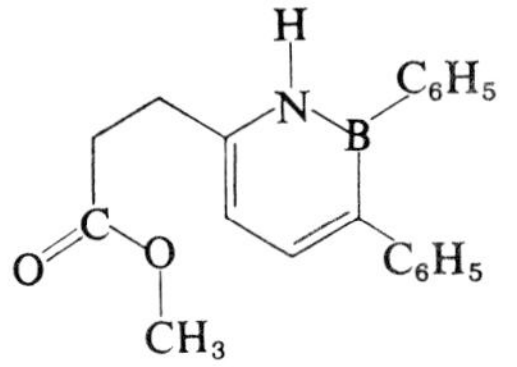 H–N–B(C_6H_5) ring with C_6H_5 and $CH_2CH_2C(=O)OCH_3$ substituents	Ethanol	220 (4.25), 275 (4.04), 306 (4.20)	39
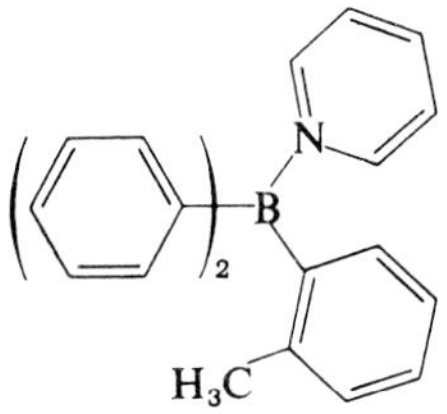 $(C_6H_5)_2B(C_6H_4CH_3\text{-}o)\cdot NC_5H_5$	Ethanol	265–280 (3.62)	78
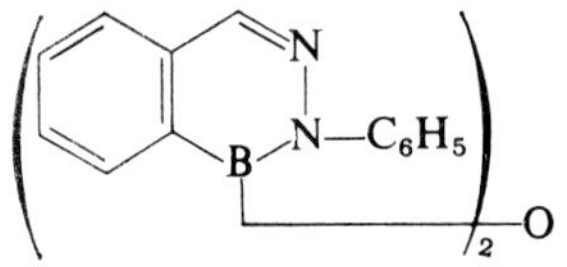 (benzo ring with N=N–C_6H_5 and B, CH_2)$_2$O	Ethanol	284 (4.45), 303 (4.38)	30
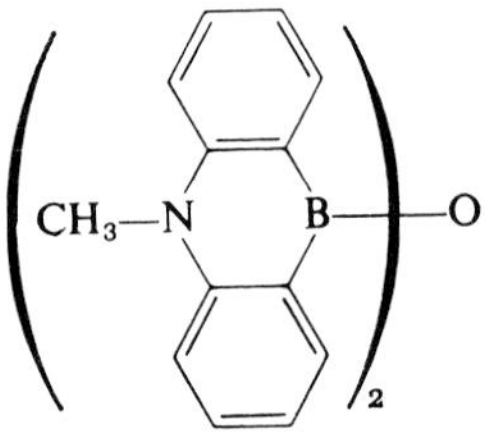 (CH_3—N(biphenyl-bridged)B)$_2$—O	Ethanol	250 (4.6), 280 (4.3), 302 (3.7), 370 (sh) (3.8), 380 (3.9)	67
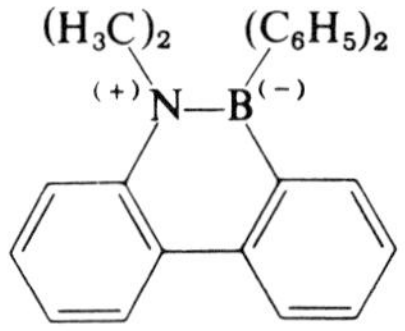 $(H_3C)_2\,^{(+)}N-B^{(-)}(C_6H_5)_2$ bridging biphenyl		(270) (brd) (4.0)	37

Compound	Solvent[a]	λ_{max}, mμ[b] (log ϵ)	Ref.[c]
C_2H_5; H_5C_6—N=C—B$(C_2H_5)_2$; $(H_5C_2)_2$B—C=N—C_6H_5 (ring)	Ethanol	230 (4.23)	14
H_5C_6—C≡C—B—N—CH_3; H_3C—N, B—C≡C—C_6H_5; H_5C_6—C≡C—B—N—CH_3 (ring)	Ether	249 (4.8), 260 (5.0), 274 (4.9), 280 (sh) (4.0)	97
CH_3C≡C—B—N—C_6H_5; C_6H_5—N, B—C≡CCH_3; CH_3C≡C—B—N—C_6H_5 (ring)	Ether	228 (4.6), 265 (3.6), 273 (3.4), 292 (2.6)	97
$(2,6\text{-}(CH_3)_2\text{-}4\text{-}CH_3C_6H_2)_3B$	MeCH	204 (4.9), 247 (3.8), 282 (3.5), 331 (4.2)	82
	CH_3CN	247 (3.5), 282 (3.2), 328 (3.9)	
$[(2,6\text{-}(CH_3)_2\text{-}4\text{-}CH_3C_6H_2)_3\bar{B}\overset{+}{Na}]_n$	Benzene	500	18
	THF	800	18
$(1\text{-}C_{10}H_7)_3B$	MeCH	221 (5.18), 263 (4.28), 286 (sh) (3.97), 353 (4.28)	82, 84
	CH_3CN	256 (4.11), 285 (4.40), 320 (3.64), 348 (3.60)	87
$(1\text{-}C_{10}H_7)_3BNH_3$	MeCH	230 (5.2), 270 (sh), 280 (sh), 289 (4.40), 300 (sh), 321 (3.48)	84

Compound	Solvent[a]	λ_{max}, mμ[b] (log ϵ)	Ref.[c]
$(1\text{-}C_{10}H_7)_3B^{\cdot-}Na^+$	THF	445 (3.7), 470 (3.6), 630 (2.8)	70
$(1\text{-}C_{10}H_7)_3B^{:-}2Na^+$	THF Ether	435 (3.8), 595 (3.5) 390 (3.8), 555 (3.7)	70
$[(C_6H_5)_3B^{\cdot-}Na^+]_2$	Ether THF	400 420	70
$[(C_6H_5)_3BNa]_2$ emission	THF THF–ether	415 (brd) (>4) 605, 645, 675	17
$(1\text{-}C_{10}H_7)_2(HO)B$—O(dioxane ring)O— $(1\text{-}C_{10}H_7)_2(HO)B$—	Ethanol	222 (4.25), 300 (3.60)	78
$C_6H_5C{\equiv}C$ $C{\equiv}CC_6H_5$ B—N $C_6H_5C{\equiv}C$—N B—$C{\equiv}CC_6H_5$ B—N $C_6H_5C{\equiv}C$ $C{\equiv}CC_6H_5$	Ether	265 (sh) (4.8), 272 (4.9), 289 (5.0)	97
$[(1\text{-}C_{10}H_7)_3\overset{(-)}{B}\cdot\overset{+}{Na}]_2$	Ether	290 (4.15), 360 (sh) (3.70), 420	70

APPENDIX C

ELECTRONIC TRANSITIONS OF GROUP IVA ORGANOMETALLOIDS

Compound	Solvent[a]	λ_{max}, mμ[b] (log ϵ)	Ref.[c]
CH_3GeBr_3		(217) (4.0)	106
CH_3GeCl_3		(182) (3.9)	106
$Br_3GeCH{=}CHGeBr_3$		(185) (4.6), (220) (sh), (235) (sh)	106
$Cl_3GeCH{=}CHGeCl_3$		(182) (sh) (4.0), (196) (4.3)	106
$Cl_3SiCH{=}CHSiCl_3$		(188) (4.0)	106
$F_3SiCH{=}CHSiF_3$		(182) (3.9)	106
$CH_2{=}CHGeBr_3$		(180) (sh) (4.3), (198) (4.2)	107
$CH_2{=}CHGeCl_3$		(180) (4.3)	106
$CH_2{=}CHSiCl_3$		(177) (4.1)	107
$H_3SiC{\equiv}CH$	Gas	190 (2.83), 195 (2.54), 198 (2.59), 204–228 (1.54 – (– 0.7))	44
$CH_2{=}CHSi(CH_3)_3$	Isooctane	202 (3.17)	125a
$(CH_3)_2GeCl_2$		(180) (sh) (3.9)	106
$(CH_3)_2SiH_2$	Gas	(110) (4.36), (135) (4.08), (143) (4.04), (160) (sh)	68a
$CH_2{=}CHCH_2GeBr_3$		(190) (4.2), (220–230) (4.0)	107
$Cl_3GeCH_2CH{=}CH_2$		(182) (4.2), 206) (3.7)	107
$Cl_3SiCH_2CH{=}CH_2$		(180) (sh) (4.0), (186) (4.0)	107
$F_3SiCH_2CH{=}CH_2$		(180) (3.7)	107
$(CH_3)GeN_3$	Isooctane	212 (2.40), 255 (1.36)	133
	Ether	211 (2.39), 253 (1.32)	133
$(CH_3)_3PbN_3$	Ether	243 (3.43)	133
	CH_3CN	243 (3.43)	133
	i-PrOH	238 (3.46)	133
	MeOH	240 (3.55)	133
$(CH_3)_3SiN_3$	Isooctane	212 (2.41), 255 (1.28)	133
	Ether	209 (2.32), 250 (1.23)	133
	CH_3CN	212 (2.35), 252 (1.36)	133

[a] CH is cyclohexane; EtOH is ethanol; *i*-PrOH is isopropanol; MeOH is methanol.
[b] Maxima enclosed in parentheses ( ) were read from small figures and are only approximate.
[c] Reference numbers refer to those listed at the end of Chapter IV.

Compound	Solvent[a]	λ_{max}, mμ[b] (log ϵ)	Ref.[c]
$(CH_3)_3SnN_3$	Ether	230 (2.50)	133
	CH_3CN	231 (2.48)	133
	i-PrOH	233 (2.51)	133
	MeOH	2.31 (2.48)	133
$(CH_3)_3SiH$	Gas	(112) (4.39), (140) (4.10), (152) (sh) (4.0), (163) (sh) (3.8)	68a
$(CH_3)_3GeNCS$	CH_2Cl_2	238	134
$(CH_3)_3SiNCS$	CH_3CN	238 (2.91), 285 (sh)	134
$(CH_3)_3SnNCS$	*i*-PrOH	238 (3.0)	134
$[(CH_3)_3Sn][CHN_2]$		280 (very strong), 396 (medium)	83
$Cl[Si(CH_3)_2]_2Cl$	CH	204 (3.51)	54
$Ge(SCH_3)_4$	Hexane	204 (3.86), 248 (3.81)	36
$(CH_3)_3Si(SCH_3)$	Hexane	204 (3.48), 224 (2.15)	36
$(CH_3)_2Si(SCH_3)_2$	Hexane	203 (3.56), 225 (3.11)	36
$(CH_3)Si(SCH_3)_3$	Hexane	204 (3.63), 226 (3.34)	36
$(CH_3S)_4Si$	Hexane	203 (3.67), 226 (3.61)	36
$(CH_3)_4Si$	Gas	115 (4.41), 138 (4.15), 167 (3.85)	68a
$(CH_3)_4Sn$	*n*-Heptane	(188) (4.3)	126
$H[Si(CH_3)_2]_2Cl$	CH	194 (3.44)	54
$(CH_3)_2SiHSiH(CH_3)_2$	CH	198 (3.31)	54
$(CH_3)_3SiC{\equiv}CH$	Gas	190 (2.83), 196 (2.42)	44
$(CH_3)_3GeCH{=}CH_2$		(182) (4.3)	107
$(CH_3)_3Si{-}O{-}C(=S){-}CH_3$	Hexane	207 (3.44), 245 (3.66), 389 (1.06)	61
$(CH_3)_3SiC(=O)CH_3$	CH	195 (3.62), 323 (1.25), 333 (1.53), 346 (1.79), 358 (2.00), 372 (2.10), 388 (1.97)	2, 15a, 21
$(CH_3)_3SiCH{=}CH_2$		(178) (4.3)	107
$(CH_3)_3SnCH{=}CH_2$		(186) (4.4)	107
$C_6H_5GeCl_3$	Heptane	269.5 (2.63), 262.9 (2.72), 256.9 (2.59), 252.5 (2.40), 247.8 (2.22)	85
$C_6H_5SnCl_3$	Heptane	269.5 (2.60), 262.8 (2.72), 256.6 (2.68), 247.1 (2.76)	85
$C_6H_5SiH_3$	CH	254 (2.26), 261 (2.42), 265 (2.46), 271 (2.37)	60
	EtOH	248, 254, 264 (3.08), 266, 271	60

Compound	Solvent[a]	λ_{max}, mμ[b] (log ε)	Ref.[c]
$(CF_2Cl)_2C{=}NSi(CH_3)_3$	Ether	213, 351	30
$(CF_3)_2C{=}NSi(CH_3)_3$		213, 351	30
$F_3CCH{=}CHSi(CH)_3$		(181) (4.1)	106
$(CH_3)_2Si[O{-}C(=S)CH_3]_2$	Hexane	222 (3.59), 243 (3.64), 389 (1.35)	61
$(CH_3)_3GeCH_2CH{=}CH_2$		(174) (4.0), (195) (4.0)	107
$(CH_3)_3SiCH_2C(=O)CH_3$	CH	283 (1.91)	92
	EtOH	276 (2.06)	92
$(CH_3)_3SiCH_2CH{=}CH_2$		(178) (sh) (3.8), (190) (4.0)	107
$(CH_3)_3SnCH_2CH{=}CH_2$		(181) (4.3), (210) (4.1)	107
$(C_2H_5)_3SiNH_2$	Isooctane	209 (3.25)	110
$[(CH_3)_3Si]_2NBr$	CH	230 (2.48), 344 (1.95)	6
$[(CH_3)_3Si]_2NCl$		295 (1.79)	6
$Cl[Si(CH_3)_2]_3Cl$	CH	219 (3.93)	52, 54
$(CH_3)_3GeGe(CH_3)_3$	*n*-Heptane	(190) (sh) (4.3)	126
$[(CH_3)_3Ge]_2S$	Hexane	213 (3.72)	36
$[(CH_3)_3Si]_2NI$	CH	344 (2.28), 408 (2.13), [?513 (1.78)]	6
$[(CH_3)_3Si]_2S$	Hexane	202 (3.46)	36
$[(CH_3)_2Sn{=}S]_3$	CH	209 (4.20), 240 (4.10)	59
$(CH_3)_3SiSi(CH_3)_3$	Gas	(140) (4.52), (163) (4.28), (192) (4.11)	68a
	CH	197–199 (3.9)	52, 126
$CH_3[(CH_3)_2Sn]_2CH_3$	Heptane	(182) (4.6), (210) (4.4)	126
$H[Si(CH_3)_2]_3Cl$	CH	218 (3.85)	54
$[(CH_3)_3Si]_2NH$	Isooctane	204 (3.46)	110
	EtOH	203	
$H[Si(CH_3)_2]_3H$	CH	218 (3.83)	54
$[(CH_3)_2SiNH]_3$	Isooctane	202 (3.79)	110
p-$CH_3OC_6H_4SiH_3$	CH	265 sh, 275 (3.09), 281 (3.02)	60
p-$CH_3C_6H_4SiH_3$	CH	258 (2.29), 263 (2.41), 270 (2.28)	60
$(CH_3)_2SiCH_2{-}C(=O){-}CH_2CH_3$	CH	279 (1.97)	92
	EtOH	276 (2.08)	92
$(CH_3)_2SiCH_2CH_2C(=O)CH_3$	CH	283 (1.38)	92
	EtOH	279 (1.60)	92
$(CH_3)_3Si(CH_2)_2CH{=}CH_2$		(175) (4.3)	107
$(CH_3)_3SiSi(CH_3)_2C(=O)CH_3$	CH	370, 384	21

Compound	Solvent[a]	λ_{max}, mμ[b] (log ϵ)	Ref.[c]
$(CH_3)_3SiOSi(CH_3)_2CH{=}CH_2$		(178–180) (4)	1
$CH_2{=}CH{-}Si(CH_3)_2Si(CH_3)_3$		223 (3.72)	120
$CH_2{=}CHSi(CH_3)[Si(CH_3)_2]_2$	*n*-Hexane	221 (sh), 238 (3.78)	122
n-$[(CH_3)_7Si_3]NH_2$	Isooctane	213 (sh) (3.81)	110
(cyclopentadienyl)$-Si(CH_3)_3$	Dioxane	241	47
$(C_2H_5O)_3SiC{\equiv}CH$	Gas	262 (0.0)	44
$(C_4H_9)_2SnCl_2$	CH	215 (3.72)	59
$CH_3C(=O){-}Si(CH_3)_2Si(CH_3)_2{-}C(=O)CH_3$	CH	331, 344, 358, 371 (max), 385	21
$CH_2{=}CHSi(OC_2H_5)_3$	Isooctane	210 (2.48)	125a
$(CH_3)_3Si(CH_2)_3CH{=}CH_2$		(175) (4.3)	107
$CH_2{=}CH[Si(CH_3)_2]_2CH{=}CH_2$	CH	225 (3.90)	52, 120
$(CH_3)_3SiCH{=}CHGe(CH_3)_3$		(195) (4.2)	106
$(CH_3)_3GeCH{=}CHGe(CH_3)_3$		195 (4.4)	106, 135
$[(CH_3)_3Ge]_2C{=}CH_2$		(190) (sh) (4.4)	106
$(CH_3CH_2)_4Pb$	EtOH	(215)	16
$(CH_3)_3SiCH{=}CHSi(CH_3)_3$		(195) (4.2)	106
$[(CH_3)_3Si]_2C{=}CH_2$		(180–190) (brd) (4.2)	106
$Cl[Si(CH_3)_2]_4Cl$	CH	235 (4.24)	52, 54
$CH_3[(CH_3)_2Ge]_3CH_3$	Heptane	(190) (sh) (4.59), (215) (4.0)	126
$CH_3[Si(CH_3)_2]_3CH_3$	CH	215 (3.96)	52
$H[Si(CH_3)_2]_4Cl$	CH	235 (4.12)	54
$(CH_3)_3SiNHSi(CH_3)_2$-$NHSi(CH_3)_3$	Isooctane	200 (3.68)	110
$H[Si(CH_3)_2]_4H$	CH	236 (4.09)	54
$[(CH_3)_2SiNH]_4$	Isooctane	201 (3.89)	110
$CH_3[Si(CH_3)_2]C_6Cl_5$	CH	216 (4.84), (280–305) (3)	57
$C_6H_5Ge(CH_3)_3$	CH	λ_0 265	94
$(CH_3)_3GeC_6H_5$	EtOH	208 (4.08), 259 (2.48)	45, 98
$(CH_3)_3SiN{=}NC_6H_5$	Ether	575	80
$(CH_3)_3SiOC_6H_5$	EtOH	218 (3.80), 273 (3.29)	97
$C_6H_5Pb(CH_3)_3$	EtOH	206 (4.43), 264 (2.90)	76
$C_6H_5Si(CH_3)_3$	EtOH	211 (3.97), 247 (1.88), 253 (2.06), 260 (2.35), 265 (2.28), 270 (2.02)	16, 45, 98
	Heptane	(188) (4.9), (210 sh) (4.0)	126
	$CHCl_3$	λ_0 269.5	28
	CH	λ_0 270	94

Compound	Solvent[a]	λ_{max}, mμ[b] (log ϵ)	Ref.[c]
$(CH_3)_3SnC_6H_5$	EtOH	209 (4.09), 252 (2.78)	98
	CH	λ_0 265	94
$(CH_3)_3SiNHC_6H_5$	Ether	244 (3.9), 289 (3.0)	80
	Isooctane	240 (4.03), 291 (3.27)	110
$(CH_3)_3SiNHNHC_6H_5$	Ether	245 (4.0), (295) (3.0)	80
4-pyridyl–$CH_2Si(CH_3)_3$	Hexane	216 (3.90), 252 (3.12), 257 (3.13)	42e
2-pyridyl–$CH_2Si(CH_3)_3$	Hexane	216 (3.88), 264 (3.46), 269 (3.48), 276 (sh) (3.32)	42e
$(CH_3)_3SiNHNHC_6H_5$	Ether	245 (4.0), (295) (3.0)	80
$CH_2{=}CH{-}[Si(CH_3)_2]_2{-}Si(CH_3)_3$	Hexane	218 (sh), 237 (3.83)	122
$[(CH_3)_3Si]_3N$	Isooctane	200.4 (3.69)	110
	EtOH	200.1	110
$(CH_3)_3GeC(=O)C_6H_5$	EtOH	252 (4.03), 412 (2.08)	139
	Heptane	197 (4.40), 248 (4.05), 419 (2.15)	2
$(CH_3)_3SiC(=O)C_6H_5$	Ethanol	252 (4.07), 402 (2.06), 413 (2.07)	26
	Heptane	198 (4.48), 251 (4.07), 281 (sh), 424 (2.00)	2
	CH	368 (sh) (1.30), 386 (1.66), 404 (1.95), 424 (2.00), 444 (1.76)	21
m-$ClC_6H_4CH_2Si(CH_3)_3$	$MeOH \cdot H_2O$	220	42a
o-$ClC_6H_4CH_2Si(CH_3)_3$	$MeOH \cdot H_2O$	218	42a
p-$ClC_6H_4CH_2Si(CH_3)_3$	$MeOH \cdot H_2O$	229	42a
p-$NO_2C_6H_4CH_2Si(CH_3)_3$	$MeOH \cdot H_2O$	305	42a
	Hexane	284	42a
o-$O_2NC_6H_4CH_2Si(CH_3)_3$	$MeOH \cdot H_2O$	269	42a
$C_6H_5CH{=}N{-}Si(CH_3)_3$	CH	207, 248, 336	79
	Ether	250, 340	30
$C_6H_5CH_2Ge(CH_3)_3$	CH	261 (2.59), 268 (2.65), 275 (2.53)	67
	EtOH	225 (3.86), 269 (3.53)	98
$(HC{\equiv}CCH_2)_2Ge(C_2H_5)_2$	EtOH	(240) (sh) (2.2)	62
$(CH_3)_3SiN{=}NC_6H_4CH_3$-$p$	Ether	575	80
p-$CH_3OC_6H_4Ge(CH_3)_3$	CH	λ_0 281	94
p-$CH_3OC_6H_4Si(CH_3)_3$		274 (3.25), 280 (3.10) 266 (sh)	42b
$C_6H_5CH_2Si(CH_3)_3$	n-Heptane	(196) (4.8), (218) (4.0), (265) (2.70)	45, 126
	EtOH	221 (3.94), 267 (2.63)	42a, 98
$(CH{\equiv}C{-}CH_2)_2Si(C_2H_5)_2$	EtOH	(232) (sh) (1.9)	62
$(CH_3)_3SnCH_2C_6H_5$	EtOH	236 (3.74), 273 (2.69)	98

Compound	Solvent[a]	λ_{max}, mμ[b] (log ε)	Ref.[c]
$(C_4H_9)_2SnCl(O_2CCH_3)$	CH	211 (3.57)	59
$(CH_3)_3SiCH{=}N$–(thiane ring, S)	CH	285 (1.85)	119
$(C_4H_9)_2Sn(OCH_3)_2$	CH	215 (3.18)	59
$Cl[Si(CH_3)_2]_5Cl$	CH	214 (4.00), 250 (4.29)	54
$CH_3[Ge(CH_3)_2]_4CH_3$	Heptane	(185) (sh) (4.7), (230) (4.2)	126
$CH_3[Si(CH_3)_2]_4CH_3$	Heptane	(186) (4.60), (235) (4.05)	52, 126
$(CH_3)_{10}Si_5$		210 (sh), 261 (3.04), 272 (2.99)	29
$H[Si(CH_3)_2]_5Cl$	CH	215 (3.94), 249 (4.21)	54
$H[Si(CH_3)_2]_5H$	CH	214 (3.97), 249 (4.13)	54
$(CH_3)_3SiNHSi(CH_3)_2NHSi(CH_3)_2$-$NHSi(CH_3)_3$	Isooctane	202 (3.82)	110
p-$(CH_3)_3SiCH_2C_6H_4CO_2^-$	H_2O	250 (4.21)	42f
m-$(CH_3)_3SiCH_2C_6H_4CO_2^-$	H_2O	206 (4.54), 230 (3.9), 284 (3.03)	42f
o-$(CH_3)_3SiCH_2C_6H_4CO_2^-$	H_2O	205 (4.42), 278 (2.97)	42f
o-$(CH_3)_3SiCH_2C_6H_4CO_2H$	Hexane	210.5 (4.55), 238.5 (3.96), 295.5 (3.45)	42f
	MeOH·HCl	208 (4.6), 232 (3.9), 290 (3.35)	42f
m-$(CH_3)_3SiCH_2C_6H_4CO_2H$	Hexane	210 (4.48), 239 (4.02), 290 (3.23)	42f
	MeOH·HCl	211 (4.5), 232 (3.95), 289 (3.17)	42f
p-$(CH_3)_3SiCH_2C_6H_4CO_2H$	Hexane	208 (4.17), 256 (4.29), 270 (sh)	42f
	MeOH·HCl	252 (4.26)	42f
$(C_6H_5)(CH_3)C{=}NSi(CH_3)_3$	Ether	222 (3.95), 346 (1.50)	30
p-$CH_3OC_6H_4CH_2Si(CH_3)_3$	MeOH·H_2O	229	42a
p-$CH_3C_6H_4CH_2Si(CH_3)_3$	MeOH·H_2O	226	42a
$(CH_3)_3SiCH_2CH_2C_6H_5$		(260) (2.48)	45
$(CH_3)_3SiSi(CH_3)_2C(=O)C_6H_5$	CH	405, 424 (max), 442	21
$C_6H_5Si(CH_3)_2Si(CH_3)_3$	Heptane or CH	(188) (4.8), 230 (3.69–4.04)	52, 66, 67, 120, 126
$(C_6H_5)_2GeCl_2$	Heptane	269.7 (2.86), 263.3 (2.98), 258.3 (2.92), 252.7 (2.93), 247.5 (2.68)	85

Compound	Solvent[a]	λ_{max}, mμ[b] (log ϵ)	Ref.[c]
$(C_6H_5)_2SnCl_2$	Heptane	269.6 (2.72), 263.0 (2.84), 258.5 (2.80), 252.4 (2.91), 246.5 (2.56)	85
$C_6H_5\overline{Si(OCH_2CH_2)_3N}$	$CHCl_3$	λ_0 269	28
$(C_2H_5)_3SiN{=}NC_6H_5$	Ether	582	80
$(C_2H_5)_3SnC_6H_5$		204 (4.26), 251 (2.93)	45
$C_6H_5Sn(C_2H_5)_3$	EtOH	204 (4.26), 206 (4.23), 208 (4.20), 251 (2.93), 256 (2.90), 264 (2.70)	16
$CH_3CH{=}C(CH_3)C{\equiv}CSi(C_2H_5)_3$	Hexane	234 (4.18)	17
p-$(CH_3)_3SiC_6H_4Si(CH_3)_3$	Hexane	226 (4.23), 230 (sh) (4.17), 258 (2.53), 264 (2.66), 270 (2.68), 276 (2.59)	3
$p(CH_3)_3SiC_6H_4Si(CH_3)_3$-$p$	Hexane	226 (4.25)	123
$C_6H_5N[Si(CH_3)_3]_2$	Isooctane	234 (3.51), 265 (2.45)	110
(pyridine ring, N)—$CH[Si(CH_3)_3]_2$	Hexane	222 (4.07), 268 (sh) (3.52), 274 (3.56), 280 (3.39)	42e
$(C_4H_9)_2Sn(O_2CCH_3)_2$	CH	210 (3.08)	59
$(C_4H_9)_2Sn(Cl)(SC_4H_9)$	CH	211 (3.91), 252 (3.40)	59
$CH_2{=}CH[Si(CH_3)_2]_4CH{=}CH_2$		244 (4.15)	52
$Cl[Si(CH_3)_2]_6Cl$	CH	225 (4.05), 259 (4.36)	52, 54
$CH_3[(CH_3)_2Ge]_5CH_3$	n-Heptane	(190) (sh) (4.9), 210 (4.5), 240 (4.3)	126
$[(CH_3)_2Ge]_6$	Heptane	(200) (sh) (4.8), (234) (3.9)	126
$CH_3[Si(CH_3)_2]_5CH_3$	CH	250 (4.26)	52
$[Si(CH_3)_2]_6$		232 (3.76), 255 (sh) (3.30)	50
$[(CH_3)_2Si]_6$		(195) (sh) (4.6), (235) (3.8)	126
$H[Si(CH_3)_2]_6Cl$	CH	225 (4.08), 258 (4.34)	54
$[(C_2H_5)_3Si]_2NH$	Isooctane	206 (3.26)	110
$H[Si(CH_3)_2]_6H$	CH	225 (4.05), 258 (4.29)	54
p-$(CH_3)_3SiC_6H_4C(=O)Si(CH_3)_3$	Hexane	256 (4.27), 285 (sh) (3.4), 427 (2.04)	15a
$C_6H_5SiCH_3[OSi(CH_3)_3]_2$	$CHCl_3$	λ_0 270.0	28
$C_6H_5[Si(CH_3)_2]_3CH_3$		221 (4.11), 240 (4.19)	120
$[(CH_3)_3Si]_2SiC_6H_5(CH_3)$	CH	243 (4.05)	53, 120
$(C_6Cl_5)_2Si(CH_3)_2$	CH	217 (4.93), (280–305) (3)	57

Compound	Solvent[a]	λ_{max}, mμ[b] (log ϵ)	Ref.[c]
$(C_6F_5)_2Si(CH_3)_2$	CH	207 (4.30), 216 (sh) (4.23), 266 (3.30)	57
N—Si—CH_3; H, CH_3	Isooctane	193 (4.50), 211 (4.53), 231 (4.42), 266 (3.90), 274 (3.92), 332 (3.79)	49
2-$C_{10}H_7C(=O)Si(CH_3)_3$	Hexane	210 (4.44), 244 (4.63), 252 (4.75), 283 (4.05), 292 (4.02), 330 (3.18), 345 (3.20), 428 (2.01)	15a
$(CH_3)_2Si(OC_6H_5)_2$	EtOH	218 (4.03), 273 (3.49)	97
$(C_6H_5)_2Si(CH_3)_2$	CH	265 (4.48)	57
$[HC{\equiv}C{-}Si(CH_3)_2]C_6H_4$	EtOH	(230) (4.2), (270) (2.9)	62
$(HC{\equiv}CCH_2)_2Sn(C_4H_9)_2$	EtOH	(220) (4.1), (250) (sh) (3.3)	62
$(CH_3)_3SiCH_2$ … $CH_2Si(CH_3)_3$	Hexane	199 (4.74), 203 (sh) (4.69), 234 (4.21), 271 (3.80), 277 (3.90), 279 (3.88), 286 (3.86)	14
p-$(CH_3)_3SiC_6H_4Si(CH_3)_2Si(CH_3)_3$	Hexane	239 (4.26)	123
$(C_2H_5)_3SiCH{=}CHSi(C_2H_5)_3$		(202) (4.2)	106
$CH_3[Si(CH_3)_2]_6CH_3$	CH	220 (sh) (4.15), 260 (4.32)	52
$(CH_3)_{14}Si_7$		217 (sh), 242 (3.32)	29
$CH_3[(CH_3)_2Sn]_6CH_3$	Heptane	(205) (5.26), (246) (4.9)	126
[n-$(CH_3)_7Si_3]_2NH$	Isooctane	218 (4.12)	110
$(C_6H_5)(CH_3)_2SiC(=O)C_6H_5$	Ethanol	255 (4.14), 403 (sh) (2.30), 415 (2.33)	26
$(C_3H_7)_3SiN{=}NC_6H_5$	Ether	587	80
p$(CH_3)_3CC_6H_4Si(CH_3)_2Si(CH_3)_3$	Hexane	233 (4.27)	123
1,3,4-$[(CH_3)_3Si]_3C_6H_3$	CH	269	94
$C_6H_5Si[OSi(CH_3)_3]_3$	$CHCl_3$	λ_0 269.5	28
$[(CH_3)_3Si]_3SiC_6H_5$	CH	241 (4.12)	53
$C_6Cl_5[Si(CH_3)_2]_2C_6Cl_5$	CH	216 (4.95), (280–305) (3)	57
Br, $Si(CH_3)_3$, NO_2	Hexane	229 (4.23), 262 (3.94), 316 (sh) (4.19), 326 (4.23)	42d
$Si(CH_3)_3$, Br, Br	Hexane	272 (sh) (4.39), 276 (4.42), 300 (4.17), 302 (4.14), 311 (4.35)	42d

Compound	Solvent[a]	λ_{max}, mμ[b] (log ε)	Ref.[c]
p-$CH_3OC_6H_4Si(CH_3)_3$	CH	λ_0 281	94
p-$CH_3OC_6H_4Sn(CH_3)_3$	CH	λ_0 282	94
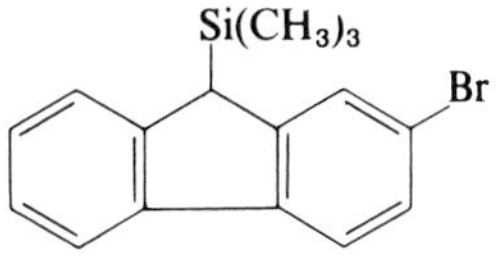	EtOH	270 (4.32), 289 (sh) (3.96), 294 (4.04), 300 (4.05), 306 (4.11)	42d
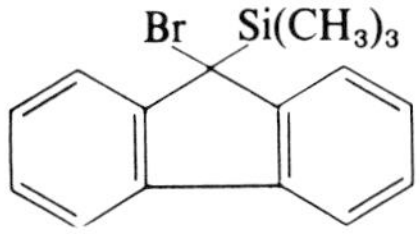	EtOH	245 (4.40), 279 (4.00)	42d
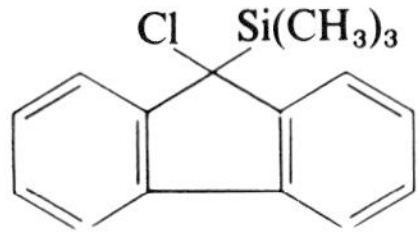	Hexane	223 (sh) (4.41), 232 (4.40), 238 (4.39), 275 (4.08), 284 (sh) (4.02)	42d
$(C_6H_5)_2C{=}NSi(CH_3)_2(CHCl_2)$	Ether	253 (4.15), 360 (1.95)	30
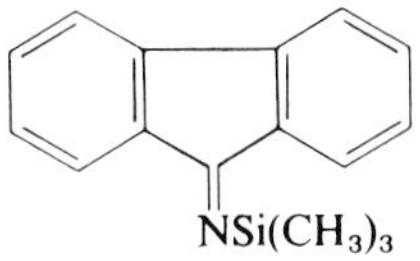	Ether	248 (4.39), 256 (4.90), 265 (3.42), 289 (3.91), 299 (3.97), 352 (2.49)	30
$(o$-$ClC_6H_4)(C_6H_5)C{=}NSi(CH_3)_3$	Ether	253 (4.20), 355 (1.91)	30
$(m$-$ClC_6H_4)(C_6H_5)C{=}NSi(CH_3)_3$	Ether	248 (4.18), 364 (1.85)	30
$(p$-$ClC_6H_4)(C_6H_4)C{=}NSi(CH_3)_3$	Ether	256 (4.26), 364 (1.95)	30
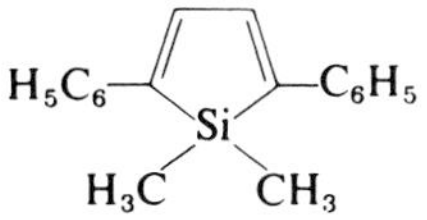	CH	230 (4.11), 370 (4.30)	5
$(C_6H_5)_2C{=}NGe(CH_3)_3$	Ether	243 (4.20), 347 (1.94)	30
$(C_6H_5)_2C{=}NSi(CH_3)_3$	Ether	244, 300, 354	79
$(C_6H_5)_2C{=}NSi(CH_3)_3$	Ether	243 (4.21), 364 (1.89)	30
$(C_6H_5)_2C{=}NSn(CH_3)_3$	Ether	233 (4.14), 361 (1.97)	30
p-$BrC_6H_4[Si(CH_3)_2]_2C_6H_4Br$-$p$	CH	244 (4.51)	57
p-$ClC_6H_4[Si(CH_3)_2]_2C_6H_4Cl$-$p$	CH	242 (4.45)	57
$(C_6H_5CH_2)_2Ge(CH_3)_2$	CH	253 (2.81), 258 (2.86), 262 (2.84), 264 (2.77), 268 (2.72)	67
$C_6H_5(CH_3)_2SiSi(C_6H_5)(CH_3)_2$	CH	238 (4.21)	67
p-$ClC_6H_4Si(CH_3)_2Si(CH_3)_2(C_6H_4Cl$-$p)$	Hexane	242.5 (4.45)	121
$C_6H_5[Si(CH_3)_2]_2C_6H_5$	CH	236 (4.26)	51, 52, 57, 66, 120
$(C_6H_5)_2(CH_3)SiSi(CH_3)_3$	CH	230 (4.22)	52
$[HC{\equiv}CCH_2Si(CH_3)_2]_2C_6H_4$	EtOH	(230) (4.2), (270) (2.9)	62
p-$(CH_3)_3SiC_6H_4[Si(CH_3)_2]_2$—$Si(CH_3)_3$-$p$	CH	245 (4.48)	57

Compound	Solvent[a]	λ_{max}, mμ[b] (log ϵ)	Ref.[c]
m-$[(CH_3)_3SiSi(CH_3)_2]C_6H_4$—$[Si(CH_3)_2Si(CH_3)_3]$	Hexane	231 (4.39)	123
p-$[(CH_3)_3SiSi(CH_3)_2]C_6H_4$—$Si(CH_3)_2Si(CH_3)_3]$	Hexane	248 (4.39)	123
$(C_4H_9)_3SnSC_4H_9$	CH	211 (3.78), 242 (3.15)	59
$(C_4H_9)_2Sn(SC_4H_9)_2$	CH	211 (3.96), 252 (3.43)	59
$(C_4H_9)_4Sn$	Hexane	202 (3.99)	36
H_5C_6 C_6H_5 Si O (ring structure)	CH	264 (3.18), 266 (3.09), 273 (2.92), 367 (sh) (2.29), 383 (2.45), 396 (2.42)	25
$(C_6H_5)_2C{=}NSi(CH_3)_2(CH{=}CH_2)$	Ether	242 (4.16), 363 (1.88)	30
H_3C $Si(CH_3)_3$ (fluorene structure)	Hexane	262 (4.16), 267 (4.18), 290 (3.95), 301 (4.07)	42d
m-$C_6H_5NHC(O)C_6H_4CH_2Si(CH_3)_3$	MeOH·H_2O	264	42a
$(p$-$CH_3OC_6H_4)(C_6H_5)C{=}NSi(CH_3)_3$	Ether	274 (4.18), 357 (2.02)	30
$(p$-$CH_3C_6H_4)(C_6H_5)C{=}NSi(CH_3)_3$	Ether	254 (4.26), 360 (1.94)	30
$(m$-$CH_3C_6H_4)(C_6H_5)C{=}NSi(CH_3)_3$	Ether	254 (4.20), 363 (1.93)	30
$(o$-$CH_3C_6H_4)(C_6H_5)C{=}NSi(CH_3)_3$	Ether	253 (4.20), 360 (1.87)	30
p-$CH_3C_6H_4[Si(CH_3)_2]_2C_6H_4Cl$-$p$	Hexane	240.5 (4.39)	121
p-$CH_3OC_6H_4[Si(CH_3)_2]_2C_6H_5$	Hexane	241 (4.41)	121
p-$CH_3C_6H_4[Si(CH_3)_2]_2C_6H_5$	Hexane	238.5 (4.32)	121
O· $(CH_3)_3C$ $Si(CH_3)_3$ $C(CH_3)_3$ (phenoxy structure)	Benzene	580	95
$(C_6H_5)_3GeCl$	Heptane	269.4 (2.82), 263.6 (2.94), 258.6 (2.95), 252.5 (2.80), 247.1 (2.57)	85
$(C_6H_5)_3SnCl$	Heptane	269.0 (2.75), 263.8 (2.93), 258.6 (2.99), 252.5 (2.92), 247.1 (2.79)	85
$(C_6H_5)_3PbI$	CH	278 (3.75)	65
$(C_6H_5)_3SnI$	CH	245 (sh) (3.82)	65
$(C_6H_5)_3SiLi$	THF	335 (~4.0)	137
$(C_6H_5)_3SiOH$	EtOH	248, 253, 259 (3.95), 264 (3.95), 266, 270	55
$C_6Cl_5[Si(CH_3)_2]_3C_6Cl_5$	CH	213 (5.02), (280–305) (3)	57

Compound	Solvent[a]	λ_{max}, mμ[b] (log ϵ)	Ref.[c]
$C_6F_5[Si(CH_3)_2]_3C_6F_5$	CH	208 (4.33), 221 (4.20), 239 (3.17)	57
$(C_6H_5)_2C{=}NSi(CH_3)_2(CH_2CH{=}CH_2)$	Ether	246 (4.16), 364 (1.95)	30
C_2H_5O $Si(CH_3)_3$ (fluorene)	Hexane	214 (4.26), 273 (4.14), 282 (4.01), 294 (3.81), 305 (3.82)	42d
$(p\text{-}CH_3C_6H_4)_2C{=}N(CH_3)_3$	Ether	258 (4.28), 361 (1.96)	30
$(o\text{-}CH_3C_6H_4)_2C{=}NSi(CH_3)_3$	Ether	355	30
$p\text{-}(CH_3)_2NC_6H_4[Si(CH_3)_2]_2\text{-}C_6H_4N(CH_3)_2\text{-}p$	CH	274 (4.70)	57
$p\text{-}CH_3OC_6H_4Si(CH_3)_2Si(CH_3)_2\text{-}C_6H_4OCH_3\text{-}p$	Hexane	242 (4.51)	121
$p\text{-}CH_3C_6H_4Si(CH_3)_2Si(CH_3)_2\text{-}C_6H_4CH_3\text{-}p$	Hexane	239.0 (4.40)	121
$p\text{-}CH_3OC_6H_4Si(CH_3)_2Si(CH_3)_2\text{-}C_6H_4CH_3\text{-}p$	Hexane	241.5 (4.45)	121
$C_6H_5[(CH_3)_2Si]_3C_6H_5$	CH	243 (4.28)	51, 53, 57
$(\text{S})_3SiI$	Hexane	209 (3.45)	42c
$CH_3[Si(CH_3)_2]_8CH_3$	CH	215 (4.46), 240 (4.26), 272 (4.59)	52
$CH_3Si(OC_6H_5)_3$	Ethanol	211 (4.24), 273 (3.45), 266 (3.49)	97
$(C_6H_5)_2C{=}NSi(CH_3)_2(OC_4H_9)$	Ether	250 (4.18), 365 (1.98)	30
$(C_6H_5)_2C{=}NSi(C_2H_5)_3$	Ether	254 (4.22), 370 (1.96)	30
$(C_6H_5)_2C{=}NSi(CH_3)_2[N(C_2H_5)_2]$	Ether	250 (4.21), 360 (sh) (2.06)	30
$(CH_3)_3Si$ $Si(CH_3)_3$ (fluorene)	Hexane	255 (sh) (4.10), 264 (4.19), 292 (4.04), 300 (4.02)	42d
$(C_6H_5)_3Ge\overset{O}{\overset{\Vert}{C}}CH_3$	Heptane	192 (4.96), 260 (3.16), 352, 366 (2.59), 380	2, 104
$(C_6H_5)_3Si\overset{O}{\overset{\Vert}{C}}CH_3$	Heptane	195 (5.04), 266 (2.85), 376 (2.61)	2
$(C_6H_5)_2(CH_3)Si\overset{O}{\overset{\Vert}{C}}C_6H_5$	EtOH	256 (4.09), 403 (sh) (232), 417 (2.38)	26
$(C_6H_5)_3Si\overset{O}{\overset{\Vert}{C}}CH_3$	CH	195 (5.04), 260 (3.16), 266 (3.15), 272 (3.01), 337 (2.01), 348 (2.29), 362 (2.56), 377 (2.64), 392 (2.53)	21, 23, 26, 104

Compound	Solvent[a]	λ_{max}, mμ[b] (log ϵ)	Ref.[c]
$(C_6H_5)_3SiOC_2H_5$	EtOH	248, 254, 260 (3.00), 264, 266, 271	55
$(C_6H_5)_3Sn\underset{\underset{O}{\|}}{C}CH_3$	Heptane	363 (sh), 375, 391	104
$[p\text{-}(CH_3)_2NC_6H_4](C_6H_5)_2SiOH$	EtOH	270 (4.25)	55
$C_6Cl_5[Si(CH_3)_2]_4C_6Cl_5$	CH	217 (5.03), (280–305) (3)	57
$NSi(CH_3)_3$ $NSi(CH_3)_3$	Ether	252 (4.57), 267 (4.30), 292 (4.40), 370 (2.39), 400 (2.50), 424 (2.51)	30
$C_6H_5[(CH_3)_2Si]_4C_6H_5$	CH	250 (4.33)	51, 57
$p\text{-}[(CH_3)_3Si]_2CHC_6H_4CH\text{-}[Si(CH_3)_3]_2$	Hexane	202 (4.80), 240 (4.33), 276 (2.91), 281 (2.98), 292 (2.86)	14
$(C_6H_5)_3Si\underset{\underset{O}{\|}}{C}CH_2CH_3$	EtOH	260 (3.20), 265 (3.17), 272 (2.99), 355 (2.37), 370 (2.48), 385 (2.39)	25
$(C_6H_5CH_2)_3SnCl$	CH	248 (3.35)	67
$(C_6H_5)_2(CH_3)SiSi(CH_3)_2C_6H_5$	CH	237 (4.33)	52
$(C_6H_5)_3SiSi(CH_3)_3$	CH	236 (4.26)	52, 66, 67
$C_6Cl_5[Si(CH_3)_2]_5C_6Cl_5$	CH	216 (5.03) (280–305) (3)	57
$(C_6H_5)_2C{=}NSi(i\text{-}C_3H_7)_3$	Ether	251 (4.21), 378 (2.00)	30
$C_6H_5[(CH_3)_2Si]_5C_6H_5$	CH	258 (4.40)	51
$C_6H_5[Si(CH_3)_2]_5C_6H_5$	CH	258 (4.40)	57
$CH_3[Si(CH_3)_2]_{10}CH_3$	CH	215 (sh) (4.45), 230 (sh) (4.32), 255 (sh) (4.39), 279 (4.63)	50
$(C_6H_5)_3GeC(O)[C(CH_3)_3]$	Heptane	363	104a
$(C_6H_5)_3SiC(O)C(CH_3)_3$	Heptane	377	104a
$(C_6H_5)_3SnC(O)[C(CH_3)_3]$	Heptane	376	104a
$(C_6F_5)_4Si$	CH	271 (3.67)	46
$(C_6F_5)_3SiC_6H_5$	CH	262 (sh) (3.49), 266 (3.60), 272 (sh) (3.57)	46
$(C_6F_5)_2Si(C_6H_5)_2$	CH	262 (sh) (3.41), 268 (3.49), 273 (3.43)	46
$C_6F_5Si(C_6H_5)_3$	CH	254 (sh) (3.08), 260 (3.25), 265 (3.31), 270 (3.24)	46
Si H_5C_6 C_6H_5		290 (weak)	64

Compound	Solvent[a]	λ_{max}, mμ[b] (log ϵ)	Ref.[c]
$(C_6H_5)_4Ge$		320 (F), 450 (P), 470 (P), 510 (P) emission	83
$(C_6H_5)_4Ge$	Heptane	268.9 (2.93), 265.4 (3.10), 262.5 (3.09), 258.8 (3.23), 252.3 (3.32)	85
	Isooctane	41,050; 40,490; 39,570; 38,620; 38,100; 37,690; 37,160 cm^{-1} (log $\epsilon \sim$ 3.0)	83
m-$HOC_6H_4Si(C_6H_5)_3$	EtOH	260, 265, 272, 285	114
p-$HOC_6H_4Si(C_6H_5)_3$	EtOH	237, 260, 265, 275, 283	114
$Si(OC_6H_5)_4$	EtOH	211 (4.39), 273 (3.78)	97
$(C_6H_5)_4Pb$		425 (P), 455 (P) emission	83
$(C_6H_5)_4Pb$	EtOH	250, 257, 263, 268 (sh)	114
	Isooctane	39,730; 38,770; 37,840 cm^{-1} (log $\epsilon \sim$ 3.2)	83
$C_6H_5[Si(C_6H_5)_2]C_6H_5$	CH	254 (4.51)	53
$(C_6H_5)_4Si$	$CHCl_3$	254 (3.04), 260 (3.11), 265 (3.15), 272 (3.04)	89
	EtOH	211 (4.65), 254, 260, 265 (3.26), 271	98, 114
	Isooctane	40,290; 39,290; 38,730; 38,330; 37,780; 37,400; 36,850 cm^{-1} (log $\epsilon \sim$ 3.2)	83
$(C_6H_5)_4Si$		310 (F), 430 (P), 460 (P) emission	83
$(C_6H_5)_4Sn$	CH	228, 257	99
	Heptane	268.4 (2.74), 264.8 (3.08), 261.9 (2.96), 258.5 (3.11), 252.4 (3.06), 246.7 (2.90)	83, 85
	EtOH	251, 258, 264, 269	114
	$CHCl_3$	301 (sh), 315 (−0.15), 322, 335, 342 (sh)	83
$(C_6H_5)_4Sn$		320 (F), 450 (P), 470 (P) emission	83
$(C_6H_5)_3GeSn(C_2H_5)_3$		242 (4.30)	35
$(p$-$(CH_3)_2NC_6H_4)_3SiOH$	EtOH	272 (5.20)	55
$[C_6H_5(CH_3)_2Si]_3SiH$	CH	237 (4.48)	53
$C_6Cl_5[Si(CH_3)_2]_6C_6Cl_5$	CH	216 (5.04), (280–305) (3)	57
$C_6H_5[(CH_3)_2Si]_6C_6H_5$	CH	265 (4.48)	51
$[(C_4H_9)_3Sn]_2O$	Hexane	206 (4.19)	36, 59
$[(C_4H_9)_3Sn]_2S$	Hexane	208 (4.32), 246 (3.67)	36

Compound	Solvent[a]	λ_{max}, mμ[b] (log ϵ)	Ref.[c]
$(C_2H_5)_{12}Sn_5$		310 (4.33)	35
$(p\text{-}ClC_6H_4)_3SiC(=O)C_6H_5$	CH	228 (4.58), 257 (sh) (4.29), 387 (sh) (2.17), 403 (2.44), 422 (2.56), 442 (2.36)	22
	EtOH	259 (4.24), 400 (2.46), 411 (2.51), 425 (2.34)	22
$p\text{-}BrC_6H_4C(=O)Si(C_6H_5)_3$	CH	271 (4.34), 390 (sh) (2.07), 407 (2.33), 428 (2.43), 448 (2.20)	22
	EtOH	271 (4.33), 410 (2.43), 421 (2.46), 442 (2.22)	22
$p\text{-}ClC_6H_4C(=O)Si(C_6H_5)_3$	CH	268 (4.26), 391 (2.07), 407 (2.34), 427 (2.44), 448 (2.20)	22
$(C_6H_5)_3GeC(=O)C_6H_4F\text{-}p$	Ether	392, 411, 435	102
	Heptane	375 (sh) (2.00), 396 (2.29), 413 (2.39), 436 (sh) (2.15)	102
$p\text{-}FC_6H_4C(=O)Si(C_6H_5)_3$	CH	260 (4.19), 384 (2.08), 400 (2.35), 418 (2.46), 440 (2.27)	22
$p\text{-}NO_2C_6H_4C(=O)Si(C_6H_5)_3$	CH	266 (4.25), 411 (sh) (2.12), 434 (2.28), 448 (2.28), 478 (1.95)	22
	EtOH	267 (4.19), 430 (2.33)	22
$(C_6H_5)_3GeC(=O)C_6H_5$	Heptane	193 (4.75), 254 (4.21), 401 (2.24), 418 (2.62), 440 (sh) (2.09)	2, 26, 102
	Ether	396 (sh), 416, 439 (sh)	102
$p\text{-}HO_2CC_6H_4Ge(C_6H_5)_3$	95% EtOH	238, (264), (269), (276), (284)	114
$m\text{-}HO_2CC_6H_4Ge(C_6H_5)_3$	95% EtOH	252, 258, 265, 276, 284	114
$C_6H_5C(=O)Si(C_6H_5)_3$	CH	195 (5.21), 257 (4.21), 388 (sh) (2.07), 405 (2.35), 424 (2.47), 440 (2.25)	2, 15a, 21, 22
	EtOH	258 (4.19), 403 (2.42), 417 (2.48)	22, 26
$(C_6H_5)_3SnC(O)C_6H_5$	Heptane	435	104a
$p\text{-}HO_2CC_6H_4Si(C_6H_5)_3$	95% EtOH	240, (260), (272)	114
$m\text{-}HO_2CC_6H_4Si(C_6H_5)_3$	95% EtOH	254, 262, 265, 272, 285	114
$p\text{-}H_2NC(=O)C_6H_4Ge(C_6H_5)_3$	95% EtOH	238, (261), (274), (284)	114

Compound	Solvent[a]	λ_{max}, mμ[b] (log ϵ)	Ref.[c]
m-$H_2NC(=O)C_6H_4Si(C_6H_5)_3$	95% EtOH	254, 260, 265, 272, 280	114
p-$H_2NC(=O)C_6H_4Si(C_6H_5)_3$	95% EtOH	240, (261), (272)	114
$(C_6H_5)_2C{=}NSi(n\text{-}C_4H_9)_3$	Ether	252 (4.22), 370 (1.97)	30
$(C_6H_5)_3GeC(=O)C_6H_4CF_3$-$p$	Ether	409 (sh), 428.2	102
	Heptane	416 (2.21), 430 (2.25), 458 (1.96)	102
$(C_6H_5)_3GeC(=O)C_6H_4OCH_3$-$p$	Ether	386 (sh), 404, 426 (sh)	102
	Heptane	390 (2.34), 406 (2.45), 424 (2.27)	102
$(C_6H_5)_3SiCH_2C(=O)C_6H_5$	Ethanol	243 (4.12), 271 (3.49), 310 (sh) (2.52)	24
p-$CH_3OC_6H_4C(=O)Si(C_6H_5)_3$	CH	290 (4.32), 298 (sh) (4.24), 380 (sh) (2.12), 396 (2.39), 414 (2.52), 434 (2.37)	22
$[(C_6H_5)_2SiO]_2[(CH_3)_2SiO]$	$CHCl_3$	λ_0 270.5	28
$(C_6H_5)_2Si(CH_3)Si(CH_3)(C_6H_5)_2$		240 (4.41)	120
$(C_6H_5)_2CH_3SiSiCH_3(C_6H_5)_2$	CH	239 (4.41)	52
p-$[(CH_3)_3Si]_3CC_6H_4C[Si(CH_3)_3]_3$	Hexane	208 (4.57), 248 (4.44), 280 (3.02), 289 (2.81)	14
$(C_6H_5)_3Si(CH_2)_2C(=O)C_6H_5$	EtOH	242 (4.08), 270 (3.24), 311 (sh) (2.12)	24
$(p\text{-}CF_3C_6H_4)_3SiC(=O)C_6H_5$	CH	266 (4.24), 384 (sh) (2.12), 401 (2.38), 418 (2.50), 438 (2.30)	22
[structure: six-membered ring of two GeI_2 and two $C_6H_5C{=}CC_6H_5$ units; labels I, I, H_5C_6, Ge, C_6H_5, H_5C_6, Ge, C_6H_5, I, I]		260 (4.38), 285 (4.32), 356 (3.67)	74
[structure: (9-bromofluorenyl)$_2$ with Br, $-Si(CH_3)_3$, subscript 2]	Hexane	227 (sh) (4.69), 261 (4.49), 266 (4.49), 278 (4.09), 290 (4.24)	42d

Compound	Solvent[a]	λ_{max}, mμ[b] (log ϵ)	Ref.[c]
(9-fluorenyl)$_2$Si(CH$_3$)$_2$	Hexane	246 (4.60), 280 (4.14)	42d
$[(C_6H_5)_2C{=}N]_2Si(CH_3)_2$	Ether	250 (4.45), 364 (2.31)	30
$(p\text{-}CH_3C_6H_4)_3SiC(=O)C_6H_5$	CH	263, 404, 423, 446 (not pure)	22
$(C_2H_5)_{14}Sn_6$		325 (4.44)	35
1-methyl-2,3,4,5-tetraphenylsilole (Si–H, Si–CH$_3$)	CH	246, 362	5
$p\text{-}(CH_3)_3CC_6H_4C(=O)Si(C_6H_5)_3$	CH	269 (4.30), 386 (sh) (2.08), 403 (2.36), 421 (4.48), 442 (4.30)	22
$(C_6H_5)_3SiSi(C_6H_5)_2Cl$		242 (4.23)	66
$(C_6H_5)_3SiSi(Cl)(C_6H_5)_2$	CH	242 (4.23), (265–280) (3.8)	67
1,1-dimethyl-2,3,4,5-tetraphenylsilole	CH	247 (4.40), 357 (4.00)	5
$(C_6H_5)_2C{=}NGe(C_6H_5)_3$	Ether	255 (4.30), 347 (2.10)	30
$(C_6H_5)_2C{=}NPb(C_6H_5)_3$	Ether	340 (sh)	30
$(C_6H_5)_2C{=}NSi(C_6H_5)_3$	Ether	259 (4.38), 366 (2.20)	30
$(C_6H_5)_2C{=}NSn(C_6H_5)_3$	Ether	222 (4.39), 358 (2.05)	30
$(C_6H_5)_3SiSi(C_6H_5)_2C(=O)CH_3$	CH	219 (4.60), 254 (sh) (4.36), 268 (sh) (3.97), 275 (sh) (3.76), 338 (sh) (2.06), 351 (sh) (2.36), 363 (2.58), 378 (2.69), 393 (2.60)	21
$(C_6H_5)_3SiSiCH_3(C_6H_5)_2$	CH	242 (4.45)	52
1,1,4,4-tetramethyl-2,3,5,6-tetraphenyl-1,4-digermacyclohexa-2,5-diene		227 (4.67), 270 (3.98)	74

Compound	Solvent[a]	λ_{max}, mμ[b] (log ϵ)	Ref.[c]
$[(C_6H_5)(CH_3)_2Si]_4Si$	CH	242 (4.59)	58
$(C_6H_5)_2C{=}NSi(CH_2C_6H_5)_3$	Ether	254 (4.27), 368 (2.06)	30
1-methyl-1,2,3,4,5-pentaphenylsilole (ring: H_5C_6, C_6H_5, H_5C_6, C_6H_5 on ring carbons; Si bearing H_5C_6 and CH_3)	CH	248 (4.39), 362 (3.95)	5
$(C_6H_5)_3GeGe(C_6H_5)_3$	CH	239 (4.48)	67
$(C_6H_5)_3GeSn(C_6H_5)_3$		243 (4.49)	35
$[(C_6H_5)_3Si]_2O$	CH	253 (3.28), 264 (3.35), 272 (3.24)	67
$[(C_6H_5)_2SiO]_3$	$CHCl_3$	λ_0 271.0	28
$(C_6H_5)_3PbPb(C_6H_5)_3$	CH	245 (sh), 294 (4.52)	40, 65, 67
$(C_6H_5)_3SiSi(C_6H_5)_3$	CH	246 (4.51), (265–280) (3.9)	50, 66, 67
$(C_6H_5)_3SnSn(C_6H_5)_3$	CH	219, 247 (4.52)	40, 65, 99
	EtOH	246 (4.52)	67
$[(C_6H_5)_3Si]_2NH$	CH	254 (3.12), 260 (3.24), 265 (3.20), 271 (3.06)	67
$(C_6H_5)_3GeC(=O)Ge(C_6H_5)_3$	CH	262–273, 451, 483, 513	23
$(C_6H_5)_3SiC(=O)Si(C_6H_5)_3$	CCl_4	262–273, 478, 524, 554	23
	EtOH	517, 545	23
$(C_6H_5)_3SiSi(C_6H_5)_2C(=O)C_6H_5$	CH	216 (4.54), 244 (4.68), 389 (2.12), 407 (2.42), 423 (2.38), 445 (2.37)	21
$[(C_6H_5)_2C{=}N]_2Si(C_6H_5)_2$	Ether	255 (4.58), 380 (2.58)	30
$[(C_6H_5)_2SiO]_3[(CH_3)_2SiO]$	$CHCl_3$	λ_0 270.5	28
hexaphenylsilole (ring: H_5C_6, C_6H_5, H_5C_6, C_6H_5 on ring carbons; Si bearing H_5C_6 and C_6H_5)	CH	249 (4.44), 365 (3.92)	5, 19
$[(C_6H_5)_2C{=}N]_3SiCH_3$	Ether	250 (4.59), 364 (2.65)	30
$(C_6H_5)_2GeSn(C_2H_5)_2Sn(C_6H_5)_3$		258 (4.52)	35
$(C_6H_5)_3GeSn(C_2H_5)_2Ge(C_6H_5)_3$		252 (4.49)	35
$[(C_6H_5)_2SiO]_3[(CH_3)_2SiO]_2$	$CHCl_3$	λ_0 270.5	28
$(p\text{-}CH_3C_6H_4)_3SnSn(C_6H_4CH_3\text{-}p)$	CH	248 (4.63)	67
$(C_6H_5CH_2)_3SnSn(CH_2C_6H_5)_3$	CH	246 (4.70)	67
$[(C_6H_5)_2C{=}N]_3SiC_6H_5$	Ether	252 (4.64), 371 (2.76)	30

Compound	Solvent[a]	λ_{max}, mμ[b] (log ϵ)	Ref.[c]
$(C_6H_5)_3Si—CH(CH_2CH_2Si(C_6H_5)_3)CH_2C(=O)C_6H_5$		263 (3.56), 270 (3.40), 314 (sh) (2.07)	24
$(C_6H_5)_3SnGe(C_6H_5)_2Sn(C_6H_5)_3$		254 (4.50)	35
$[(C_6H_5)_2Ge]_4$		(265)	100, 101
$[(C_6H_5)_2SiO]_4$	$CHCl_3$	260 (2.51), 265 (2.56), 271 (2.45)	28
$[C_6H_5SiO_{3/2}]$	$CHCl_3$	257 (2.53), 263 (2.62), 270 (2.45)	28
$C_6H_5[Si(C_6H_5)_2]_3C_6H_5$		255 (4.51)	50
$[(C_6H_5)_2Si]_4$		(241) (sh), (275)	100, 101
$[Si(C_6H_5)_2]_4$		234 (4.81), 270 (4.54)	50
$(C_6H_5)_3SnSn(C_6H_5)_2Sn(C_6H_5)_3$	CH	215, 247, 275	99
$C_2H_5Sn[Ge(C_6H_5)_3]_3$		262 (5.09), 269 (5.09)	35
$(C_6H_5)_3Ge[Sn(C_2H_5)_2]_3Ge(C_6H_5)_3$		293 (sh) (4.36)	35
cyclo-$Si(CH_3)_2$—$Si(C_6H_5)_2$—$Si(C_6H_5)_2$—$Si(C_6H_5)_2$—$Si(C_6H_5)_2$ (five-membered ring: H_3C, CH_3 on Si; $(C_6H_5)_2$—Si, Si—$(C_6H_5)_2$, $(C_6H_5)_2$—Si—Si—$(C_6H_5)_2$)		249 (sh) (4.61)	50
$[(C_6H_5)_2SiO]_4[(CH_3)_2SiO]_2$	$CHCl_3$	λ_0 271.0	28
$[(C_6H_5)_3Ge]_3SiH$	CH	240 (sh) (4.69)	53
$[(C_6H_5)_3Si]_3SiH$	CH	240 (sh) (4.62)	53
$(C_6H_5)_3GeSn(C_2H_5)_2Sn(C_6H_5)_2$-$Sn(C_2H_5)_2Ge(C_6H_5)_3$		303 (4.75)	35
$(C_6H_5)_2GeSn(C_2H_5)_2Ge(C_6H_5)_2$-$Sn(C_2H_5)_2Ge(C_6H_5)_3$		298 (4.47)	35
$[(C_6H_5)_2Ge]_5$		(250)	100
$(C_6H_5SiO_{3/2})_{10}$	$CHCl_3$	λ_0 269.7	28
$C_6H_5[Si(C_6H_5)_2]_4C_6H_5$		255 (sh) (4.52), 288 (4.36)	50
$[Si(C_6H_5)_2]_5$		251–255 (sh) (4.78)	50, 100, 101
$(\text{S-ring})_3$PbPb$(\text{S-ring})_3$	CH	254 (4.50), 320 (sh) (4.2)	40
cyclo-$Si(CH_3)_2$—$[Si(C_6H_5)_2]_5$ (six-membered ring: H_3C, CH_3 on Si; $(C_6H_5)_2$—Si, Si—$(C_6H_5)_2$, $(C_6H_5)_2$—Si, Si—$(C_6H_5)_2$, Si—$(C_6H_5)_2$)		250 (sh) (4.75)	50

Compound	Solvent[a]	λ_{max}, mμ[b] (log ϵ)	Ref.[c]
Si $(C_6H_5)_2$—Si Si—$(C_6H_5)_2$ $(C_6H_5)_2$—Si Si—$(C_6H_5)_2$ Si $(C_6H_5)_2$		250 (sh) (4.72)	50
$(C_6H_5)_3GeSn(C_2H_5)_2Sn(C_6H_5)_2Sn$-$(C_6H_5)_2Sn(C_2H_5)_2Ge(C_6H_5)_3$		302 (sh) (4.41)	35
$[(C_6H_5)_3Pb]_4Ge$	$CHCl_3$	328 (4.80)	40
$[(C_6H_5)_3Sn]_4Ge$	$CHCl_3$	276 (4.86)	40
$[(C_6H_5)_2Ge]_6$		(250)	100, 101
$[C_6H_5SiO_{3/2}]_{12}$	$CHCl_3$	λ_0 270.0	28
$[(C_6H_5)_3Sn]_4Pb$	$CHCl_3$	298 (4.77)	40
$[(C_6H_5)_3Pb]_4Sn$	C_6H_6	319 (4.83)	40
$[(C_6H_5)_3Pb]_4Pb$	C_6H_6	358 (4.75), 444 (4.46)	40
$C_6H_5[Si(C_6H_5)_2]_5C_6H_5$		250 (sh) (4.57), 296.5 (4.44)	50
$[Si(C_6H_5)_2]_6$		255 (sh)	100, 101
$[(C_6H_5)_3Sn]_4Sn$	$CHCl_3$	277 (4.90)	40
$[(C_6H_5)_2Sn]_6$	CH	(215), 260 (sh)	99
$C_6H_5[Si(C_6H_5)_2]_6C_6H_5$		255 (sh) (4.58), 312 (4.49)	50
$[C_6H_5Si(C_6H_5)_2]_7C_6H_5$		255 (sh) (4.68), 324 (4.58)	50
$(CH_3)_3SiO[(C_6H_5)_2SiO]_nSi(CH_3)_3$, $n = 1, 2, 3$	$CHCl_3$	λ_0 270.0	28
$(C_6H_5SiO_{3/2})_n$, $n = 24, 84, 1300$	$CHCl_3$	λ_0 270.0	28
$[(C_6H_5)_2SiO_2(CH_3)_2SiO]_n$, $\bar{n} > 500$	$CHCl_3$	λ_0 271.0	28

APPENDIX D

ELECTRONIC TRANSITIONS OF GROUP VA ORGANOMETALLOIDS

Compound	Solvent[a]	λ_{max}, mμ[b] (log ϵ)	Ref.[c]
CH_3PH_2	Gas	187 (2.11), 196 (weak), 201 (3.18)	42
$CF_3PHPHCF_3$	Gas	204 (3.87)	56
$CH_2{=}CHPOCl_2$		178 (4.07)	51
$(CH_3O)_2PHO$	Gas	(192) (sh) (2.2), (202) (sh) (1.5)	42
$(CH_3)_2PH$	Gas	189 (3.80)	42
$CF_3PHP(CF_3)PHCF_3$	Gas	224 (3.74)	56
$(CH_3O)_3P$	Gas	190 (brd) (4.8)	42
$(CH_3)_3P$	Gas	201 (4.27)	42
cyclo-$(CF_3As)_4$		197 (3.59), 224 (3.59)	19
$(CF_3)_2PP(CF_3)_2$	Gas	216 (3.89)	56
cyclo-$(CF_3P)_4$	Gas	221 (3.48), 239 (3.51), 259 (3.49)	56
$(CH_3)_2AsCH{=}CH_2$	Isooctane	231 (3.49)	81
	CH_3OH	230 (3.48)	81
	Gas	225	81
(2-thienyl)$_3$PO	EtOH	238 (4.52), 250 (sh) (4.45)	39
$(CH_3CH_2)_2PCl$	Gas	194 (3.5), 240 (2.8)	42
cyclo-$(CF_3P)_5$	Gas	240 (3.78), 268 (sh) (3.24)	56
$CH_3As(CH{=}CH_2)_2$	Isooctane	228 (3.65)	11
	CH_3OH	227 (3.65)	11
	Gas	223	11
$(CH_3)_2P(S)C{\equiv}CCH_3$		221 (3.65)	10
$C_2H_5C(O)P(O)(OC_2H_5)$	CH	340 (1.85)	5
cyclo-$(CH_3P)_5$	Gas	295	43
$[p\text{-}NO_2C_6H_4PO_3]^{2-}$	H_2O	216 (3.75), 278 (4.01)	49
$(C_6H_5AsO_3)^{2-}$	H_2O	245 (2.35), 251 (2.47), 256 (2.57), 261 (2.60), 268 (2.47)	47
$C_6H_5PCl_2$	96% EtOH	264 (3.23)	75
$C_6H_5N_2P(O)(OK)_2$	EtOH	286 (4.05), 433 (1.81)	9
$[p\text{-}NO_2C_6H_4PO_3H]^-$	H_2O	272 (4.03)	49

[a] CH is cyclohexane; EtOH is ethanol; *i*-PrOH is isopropanol; MeOH is methanol.

[b] Maxima enclosed in parentheses ( ) were read from small figures and are only approximate.

[c] Reference numbers refer to those listed at the end of Chapter V.

Compound	Solvent[a]	λ_{max}, mμ[b] (log ϵ)	Ref.[c]
$[C_6H_5PO_3]^{2-}$	H_2O	258 (2.38)	49
$(C_6H_5AsO_3H)^-$	H_2O	250 (sh) (2.52), 256 (2.68), 262 (2.78), 268 (2.69)	47
o-$BrC_6H_4PO_3H_2$	EtOH	270 (2.93)	48
m-$ClC_6H_4PO_3H_2$	EtOH	215 (3.91), 271 (2.81)	49
o-$ClC_6H_4PO_3H_2$	EtOH	217 (4.09), 270 (2.90)	49
p-$ClC_6H_4PO_3H_2$	EtOH	224 (4.16), 264 (2.47)	49
m-$NO_2C_6H_4PO_3H_2$	EtOH	263 (3.81)	49
p-$NO_2C_6H_4PO_3H_2$	EtOH	213 (3.77), 270 (4.02)	49
$[C_6H_5PO_2H]^-$	H_2O	264 (2.67)	49
$[C_6H_5PO_3H]^-$	H_2O	263 (2.58)	49
$C_6H_5AsO_3H_2$	EtOH	214 (3.83), 245 (sh) (2.39), 251 (sh) (2.59), 256 (2.75), 262 (2.87), 269 (2.79)	47
$C_6H_5PO_2H_2$	EtOH	216 (3.85), 264 (2.79)	49
$C_6H_5PO_3H_2$	EtOH	246 (sh), 252 (sh), 257 (2.59), 263 (2.73), 270 (2.62)	47, 49
$C_6H_5PH_2$		258, 265 (2.77), 282	6
	96% EtOH	(235) (sh) (3.6), (267) (2.9)	75
$(CH_2{=}CH{-})_3As$	Isooctane	227 (3.70)	11
	Methanol	227 (3.67)	11
	Gas	222	11
$(CH_2{=}CH{-})_3P$	Isooctane	235 (3.81)	11
	Methanol	234 (3.81)	11
	Gas	229	11
$C_2H_5As(CH{=}CH_2)_2$	Isooctane	230 (3.63)	11
	Methanol	229 (3.60)	11
	Gas	225	
$C_2H_5P(CH{=}CH_2)_2$	Isooctane	237 (3.72)	11
	Methanol	236 (3.72)	11
	Gas	231	11
$(n$-$C_4H_9)P(O)(CH{=}CH_2)$	Isooctane	(200) (sh)	11
$(C_2H_5)_2AsCH{=}CH_2$	Isooctane	235 (3.45)	11
	MeOH	234 (3.45)	11
	Gas	230	11
$CH_3C(O)P(O)(OC_2H_5)_2$	CH	334 (1.72)	5
$(C_2H_5)_2PCH{=}CH_2$	Isooctane	244 (3.43)	11
	Methanol	242 (3.44)	11
	Gas	239	11
$(C_2H_5)_3As$	Isooctane	208 (4.07)	11
	MeOH	207 (4.02)	11
	Gas	206	11

Compound	Solvent[a]	λ_{max}, mμ[b] (log ϵ)	Ref.[c]
$CH_2{=}CHP(O)[N(CH_3)_2]_2$		174 (4.22)	51
n-$C_3F_7As(CH{=}CH_2)_2$	Isooctane	(215–220) (sh)	11
p-$CH_3OC_6H_4PH_2$		270, 276 (3.15), 284	6
p-$CH_3C_6H_4PH_2$	96% EtOH	(225) (3.75), (270) (2.6)	75
$CH_3P(S)(C{\equiv}CCH_3)_2$		234 (3.53)	10
$CH_2{=}CHP(O)(CH_3)(OC_4H_9)$		176 (4.10)	51
$C_6H_5As(CF_3)_2$	CH	224 (3.76), 253, 259, 265, 272 (2.81)	25
$C_6H_5As(CF_3)(CH_3)$	CH	221 (3.70), 250, 256, 263, 270 (2.7)	25
(o-$BrC_6H_4)(C_2H_5)PO_2H$	EtOH	271 (3.01)	48
(p-$O_2NC_6H_4)(C_2H_5)PO_2H$	EtOH	266 (3.91)	48
$C_6H_5As(CH_3)_2$	EtOH	242 (3.83)	11
	CH	239 (3.83)	25
$C_6H_5Sb(CH_3)_2Br_2$	EtOH	265 (2.78), 270 (2.63)	11
$C_6H_5N_2P(O)(CH_3)_2$	Dioxane	(512) (1.19)	9
$C_6H_5N_2P(O)(OCH_3)_2$	Dioxane	293 (4.08), 488 (1.89)	9
$C_6H_5P(CH_3)_2$	CH	251.0 (3.54)	11
	EtOH	266 (3.35)	75
p-$C_2H_5C_6H_4PH_2$	EtOH	(230) (3.8), (270) (sh) (2.9)	75
$C_6H_5Sb(CH_3)_2$	EtOH	250 (3.57)	11
n-$C_4H_9As(CH{=}CH_2)_2$	Isooctane	231 (3.60)	11
	Methanol	230 (3.61)	11
n-$C_4H_9P(CH{=}CH_2)_2$	Isooctane	238 (3.71)	11
	Methanol	237 (3.72)	11
$[(CH_3)_2N]_2P(O)N_2P(O)[N(CH_3)_2]_2$	Dioxane	(287) (3.48), 562 (1.11)	9
o-$H_2O_2AsC_6H_4N{=}NCH(CN)CO_2H$		240 (4.0), 326 (4.40)	32
$(CH_3C{\equiv}C)_3PS$		250 (3.65)	10
$C_6H_5As^+(CH_3)_3I^-$	EtOH	250 (2.70), 256 (2.78), 262 (2.89), 269 (2.83)	11
$C_6H_5Sb^+(CH_3)_3I^-$	EtOH	256 (2.68), 262 (2.72), 267 (2.60)	11
$C_6H_5P^+(CH_3)_3I^-$	EtOH	251 (2.65), 260 (2.79), 265 (2.96), 272 (2.91)	11
$(CH_3)_2P(S)(C{\equiv}CC_6H_5)$		246 (4.31)	10
p-$(CH_3)_2NC_6H_4P(O)(CH_3)_2$	CH_3OH	274 (4.04)	74
p-$(CH_3)_2NC_6H_4P(CH_3)_2$	CH_3OH	272	74
	CH	268	74
$C_6H_5N_2P(O)[N(CH_3)_2]_2$	Dioxane	288 (3.98), 505 (1.97)	9
(cyclopentyl)$-C(=O)-P(=O)(OC_2H_5)_2$	CH	340 (1.79)	5

Compound	Solvent[a]	λ_{max}, mμ[a] (log ϵ)	Ref.[c]
$CH_2{=}CHP(O)(C_4H_9)_2$		179 (4.17)	51
$CH_2{=}CHP(OC_4H_9)_2$		176 (4.08)	51
$CH_2{=}CHP(O)(OC_4H_9)_2$		177 (4.13)	51
$(n\text{-}C_4H_9)_2AsCH{=}CH_2$	Isooctane	236 (3.39)	11
$(n\text{-}C_4H_9)_2PCH{=}CH_2$	Isooctane	246 (3.42)	11
	MeOH	245 (3.42)	11
		182 (4.23)	51
$o\text{-}H_2O_3AsC_6H_4N{=}NCH(CN)\text{-}(CH_2C_2H_5)$		247 (3.87), 355 (4.31)	32
$p\text{-}ClC_6H_4C(O)P(O)(OC_2H_5)_2$	CH	268 (4.10)	11
$C_6H_5C(O)P(O)(OC_2H_5)_2$	CH	258 (4.05)	5
S–C(=O)–P(=O)–$(OC_2H_5)_2$	CH	345 (1.73)	5
$CH_2{=}CHCH_2P(O)(OC_4H_9)_2$		182 (4.04)	51
$(p\text{-}NO_2C_6H_4)_2PO_2^-$	H_2O	272 (4.33)	49
As(=O)OH	EtOH	226 (4.41), 232 (4.47), 240 (4.45), 276 (3.91), 288 (3.84)	34
$(o\text{-}BrC_6H_4)_2PO_2H$	EtOH	275 (3.89)	48
$(p\text{-}ClC_6H_4)_2PO_2H$	EtOH	234 (4.34), 265 (3.00)	49
$(m\text{-}ClC_6H_4)_2PO_2H$	EtOH	273 (1.14)	49
$(o\text{-}ClC_6H_4)_2PO_2H$	EtOH	274 (3.24)	49
$(p\text{-}NO_2C_6H_4)(C_6H_5PO_2^-)$	EtOH	272 (4.06)	49
$(m\text{-}NO_2(C_6H_4)_2PO_2H$	EtOH	218 (4.41), 263 (4.16)	49
$(p\text{-}NO_2C_6H_4)_2PO_2H$	EtOH	274 (4.32)	49
$(C_4H_3O)_3P{=}O$	EtOH	238 (4.53)	39
$(o\text{-}BrC_6H_4)(C_6H_5)PO_2H$	EtOH	274 (3.16)	48
$(m\text{-}ClC_6H_4)(C_6H_5)PO_2H$	EtOH	272 (3.10)	49
$(KO)(C_6H_5O)P(O)N_2P(O)\text{-}(OC_6H_5)(OK)$	Dioxane	255 (3.08), 260, 268, 538 (1.15)	9
N(H)–P(=O)(H)	EtOH	(222) (4.4), (258) (4.07), (271) (4.0), (317) (3.8)	28a
$(m\text{-}NO_2C_6H_4)(C_6H_5)PO_2H$	EtOH	220 (4.29), 263 (3.88)	49

Compound	Solvent[a]	λ_{max}, mμ[b] (log ϵ)	Ref.[c]
$(C_6H_5)_2PO_2^-$	H_2O	222 (4.08), 264 (2.95)	49
o-$(C_6H_4)C_6H_5AsO_3H_2$	EtOH	239 (3.93), 276 (3.46)	34
$(C_6H_5)_2PO_2H$	EtOH	224 (4.12), 265 (3.08)	49
p-$(C_6H_5)C_6H_4PO_3H_2$	EtOH	255 (4.34)	34
o-$(C_6H_4)C_6H_5PO_3H_2$	EtOH	237 (3.91), 274 (3.31)	34
$(C_6H_5)_2PH$	EtOH	(235) (sh) (3.9), (270) (sh) (3.3)	75
$(\text{N})_3$—P=O	EtOH	238 (4.06), 240 (sh) (4.06)	39
$C_6H_5P(O)(CH_2CH_2CN)_2$	EtOH	218 (4.13), 259 (2.81), 265 (2.93), 272 (2.86)	75
$C_6H_5P(CH_2CH(CN)_2$	EtOH	246 (3.51)	75
p-$H_2O_3PC_6H_9$—$C_6H_4PO_3H_2$	EtOH	262 (4.37)	34
PO_3H_2 … PO_3H_2	EtOH	250 (4.19)	34
$C_6H_5P(O)(CH_2CH_2CO_2H)_2$	EtOH	218 (3.51), 258 (2.49), 264 (2.63), 271 (2.56)	75
$(C_2H_5)_2PC{\equiv}CC_6H_5$	EtOH	220 (4.08), 237 (4.08), 247 (4.11), 258 (4.08), 282 (3.87)	18
p-$CH_3OC_6H_4C(O)P(O)(OC_2H_5)_2$	CH	(228) (3.7), 295 (4.01)	5
—$C{\equiv}C$—$P(C_2H_5)_2$	EtOH	219 (4.08), 229 (4.06), 240 (4.02), 252 (4.95)	17
$O^{(-)}$, Cl, Cl, —$N{\equiv}NP^+[N(CH_3)_2]_3$	MeOH	516 (3.08)	70
$O^{(-)}$, Cl, —$N{\equiv}NP^+[$—$N(CH_3)_2]_3$	MeOH	504 (3.08)	70
o-$O^-C_6H_4N{=}NP^+[N(CH_3)_2]_3$	MeOH	494 (3.01)	70
$[p$-$(CH_3)_2NC_6H_4P^{(+)}(CH_3)_4]I^{(-)}$	MeOH	282 (4.45)	74
$(n$-$C_4H_9)_3P$	Isooctane	204 (4.05)	11
	MeOH	203 (4.06)	11

Compound	Solvent[a]	λ_{max}, mμ[b] (log ϵ)	Ref.[c]
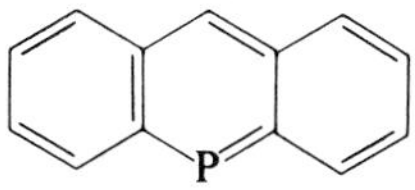 P	Toluene	(385), (400), 429 (max)	26
$(C_6H_5)_2AsCF_3$	CH	224 (2.95), 237 (3.8), 258, 269, 272 (2.96)	25
$(C_6H_5)_2AsCH_3$	CH	240 (4.16), 264 (3.23), 272 (3.26)	25
$(C_6H_5)_2(CH_3)PO$	EtOH	223 (4.22), 259 (2.96), 264 (3.10), 271 (3.13)	75
o-$Na_2O_3AsC_6H_4N{=}NC(CN){=}NNHC_6H_5$		456 (3.77)	32
o-$Na_2O_3AsC_6H_4N{=}NC(CN){=}NNHC_6H_4OH$-$o$		470 (3.70)	32
o-$KHO_3AsC_6H_4N{=}NC(CN){=}N{-}NHC_6H_3$-2-(OH)-3-$(SO_3K)$		482 (3.74), 512 (3.76)	32
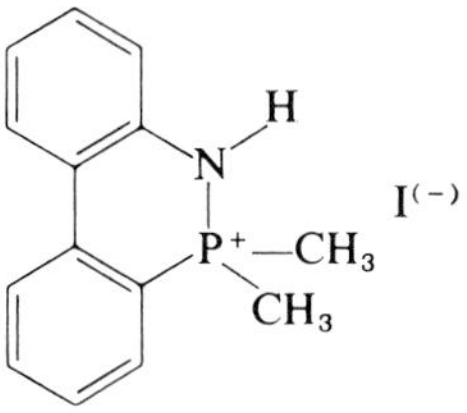 H, N, $P^+{-}CH_3$, CH_3, $I^{(-)}$		(205), (4.7), (263) (3.6), (272) (3.6), (300) (3.5)	28a
$C_6H_5P(n\text{-}C_4H_9)_2$	EtOH	251 (3.44)	75
$C_6H_5P(i\text{-}C_4H_9)_2$	EtOH	253 (3.39)	75
o-$NaHO_3AsC_6H_4N{=}NC(CN){=}NNHC_6H_4CO_2Na$-$o$		468 (3.75)	32
$(C_6H_5)_2PCH_2CH_2CN$	EtOH	247 (3.85)	75
$(C_6H_5)_2PCH_2CH_2CO_2H$	EtOH	248 (3.79)	75
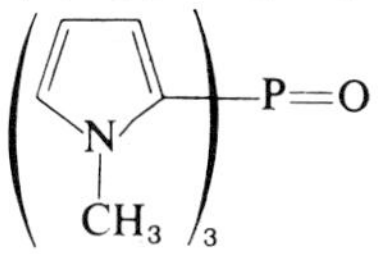 N, CH_3, P=O, 3	EtOH	243 (4.11)	39
p-$(t\text{-}C_4H_9)C_6H_4C(O)P(O)(OC_2H_5)_2$	CH	271 (4.06)	5
p-$CH_3C_6H_4P(C_4H_9\text{-}i)_2$	EtOH	251 (3.45)	75
$(C_6H_5)_2PC{\equiv}C{-}C{\equiv}CH$	EtOH	236 (4.58), 247 (4.52), 259 (4.43), 275 (4.30), 291 (4.23), 307 (4.08)	17
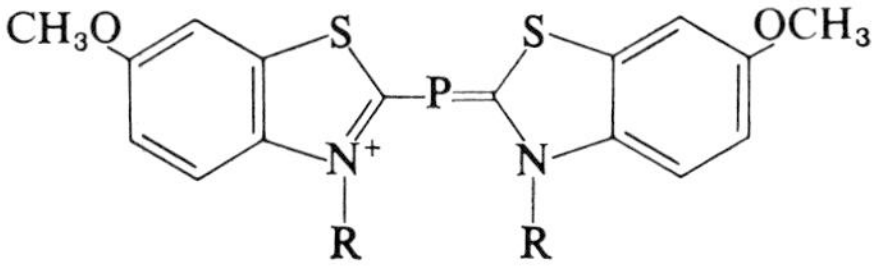 CH_3O, S, P, S, OCH_3, N^+, N, R, R; R = H, C_2H_5		485 (4.64)	29

Compound	Solvent[a]	λ_{max}, mμ[b] (log ϵ)	Ref.[c]
[structure: two benzothiazole rings joined via P; S, S, —P=, N⁺, N, R, R] R = M_2, Et		472 (4.64)	29
$(C_6H_5CH{=}CH)_2PS$		277 (4.66)	10
o-$H_2O_3AsC_6H_4N{=}N{-}C(CN){=}$ $N{-}NHC_6H_2$-2-(OH)-3,5-$(CH_3)_2$		486 (3.84)	32
cyclo-(*i*-$C_4H_9P)_4$	Gas	(277), (282), (292)	43
$CH_3(C{\equiv}C)_2P(O){-}(C_6H_5)_2$	EtOH	225 (4.45), 247 (4.62), 261 (4.40), 274 (4.23)	17
$CH_3C{\equiv}C{-}C{\equiv}C{-}P(C_6H_5)_2$	EtOH	228 (4.43), 250 (4.00), 266 (3.8), 283 (3.5)	17
$CH_3P(S)(C{\equiv}CC_6H_5)_2$		253 (4.58)	10
[structure: cyclopentenyl]$-C{\equiv}C{-}P(O)(C_6H_5)_2$	EtOH	228 (4.24), 246 (4.16), 255 (4.15)	17
$(C_6H_5)_3PO$	CH_3OH	275 (3.42)	80
$(C_6H_5)_3P$	CH_3OH	253 (4.02)	80
(*p*-$O^{(-)}C_6H_4)_3PO$	H_2O	266 (primary band)	73
(*p*-$BrC_6H_4)(C_6H_5)_2P$	Dioxane	214 (4.56), 263 (4.14)	36
(*p*-$ClC_6H_4)(C_6H_5)_2PO$	Dioxane	215 (4.34), 225 (4.44), 229 (4.45), 266 (3.28)	36
(*p*-$ClC_6H_4)(C_6H_5)_2P$	Dioxane	219 (4.42), (230) (sh), 265 (4.10)	36
(*p*-$ClC_6H_4)_2(C_6H_5)_2P{=}S$	Dioxane	218 (4.46), 228 (sh), 260 (3.71)	36
[structure: dibenzo N(H)–P(O)(C_6H_5) ring; H, N, O, P, C_6H_5]	EtOH	(235) (4.5), (260) (4.2), (320) (3.8)	28a
[structure: dibenzo N(H)–P(O)(OC_6H_5) ring; H, N, O, P, OC_6H_5]	EtOH	(235) (4.5), (260) (4.2), (320) (3.8)	28a

Compound	Solvent[a]	λ_{max}, mμ[b] (log ϵ)	Ref.[c]
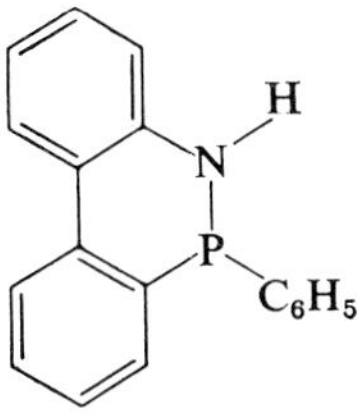	EtOH	(230) (brd) (4.2), (337) (3.5)	28a
$(p\text{-}O^-C_6H_4)(C_6H_5)_2PO$	50% EtOH–NaOH (1 *M*)	268	65
$(C_6H_5)_3As$	EtOH	248 (4.12)	47, 72
	CH_3OH	246 (4.12), (273) (sh), (277) (sh)	24
	CH	248 (4.12), 261, 267, 273 (3.3)	25
$(C_6H_5)_3AsO$	EtOH	221 (4.33), 254 (sh) (3.02), 258 (3.15), 263 (3.24), 270 (3.14)	47
$(C_6H_5)_3Bi$	EtOH	248 (sh) (4.11), 280 (sh) (3.63)	47, 72
$(C_6H_5)_2SbCl_2$	EtOH	218 (4.46), 260 (3.11), 264 (3.16), 270 (3.03)	47, 72
$C_6H_5N_2P(O)(C_6H_5)_2$	Dioxane	296 (4.14), 501 (2.00)	9
$C_6H_5N_2P(O)(OC_6H_5)_2$	Dioxane	300 (4.13), 493 (1.97)	9
$C_6H_5N_2P(S)(C_6H_5)_2$	Dioxane	294 (4.13), 493 (2.16)	9
$(C_6H_5)_3PO$	CH	224 (4.36)	2
	EtOH	225 (4.30), 259 (3.13), 264 (3.22), 271 (3.13)	2, 49, 65, 75
	Dioxane	224 (4.37), 266 (3.28)	36
	MeOH	(266) (3.5)	44a
	H_2O	222 (primary band)	73
$(p\text{-}HOC_6H_4)(C_6H_5)_2PO$	50% EtOH–HCl (1 *M*)	240	65a
$(p\text{-}HOC_6H_4)_3PO$	H_2O	242 (primary band)	73
$(C_6H_5)_3P$	Dioxane	215 (4.46), (226) (sh), 264 (4.15)	36
	CH	262 (3.90)	75
	EtOH	260 (4.06)	49, 72, 75
	MeOH	(262) (4.05)	44a
$(C_6H_5)_3P{=}S$	Dioxane	221 (4.46), 260 (3.70)	36
	MeOH	(262) (sh) (3.8)	44a
$(C_6H_5)_3Sb$	EtOH	256 (4.07)	47, 72

Compound	Solvent[a]	λ_{max}, mμ[b] (log ε)	Ref.[c]
$(p\text{-}H_2NC_6H_4)_3PO$	H_2O	263 (primary band)	73
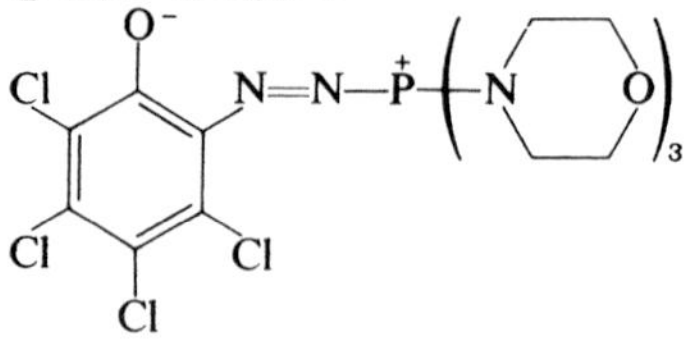 2-O^--3,4,5,6-Cl-substituted phenyl: O^-; Cl, Cl, Cl, Cl; $N{=}N{-}\overset{+}{P}{-}(N\ \ O)_3$	MeOH	(350) (3.3), (488) (3.0)	70
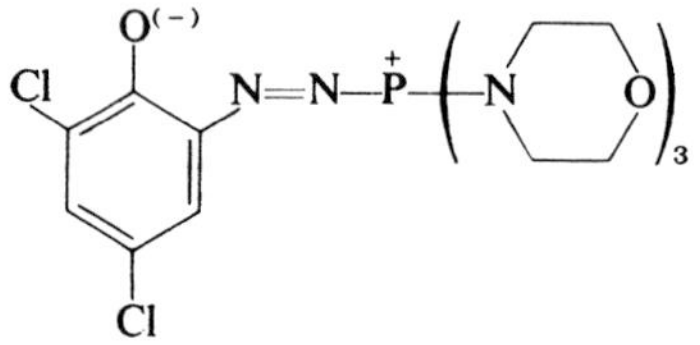 $O^{(-)}$; Cl; Cl; $N{=}N{-}\overset{+}{P}{-}(N\ \ O)_3$	MeOH	(350) (2.8), 519 (3.10)	70
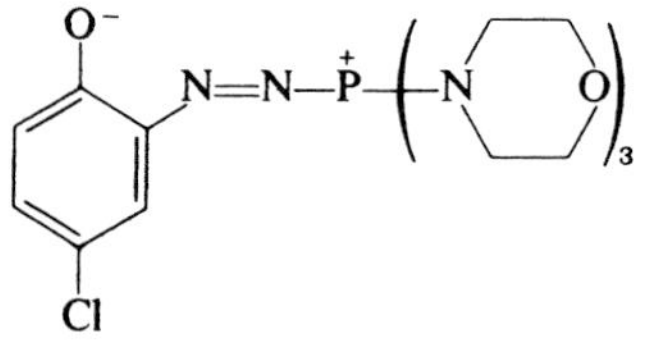 O^-; Cl; $N{=}N{-}\overset{+}{P}{-}(N\ \ O)_3$	MeOH	(343) (2.8), 507 (3.10)	70
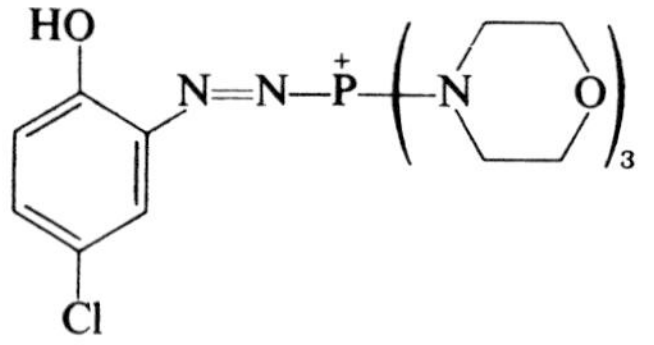 HO; Cl; $N{=}N{-}\overset{+}{P}{-}(N\ \ O)_3$	MeOH	(313) (3.0), (427) (2.8)	70
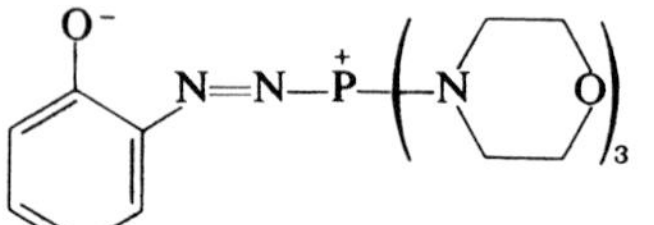 O^-; $N{=}N{-}\overset{+}{P}{-}(N\ \ O)_3$	MeOH	(345) (2.95), 498 (3.07)	70
$CH_3(C{\equiv}C)_3P(C_6H_5)_2$	EtOH	228 (4.78), 253 (4.51), 272 (4.18), 297 (3.40), 316 (3.34), 338 (3.08)	17
$n\text{-}C_3H_7C{\equiv}C{-}C{\equiv}C{-}P{-}(C_6H_5)_2$	EtOH	229 (4.46), 252 (4.00), 267 (3.8), 284 (3.5)	17
$(p\text{-}ClC_6H_4)_3P^+CH_3(I)^-$	H_2O	235 (primary band)	73
$(C_6H_5)_2P(O)N_2C(O)(C_6H_5)$	Dioxane	509 (1.58)	9
$(C_6H_5)_2P{-}\overset{O}{\overset{\Vert}{C}}C_6H_5$	EtOH	265 (4.08), 377 (2.39), 390 (2.43), 404 (2.24)	14
$(p\text{-}O^{(-)}C_6H_4)_3P^+CH_3$	H_2O	271 (primary band)	73

Compound	Solvent[a]	λ_{max}, mμ[b] (log ϵ)	Ref.[c]
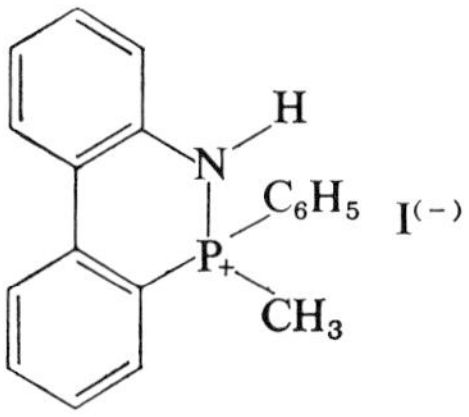 H N C_6H_5 $I^{(-)}$ P_+ CH_3		(220) (4.7), (262) (3.9), (310) (3.6)	28a
$n\text{-}C_3H_7(C{\equiv}C)_2P(C_6H_5)_2$ (P=O)	EtOH	226 (4.54), 248 (3.57), 262 (3.46), 274 (3.26)	17
$(p\text{-}CH_3C_6H_4)(C_6H_5)_2PO$	50% EtOH	231	65
$(p\text{-}CH_3OC_6H_4)(C_6H_5)_2P$	Dioxane	216 (4.39), 235 (sh) (4.18), 261 (4.10)	36
$(p\text{-}CH_3OC_6H_4)(C_6H_5)_2P{=}S$	Dioxane	216 (4.42), 239 (4.28), 270 (sh)	36
$(p\text{-}CH_3OC_6H_4)(C_6H_5)_2PO$	EtOH	224 (4.32), 241 (sh) (4.19), 265 (3.4)	36, 65
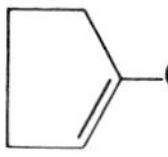 $-C{\equiv}CP(C_6H_5)_2$	EtOH	225 (4.26), 231 (4.23), 246 (4.20), 256 (4.13)	17
$(C_6H_5)_3P^+CH_2BrBr^-$	EtOH	228 (4.31)	31
$(p\text{-}HOC_6H_4)_3P^+CH_3Cl^{(-)}$	H_2O	246 (primary band)	73
$(C_6H_5)_3P^+CH_3I^{(-)}$	MeOH	254, 260, 266, 274	74
$(C_6H_5)_2P(O)C{\equiv}CC_6H_5$	EtOH	230 (4.26), 244 (4.26), 253 (4.36), 264 (4.28), 286 (3.25)	18
$(C_6H_5)_2PC{\equiv}CC_6H_5$	EtOH	230 (4.20), 246 (4.20), 254 (4.28), 265 (4.20), 287 (3.56)	18
$(C_6H_5)_2P(S)C{\equiv}CC_6H_5$	EtOH	229 (4.32), 245 (4.32), 252 (4.36), 264 (4.26), 285 (3.9)	18
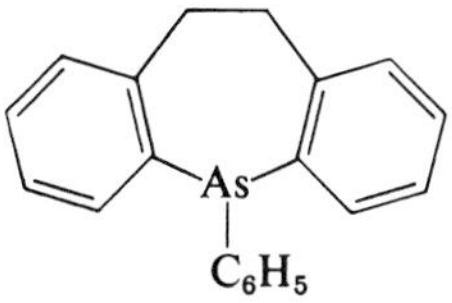 As C_6H_5	EtOH	247 (4.03)	59
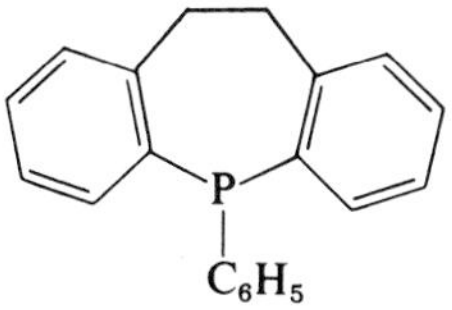 P C_6H_5	EtOH	256 (4.04)	59
$(C_6H_5)_2P(S)CH{=}CHC_6H_5$		264 (4.48)	10

Compound	Solvent[a]	λ_{max}, mμ[b] (log ϵ)	Ref.[c]
$-C\equiv C-P(O)(C_6H_5)_2$	EtOH	223 (3.99), 229 (4.0), 242 (3.95), 252 (3.95)	18
$-C\equiv CP(C_6H_5)_2$	EtOH	225 (4.06), 231 (4.36), 243 (3.95), 255 (3.9)	18
$(C_6H_5)_2P(S)C\equiv C-$	EtOH	228 (4.36), 241 (4.0), 252 (3.9)	18
p-$(CH_3)_2NC_6H_4P(O)(C_6H_5)_2$	H_2O	286 (primary band)	73
	EtOH	282	65
	Dioxane	215 (4.46), 282 (4.42)	36
	CH_3OH	251, 258, 271, 284 (4.44)	74
	CH	277 (4.44)	74
$[p$-$(CH_3)_2NC_6H_4](C_6H_5)_2P$	Dioxane	215 (4.48), 284 (4.42)	36
p-$(CH_3)_2NC_6H_4P(C_6H_5)_2$	MeOH	282 (4.44)	74
	CH	279 (4.44)	74
$[p$-$(CH_3)_2NC_6H_4]P(S)(C_6H_5)_2$	MeOH	252, 286 (4.47)	74
	CH	279	74
	Dioxane	216 (4.55), 288 (4.43)	36
cyclo-$[(C_2H_5)_2CHP]_4$	Gas	(275), (282), (289)	43
$(m$-$CF_3C_6H_4)_3As$	MeOH	246 (4.03), (260) (sh), (266) (sh), (271) (sh)	24
$(p$-$CF_3C_6H_4)_3As$	MeOH	252 (4.16), (265) (sh), (273) (sh)	24
$(C_6H_5)_3P{=}C(CN)_2$		(226) (4.4), (270) (3.6), (294) (sh) (3.1)	44a
P, H_5C_6, C_6H_5	Benzene	362, 420	61
$(C_6H_5)_2P^+(CH_3)(C\equiv CC_6H_5)$ $I^{(-)}$	EtOH	237 (4.23), 248 (4.24), 259 (4.34), 268 (4.26), 290 (3.26)	17
As^+, H_3C, C_6H_5, $I^{(-)}$	EtOH	(220) (3.7), 270 (3.43)	59
P^+, H_3C, C_6H_5, $I^{(-)}$	EtOH	(220) (4.5), 275 (3.55)	59

Compound	Solvent[a]	λ_{max}, mμ[b] (log ϵ)	Ref.[c]
$(p\text{-}CH_3C_6H_4)_3As$	MeOH	248 (4.31), (268) (sh), (275) (sh)	24
$(o\text{-}CH_3C_6H_4)_3As$	MeOH	255 (4.12), (269) (sh), (276) (sh)	24
$(m\text{-}CH_3C_6H_4)_3As$	MeOH	249 (4.12), (270) (sh), 276 (sh)	24
$(p\text{-}CH_3C_6H_4)_3PO$	CH	231 (4.52)	2, 73
	H_2O or EtOH	232 (4.55)	2, 73
$(C_6H_5CH_2)_3PO$	EtOH	219 (4.26), 255 (2.67), 261 (2.78), 267 (2.66)	75
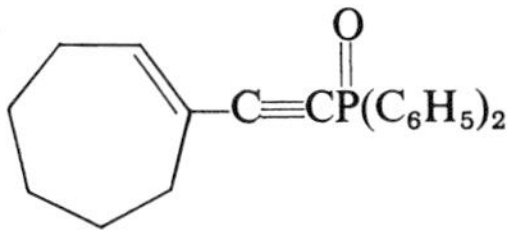 1-cycloheptenyl—C≡CP(O)$(C_6H_5)_2$	EtOH	223 (4.23), 228 (4.26), 245 (4.15), 254 (4.17)	17
$(p\text{-}CH_3OC_6H_4)_3PS$	CH	240 (4.55)	2
	EtOH	242 (4.55)	2
$(p\text{-}CH_3OC_6H_4)_3PO$	CH	244 (4.64)	2
	EtOH	246 (4.65)	2
$(p\text{-}CH_3OC_6H_4)_3PO$	H_2O	244 (primary band)	73
$(C_6H_5CH_2)_3P$	EtOH	247 (3.46)	75
$(p\text{-}CH_3C_6H_4)_3PS$	CH	227 (4.49)	2
	EtOH	227 (4.48)	2
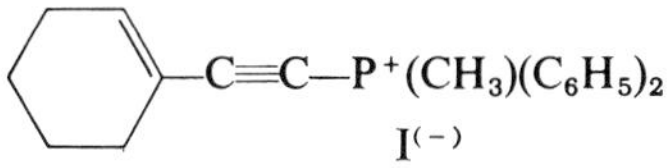 1-cyclohexenyl—C≡C—$P^+(CH_3)(C_6H_5)_2$ $I^{(-)}$	EtOH	225 (4.48), 238 (4.26), 250 (4.20), 256 (4.23)	17
$p\text{-}(CH_3)_2NC_6H_4P^+(CH_3)(C_6H_5)_2$ $I^{(-)}$	H_2O	295 (primary band)	73
	MeOH	265, 272, 300 (4.42)	74
HC≡C—C≡C—P(O)$(C_6H_5)_3$	EtOH	237 (4.75), 248 (4.48), 259 (4.00), 273 (4.26), 289 (4.40)	17
$(C_6H_5)_3P{=}C(CO_2CH_3)(CN)$		(226) (4.5), (266) (3.6), (294) (sh) (3.1)	44a
$C_6H_5P(S)(CH{=}CHC_6H_5)_2$		267 (4.67)	10
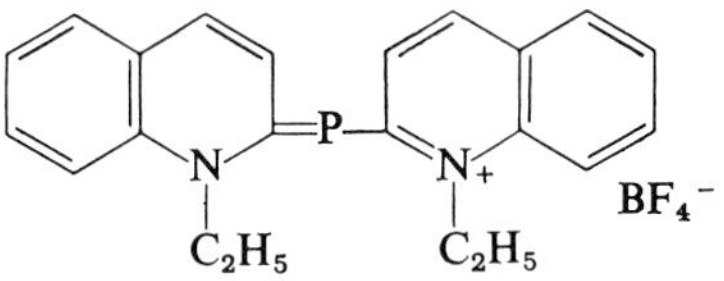 (1-ethyl-2(1H)-quinolylidene)=P—(1-ethyl-2-quinolinium) BF_4^-		592 (4.63)	29
$(p\text{-}CH_3OC_6H_4)_3P^+CH_3I^{(-)}$	H_2O	247 (primary band)	73
$(p\text{-}CH_3C_6H_4)_3P^+CH_3$	H_2O	232 (primary band)	73
$[(p\text{-}CH_3C_6H_4)_3P^+CH_3]I^-$	CH_3OH	233 (4.6), 264 (3.2), 275 (2.7)	4

Compound	Solvent[a]	λ_{max}, mμ[b] (log ϵ)	Ref.[c]
$[p\text{-}(CH_3)_2NC_6H_4]_2P(O)C_6H_5$	MeOH	284 (4.62)	74
	CH	276	74
$[p(CH_3)_2NC_6H_4]_2PC_6H_5$	CH_3OH	284 (4.59)	74
	CH	281 (4.59)	74
$[p\text{-}(CH_3)_2NC_6H_4]_2P(S)(C_6H_5)$	CH_3OH	285	74
	CH	280	74
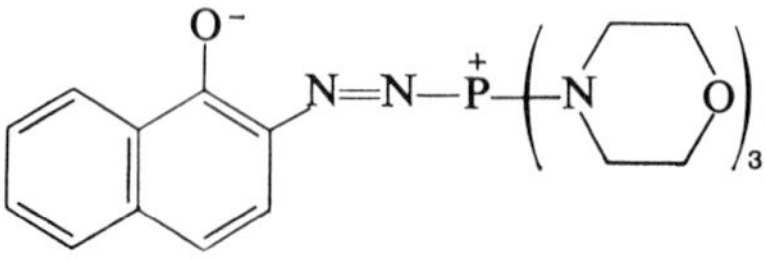 O^- ; $N{=}N{-}\overset{+}{P}{-}(N\ \ O)_3$	MeOH	(345), (467) (3.3), (500) (sh) (3.2)	70
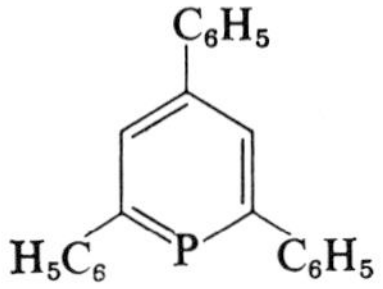 C_6H_5 ; H_5C_6 P C_6H_5	MeOH	278 (4.61)	60, 62
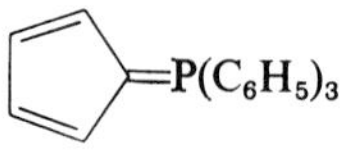 $=P(C_6H_5)_3$	CH_3CN	222 (4.59), 250 (4.33), 295 (sh) (3.77)	69
	EtOH	(222), (290) (sh) (3.7)	69
$[(C_6H_5)_3P^+(C_5H_9)\text{-}cyclo]Br^-$	EtOH	226 (4.40), 262 (4.60), 268 (4.68), 271 (4.59)	69
$[(p\text{-}CH_3C_6H_4)_2P^+C_2H_5]I^-$	MeOH	233 (4.4), 263 (3.3), 274 (3.2)	4
$(C_6H_5C{\equiv}C)_3PS$		256 (4.69)	10
S ; $\overset{O}{\overset{\|}{C}}{-}\overset{I}{\overset{\|}{C}}{=}P(C_6H_5)_3$	EtOH	250 (4.19), 335 (3.81)	40
$(C_6H_5)_3P{=}NC_6H_4Cl\text{-}p$	MeOH	(256) (4.2), (329) (sh) (3.7)	44
$o\text{-}O_2NC_6H_4N{=}P(C_6H_5)_3$	CH_3OH	(380) (3.5)	44
$m\text{-}O_2NC_6H_4N{=}P(C_6H_5)_3$	CH_3OH	(260) (sh) (3.4), (377) (3.2)	44
$p\text{-}O_2NC_6H_4N{=}P(C_6H_5)_3$	CH_3OH	(267) (sh) (3.6), (384) (4.4)	44
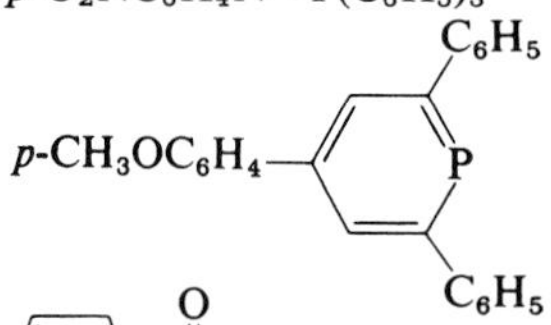 $p\text{-}CH_3OC_6H_4$— P ; C_6H_5 ; C_6H_5	EtOH	283 (4.56)	62
S ; $\overset{O}{\overset{\|}{C}}CH{=}P(C_6H_5)_3$	EtOH	248 (4.11), 253 (4.12), 325 (4.16)	77
O ; $\overset{O}{\overset{\|}{C}}CH{=}P(C_6H_5)_3$	EtOH	255 (4.23), 260 (4.19), 320 (4.27)	77
$(C_6H_5)_4As^+Cl^-$	EtOH	256 (sh), 260, 266, 273	72

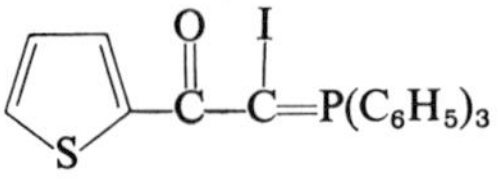

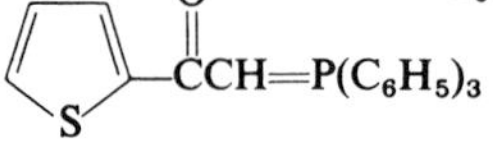

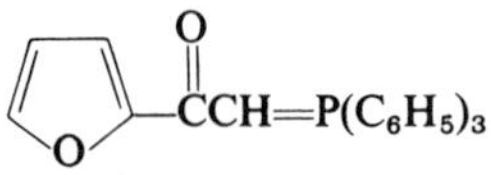

Compound	Solvent[a]	λ_{max}, mμ[b] (log ϵ)	Ref.[c]
$(C_6H_5)_4P^+Br^-$	EtOH	257 (sh), 263, 269, 277	72
	MeOH	(268) (3.7)	44
$(C_6H_5)_4SbBr$	EtOH	252, 258, 263, 270	72
$(C_6H_5)_3P{=}NC_6H_5$	MeOH	(285) (broad shoulder) (3.6)	44
$(C_6H_5)_2P(O)N_2P(O)(C_6H_5)_2$	Dioxane	255 (4.11), 568 (1.36)	9
$(C_6H_5O)_2P(O)N_2P(O)(C_6H_5O)_2$	Dioxane	(255) (3.36), (260), (268), 555 (1.20)	9
$(C_6H_5)_2P(S)N_2P(S)(C_6H_5)_2$	EtOH	538	9
$[(C_6H_5)_3P^+NHC_6H_5]Br^-$	MeOH	(268) (3.8)	44
$(C_6H_5)_2P(O)NHNHP(O)(C_6H_5)_2$	Dioxane	(286) (4.06), (562) (3.26)	9
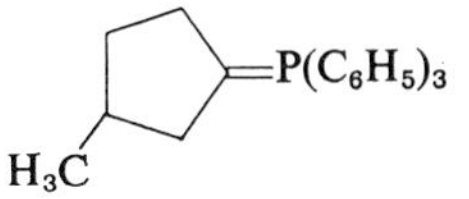 H_3C (cyclopentylidene) $=P(C_6H_5)_3$	Isooctane	416.5 (3.34)	63a
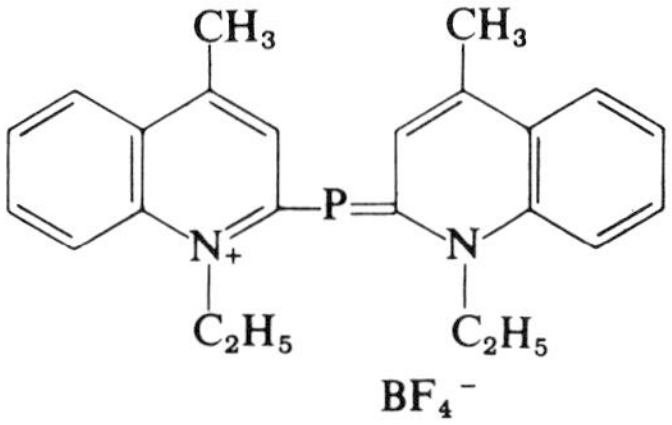 CH_3, CH_3; N+, —P=, N; C_2H_5, C_2H_5; BF_4^-		587 (4.61)	29
$(2,5\text{-}(CH_3)_2C_6H_3)_3As$	CH_3OH	256 (4.20), (266) (sh), (273), (283)	24
$[(p\text{-}CH_3C_6H_4)_3P^+C_3H_7\text{-}n]I^-$	MeOH	233 (4.5), 263 (3.7), 276 (3.5)	4
$[p(CH_3)_2NC_6H_4]_3P{=}O$	MeOH	284 (4.88)	74
	H_2O	288 (primary band)	73
$[p\text{-}(CH_3)_2NC_6H_4]_3P$	MeOH	385 (4.85)	74
	$CHCl_3$	383	74
$[p\text{-}(CH_3)_2NC_6H_4]_3P(S)$	MeOH	286	74
	CH	279	74
cyclo-$(C_6H_{11}P)_4$	Gas	(275) (sh), (282), (290)	43
$p\text{-}O_2NC_6H_4N{=}P(C_6H_5)_2C_6H_4CF_3\text{-}p$	EtOH	(375) (4.2)	82
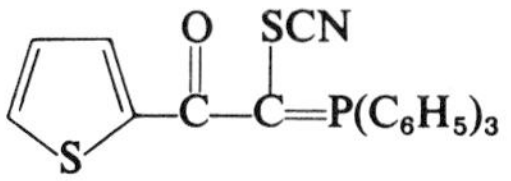 (2-thienyl)—C(=O)—C(SCN)$=P(C_6H_5)_3$	EtOH	260 (4.08), 275 (4.03), 310 (4.03)	40
$C_6H_5N{=}P(C_6H_5)_2(C_6H_4CF_3\text{-}p)$	EtOH	(260) (sh) (3.7)	82
$p\text{-}NCC_6H_4N{=}P(C_6H_5)_3$	MeOH	(326) (brd) (4.2)	44
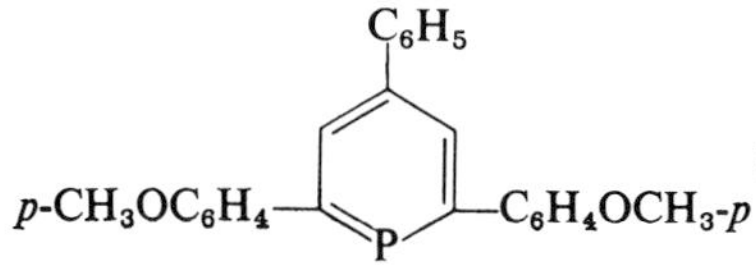 $p\text{-}CH_3OC_6H_4$—(phosphorin, 4-C_6H_5)—$C_6H_4OCH_3\text{-}p$	EtOH	292 (4.60)	62

Compound	Solvent[a]	λ_{max}, mμ[b] (log ϵ)	Ref.[c]
2,6-bis(p-$CH_3C_6H_4$)-4-C_6H_5-phosphorin (p-$CH_3C_6H_4$, C_6H_5, $C_6H_4CH_3$-p substituents on P-ring)	EtOH	283 (4.61)	62
$(C_6H_5)_3P{=}C(CO_2C_2H_5)_2$		(230) (4.4), (270) (sh) (3.5), (294) (sh) (2.7)	44a
$[(p\text{-}CH_3C_6H_4)_3P^+(C_4H_9\text{-}n)]I^-$	MeOH	234 (4.8), 264 (3.5), 276 (3.3)	4
$(p\text{-}CH_3)_2NC_6H_4)_3P^+CH_3I^{(-)}$	H_2O	298 (primary band)	73
$(n\text{-}C_4H_9)_3P{-}N{=}N{-}$(fluorenylidene)	Dioxane	215 (4.23), 249 (4.78), 295 (3.79), 306 (3.82), 366 (4.32), 369 (4.33)	35
p-$BrC_6H_4C(O)C(I){=}P(C_6H_5)_3$	EtOH	320 (3.68)	40
p-$ClC_6H_4C(O)C(I){=}P(C_6H_5)_3$	EtOH	269 (4.16), 310 (3.94)	40
p-$O_2NC_6H_4C(O)C(I){=}P(C_6H_5)_3$	EtOH	268 (4.18), 360 (3.46)	40
$(C_6H_5)_2P{-}C{\equiv}C{-}B(C_6H_5)_2$	EtOH	213 (4.40), 223 (4.30)	17
$(C_6H_5)_3P{=}CHC(O)C_6H_4Br$-$p$	EtOH	278 (3.77), 325 (4.12)	77
$(C_6H_5)_3P{=}CHC(O)C_6H_4Cl$-$p$	EtOH	240 (4.34), 274 (3.93), 320 (4.21)	77
$C_6H_5C(O)C(I){=}P(C_6H_5)_3$	EtOH	267 (3.83), 308 (3.65), 318 (3.67)	40
$(C_6H_5)_3P{=}CHC(O)C_6H_4NO_2$-$p$	EtOH	262 (4.23), 268 (4.25), 360 (4.92)	77
$(C_6H_5)_3P{=}CHC(O)C_6H_5$	EtOH	268 (3.88), 275 (3.89), 317 (4.13)	77
$(C_6H_5)_3P{=}C(CH_3)P(O)(C_6H_4Br$-$p)$	EtOH	275 (3.85), 315 (3.85)	77
2,4,6-tris(p-$CH_3OC_6H_4$)phosphorin (p-$CH_3OC_6H_4$, p-$CH_3OC_6H_4$, $C_6H_4OCH_3$-p substituents on P-ring)	EtOH	299 (4.67)	62
p-$O_2NC_6H_4N{=}P(C_6H_5)_2$-$[C_6H_4N(CH_3)_2$-$p]$	EtOH	(288) (4.4), (395) (4.4)	82
p-$(CH_3)_2NC_6H_4N{=}P(C_6H_5)_3$	EtOH	(265) (4.5)	82

Compound	Solvent[a]	λ_{max}, mμ[b] (log ϵ)	Ref.[c]
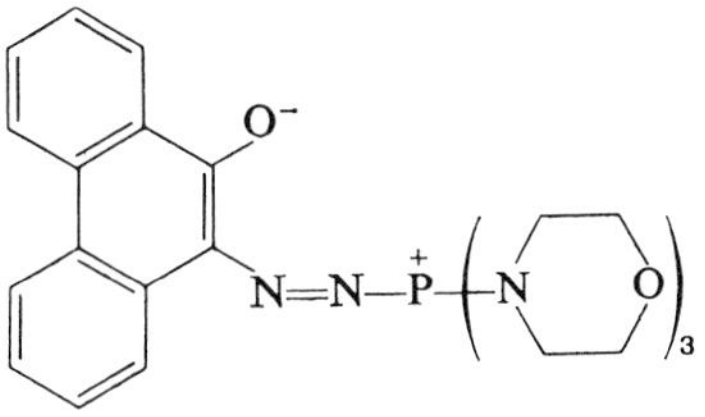 O^-; $N{=}N{-}\overset{+}{P}{-}(N\ \ O)_3$	MeOH	(375) (sh) (2.9), (427) (3.1)	70
$[p\text{-}(CH_3)_2NC_6H_4]_3P^+C_2H_5Br^{(-)}$	MeOH	301 (4.91)	74
$p\text{-}O_2NC_6H_4N{=}P(C_6H_4CF_3\text{-}p)_3$	EtOH	(370) (4.2)	82
$C_6H_5N{=}P(C_6H_4CF_3\text{-}p)_3$	EtOH	(260) (sh) (3.7)	82
$p\text{-}ClC_6H_4C(O)C(SCN){=}P(C_6H_5)_3$	EtOH	262 (3.92), 276 (3.97)	40
$C_6H_5C(O)C(SCN){=}P(C_6H_5)_3$	EtOH	246 (4.08), 275 (3.97)	40
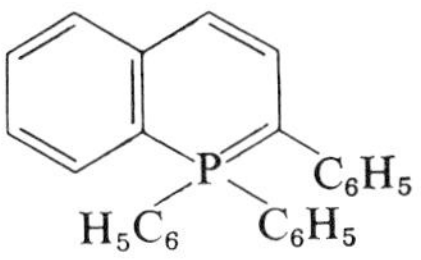 P; C_6H_5; H_5C_6; C_6H_5	Benzene	387, 426, 488	61
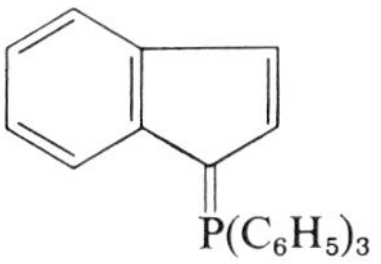 $P(C_6H_5)_3$	EtOH	202 (4.90), 254 (4.22), 331 (3.66)	23
$(C_6H_5)_3P{=}C(CH_3)C(O)C_6H_4Cl\text{-}p$	EtOH	275 (3.81), 315 (3.78)	77
$p\text{-}CH_3C_6H_4C(O)CI{=}P(C_6H_5)_3$	EtOH	265 (3.89), 320 (3.75)	40
$p\text{-}CH_3OC_6H_4C(O)C(I){=}P(C_6H_5)_3$	EtOH	249 (4.13), 302 (3.82), 310 (3.76)	40
$(C_6H_5)_3P{=}C(CH_3)C(O)C_6H_5$	EtOH	265 (3.81), 310 (3.76), 320 (3.77)	77
$(C_6H_5)_2P{=}CH\underset{\underset{O}{\Vert}}{C}C_6H_4CH_3\text{-}p$	EtOH	275 (3.96), 322 (4.18)	77
$(C_6H_5)_3P{=}CH\underset{\underset{O}{\Vert}}{C}C_6H_4OCH_3\text{-}p$	EtOH	248 (4.14), 320 (4.26)	77
$p\text{-}(CH_3)_2NC_6H_4N{=}P(C_6H_5)_2\text{-}(C_6H_4CF_3\text{-}p)$	EtOH	(265) (4.5)	82
$[p\text{-}(CH_3)_2NC_6H_4P^+(C_6H_5)_2\text{-}(CH_2C_6H_5)]Cl^-$	MeOH	268, 274, 303 (4.42)	74
$(p\text{-}(CH_3)_2NC_6H_4)(C_6H_5)_2\text{-}PN{=}\ddot{N}C(C_6H_4Cl\text{-}p)_2$	Dioxane	240 (4.44), 245 (4.61), 248 (4.61), 287 (4.48), 355 (4.18)	35
$[(p\text{-}(CH_3)_2NC_6H_4)_2P^+(C_6H_5)\text{-}(CH_3)]I^{(-)}$	MeOH	304 (4.65)	74

Compound	Solvent[a]	λ_{max}, mμ[b] (log ϵ)	Ref.[c]
$(2,4,6\text{-}(CH_3)_3(C_6H_2)_3As$	CH_3OH	274 (4.21), (285) (sh), (291) (sh)	24
$[(p\text{-}CH_3C_6H_4)_3P^+(C_6H_{13}\text{-}n)]I^-$	CH_3OH	233 (4.6), 264 (3.5), 277 (3.4)	4
H_5C_6 C_6H_5 H_5C_6 As C_6H_5 H	CH	244 (4.50), 360 (3.94) 480 fluorescence	13 13
$p\text{-}CH_3OC_6H_4C(O)C(SCN){=}P(C_6H_5)_3$	EtOH	255 (4.15), 276 (4.09), 290 (4.06)	40
$p\text{-}O^-C_6H_4CH{=}CH{-}CH{=}CHP^+\text{-}(C_6H_5)_3$		510 (4.25), 375 (3.90), 267 (3.97)	66
$(C_6H_5)_3P{=}CHC(O)C_6H_3(CH_3)_2\text{-}2,4$	EtOH	275 (4.02), 302 (4.07)	77
$(C_6H_5)_3P{=}C(CH_3)C(O)C_6H_4CH_3\text{-}p$	EtOH	265 (3.88), 275 (3.85), 320 (3.80)	77
$(C_6H_5)_3P{=}C(CH_3)C(O)\text{-}(C_6H_4OCH_3\text{-}p)$	EtOH	275 (3.96), 320 (3.79)	77
$(n\text{-}C_5H_{11})_3PN{=}N{-}$(fluorenylidene) 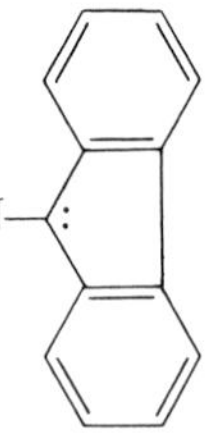	Dioxane	215 (4.16), 248 (4.77), 295 (3.81), 306 (3.80), 369 (4.32)	35
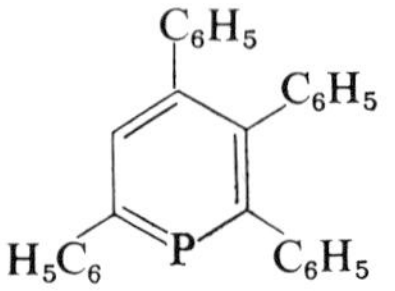 C_6H_5 C_6H_5 H_5C_6 P C_6H_5	EtOH	270 (4.45)	62
$p\text{-}(CH_3)_2NC_6H_4N{=}P(C_6H_4CF_3\text{-}p)_3$	EtOH	(260) (4.6)	82
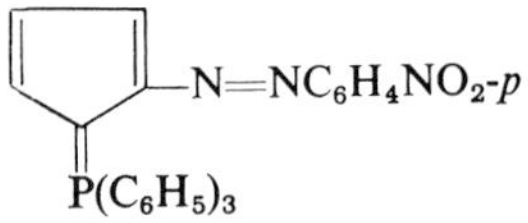 $-N{=}NC_6H_4NO_2\text{-}p$ $P(C_6H_5)_3$	EtOH	205 (4.74), 529 (4.46)	23
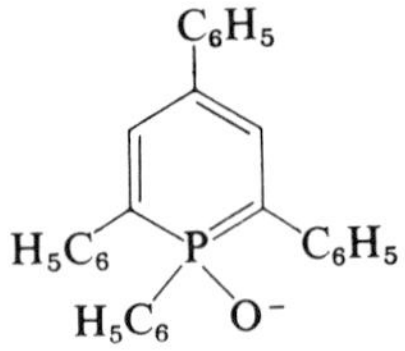 C_6H_5 H_5C_6 P C_6H_5 H_5C_6 O^-	Benzene	348, 365, 475	63

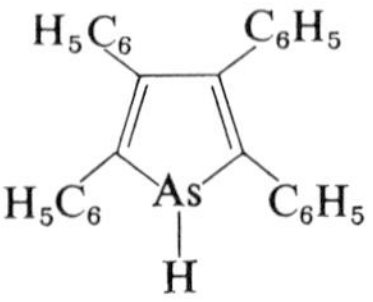

Compound	Solvent[a]	λ_{max}, mμ[b] (log ϵ)	Ref.[c]
C_6H_5; H; H_5C_6 P C_6H_5; H_5C_6 O	Benzene	332 (3.8)	63
	CH_3OH	340	63
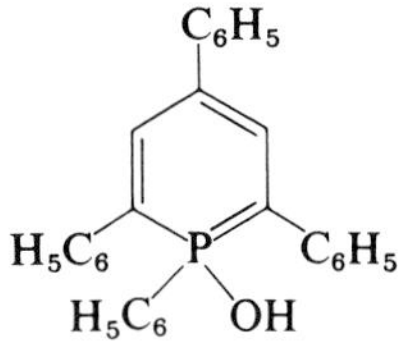 C_6H_5; H_5C_6 P C_6H_5; H_5C_6 OH	MeOH	432	63
$2\text{-}C_{10}H_7C(=O)\text{—}C(I)=P(C_6H_5)_3$	EtOH	267 (4.16), 320 (3.79)	40
$2\text{-}C_{10}H_7C(=O)CH=P(C_6H_5)_3$	EtOH	268 (4.25), 275 (4.23), 317 (4.27), 324 (4.31)	77
$(C_6H_5)_2PC_6H_4P(C_6H_4)_2$	EtOH	272 (4.70)	75
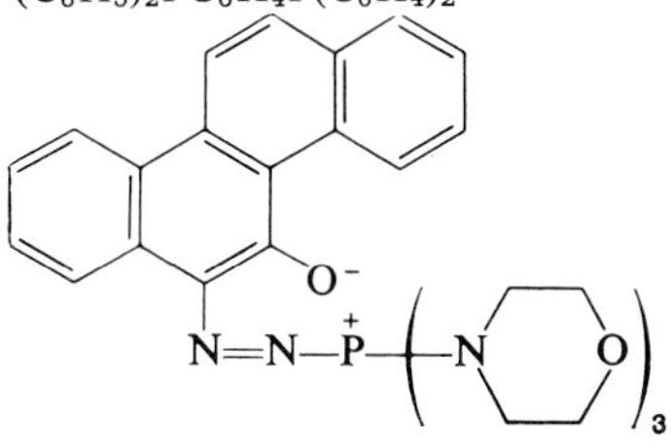 O^-; $N=N\text{—}\overset{+}{P}\text{—}(N\;\;O)_3$	MeOH	(401) (3.25), (425) (sh) (3.05)	70
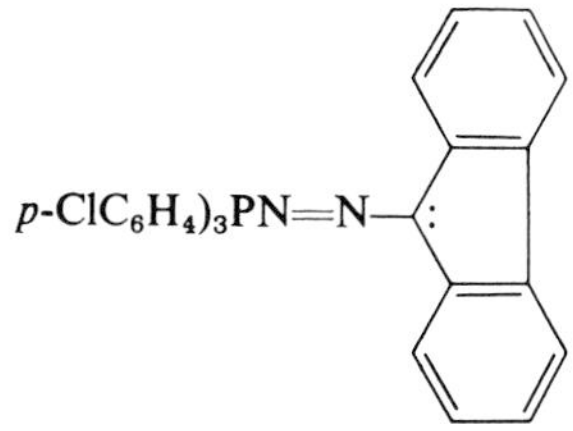 $p\text{-ClC}_6H_4)_3PN=N\text{—}$:	Dioxane	212 (4.72), 240 (4.87), 298 (3.92), 310 (3.92), 366 (4.29), 372 (4.29), 379 (4.30)	35
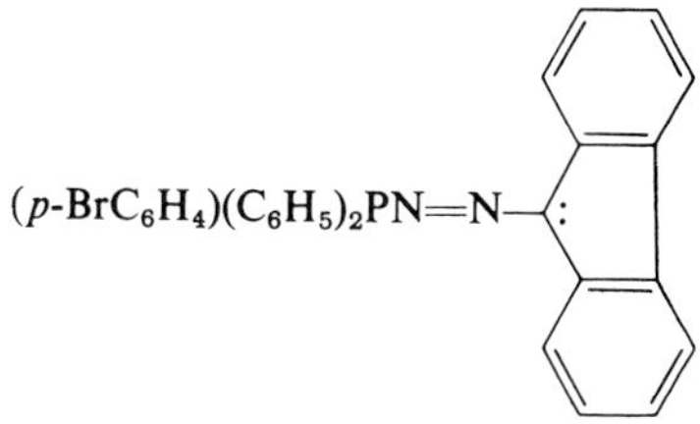 $(p\text{-BrC}_6H_4)(C_6H_5)_2PN=N\text{—}$:	Dioxane	214 (4.66), 246 (4.80), 297 (3.90), 308 (3.90), 378 (4.33)	35

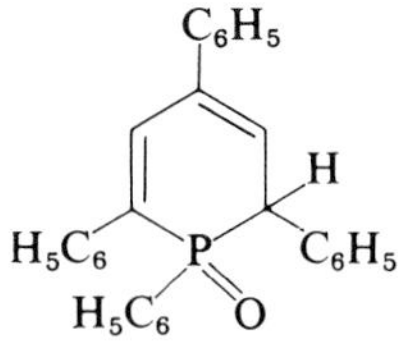

Compound	Solvent[a]	λ_{max}, mμ[b] (log ϵ)	Ref.[c]
p-ClC$_6$H$_4$)(C$_6$H$_5$)$_2$P—N=N— 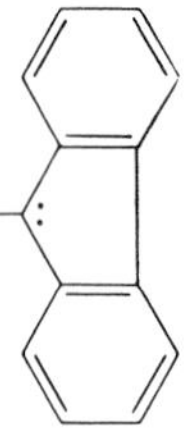 :	Dioxane	240 (4.76), 247 (4.77), 297 (3.90), 308 (3.90), 372 (4.34)	35
C$_6$H$_5$)$_3$P—N=N$\ddot{C}$(C$_6$H$_4$Cl-p)$_2$	Dioxane	240 (4.42), 245 (4.48), 350 (4.12)	35
(C$_6$H$_5$)$_3$P—N=N— 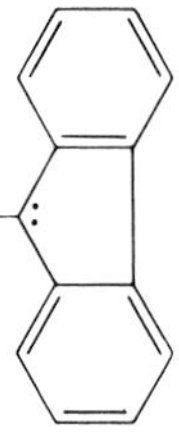 :	Dioxane	227 (4.64), 247 (4.83), 337 (3.91), 308 (3.93), 375 (4.47), 388 (4.47)	35
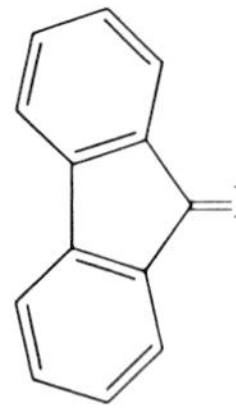 =P(C$_6$H$_5$)$_3$	CHCl$_3$	382 (4.60)	33
(C$_6$H$_5$)$_3$PN=N—$\ddot{C}$(C$_6$H$_5$)-(C$_6$H$_4$—NO$_2$-p)	Dioxane	(310) (~3.8), 422 (4.22)	35
(C$_6$H$_5$)$_3$P=C(CH$_3$)C(O)(2-C$_{10}$H$_7$)	EtOH	275 (4.16), 315 (3.89)	77
(p-CH$_3$OC$_6$H$_4$)-(C$_6$H$_5$)$_2$—PN=N— 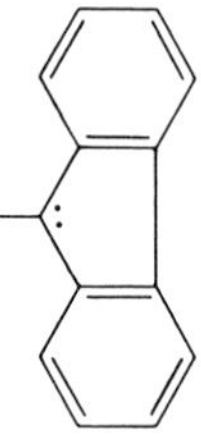 :	Dioxane	214 (4.65), 248 (4.85), 296 (3.91), 308 (3.91), 380 (4.36), 394 (4.36)	35
(p-CH$_3$OC$_6$H$_4$)(C$_6$H$_5$)$_2$PN=N—$\ddot{C}$(C$_6$H$_5$)(C$_6$H$_4$NO$_2$-p)	Dioxane	241 (4.44), 243 (4.44), (310) (~3.8), 427 (4.25)	35
(p-CH$_3$OC$_6$H$_4$)(C$_6$H$_5$)$_2$P—N=N—$\ddot{C}$(C$_6$H$_5$)$_2$	Dioxane	240 (4.68), 244 (4.83), 342 (4.08)	35
[p-(CH$_3$)$_2$NC$_6$H$_4$]$_4$P$^{(+)}$I$^{(-)}$		301 (5.01)	74
cyclo-(n-C$_8$H$_{17}$P)$_4$	Gas	(275), (281), (290)	43

Compound	Solvent[a]	λ_{max}, mμ[b] (log ϵ)	Ref.[c]
(9-fluorenylidene)$=CHCH=P(C_6H_5)_3$	CH_2Cl_2	445 (4.34)	33
$(p\text{-}(CH_3)_2NC_6H_4)(C_6H_5)_2PN=N—\ddot{C}(C_6H_4Cl\text{-}p)(C_6H_5)$	Dioxane	245 (4.70), 247 (4.68), 286 (4.43), 348 (4.13)	35
$[p\text{-}(CH_3)_2NC_6H_4]\text{-}(C_6H_5)_2P—N=N—$(9-fluorenylidene carbene):	Dioxane	215 (4.59), 248 (4.68), 392 (4.28), 398 (4.28)	35
$(p\text{-}(CH_3)_2NC_6H_4)(C_6H_5)_2P—N=N\ddot{C}(C_6H_5)(C_6H_4NO_2\text{-}p)$	Dioxane	246 (4.23), 251 (4.32), 255 (4.33), 283 (4.44), (314) (~3.7), 416 (4.26)	35
1,2,3,4,5-pentaphenylphosphole oxide (H_5C_6, C_6H_5, H_5C_6, P, C_6H_5, H_5C_6, O)	CH	260 (4.88), 272 (4.92), 385 (4.60)	13
		528 fluorescence	13
1,2,3,4,5-pentaphenylphosphole (H_5C_6, C_6H_5, H_5C_6, P, C_6H_5, C_6H_5)	CH	248 (4.51), 320 (3.94), 358 (3.94)	13
		480 fluorescence	13
2,3,4,5,6-pentaphenylphosphabenzene (C_6H_5, H_5C_6, C_6H_5, H_5C_6, P, C_6H_5)	EtOH	258 (4.51), 283 (sh) (4.41)	62
(9-fluorenylidene)$=CHCH=CHCH$ $\Vert$ $(C_6H_5)_3P$	CH_2Cl_2	531 (4.47)	33

Compound	Solvent[a]	λ_{max}, mμ[b] (log ε)	Ref.[c]
9-fluorenylidene=CHCH=CH–CH$_2$–$P^+(C_6H_5)_3$ ClO_4^- (structure)	CH_2Cl_2	355 (4.55)	33
$(p\text{-}ClC_6H_4)(C_6H_5)_3PN{=}N{-}\ddot{C}{-}(C_6H_5)(C_6H_4NO_2\text{-}p)$	Dioxane	(310) (~3.9), 417 (4.22)	35
$(C_6H_5)_3P{=}C{=}P(C_6H_5)_3$	CH	225 (4.5), 258 (3.8), 325 (3.85)	68
$(C_6H_5)_3P^+{-}CH{=}P(C_6H_5)_3$ $Br^{(-)}$		268 (3.90)	66
$(C_6H_5)_3P^+{-}CH{=}P(C_6H_5)_3$ $I^{(-)}$		267 (4.03)	66
$[(C_6H_5)_3P^+]_2\ddot{C}HBr^{(-)}$	EtOH	268 (3.90)	31
$(C_6H_5)_3P^+{-}CH{=}CH{-}CH{=}P(C_6H_5)_3$ $I^{(-)}$		267 (4.34), 305 (4.15), 345 (4.11)	66
$(C_6H_5)_3P^+{-}CH{=}CH{-}CH{=}P(C_6H_5)_3$ $Br^{(-)}$		267 (4.12), 272 (4.12), 290 (4.01), 295 (4.01)	66
1,1,2,4,4,5-hexaphenyl-1,4-diphosphoniacyclohexa-2,5-diene $2Br^-$ (structure: $(H_5C_6)_2P^+$, C_6H_5, H_5C_6, $P^+(C_6H_5)_2$)	MeOH	267 (4.27)	1
1,1,2,4,4,5-hexaphenyl-1,4-diphosphacyclohexane (structure: $(H_5C_6)_2P$, C_6H_5, H_5C_6, $P(C_6H_5)_2$)	MeOH	269 (3.41)	1
$p\text{-}CH_3C_6H_4{-}P(n\text{-}C_6H_{13}){-}C_{19}H_{38}{-}P(n\text{-}C_6H_{13}){-}C_6H_4CH_3\text{-}p$	EtOH	256 (3.73)	75

Compound	Solvent[a]	λ_{max}, mμ[b] (log ϵ)	Ref.[c]
(fluorenylidene=CH)$_2$C=P$(C_6H_5)_3$	CH_2Cl_2	574 (4.50)	33
tetraphenylcyclopentadienide–As$^+(C_6H_5)_3$	$CHCl_3$	299 (4.42), 335 (sh)	55
tetraphenylcyclopentadienide–Sb$^+(C_6H_5)_3$	$CHCl_3$	349 (4.62)	54
(fluorenylidene=CH)$_2$C=CHCH=P$(C_6H_5)_3$	CH_2Cl_2	554 (4.61)	33
(fluorenylidene=CH)$_2$C=CHCH$_2$P$^+(C_6H_5)_3$	CH_2Cl_2	342 (4.43)	33
$[(C_6H_5)_3P^+]_2\ddot{C}—B(C_6H_5)_3$	CH_3CN	267 (3.97)	31
$(KO)_2P(O)N_2P(O)(OK)_2$	H_2O	(250) (sh) (2.5), 320 (sh) (0.6), 532 (1.04)	9

Addendum

Recent studies [43] of the temperature dependence of the NMR spectrum of allyllithium in THF or ether are not consistent with rapid equilibration of the type $LiC^*H_2CH{=}CH_2 \rightleftarrows C^*H_2{=}CH{-}CH_2Li$. Rather the results suggest a structure of the type **A1a**, in which case the observed absorption at 310–315 mμ would be characteristic of the free anion. A structure of the type **A1b** however also seems possible to this author.

Häfelinger and Streitwieser [24] have examined the agreement for 20 carbanions between observed transition energies for cesium salts in cyclohexylamine, and those predicted by simple Hückel molecular orbital calculations. A reasonably good linear relationship is found, $\Delta E = 20{,}270\ \Delta m + 2520/\text{cm}$, where Δm is the Hückel molecular orbital energy difference in units of β. Only the ions **A2** and **A3** deviate significantly from the regression line and the assumption that the transition energies of the solvent separated ions in cyclohexylamine closely approximate, or at least parallel, those of the free carbanion seem justified.

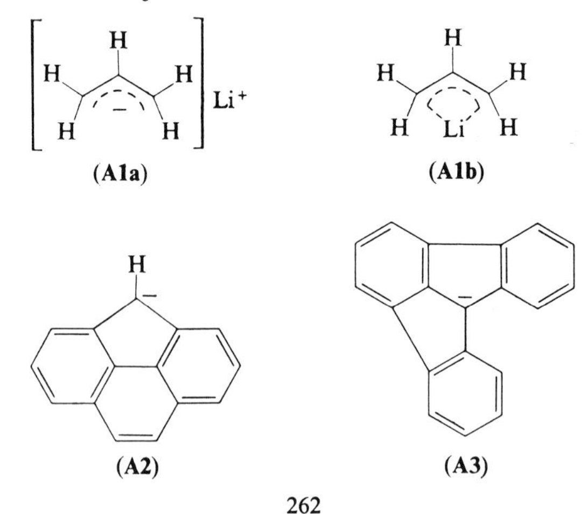

Hammons [26] has studied the effect of substituent on the ultraviolet spectra of the lithium salts (in cyclohexylamine) of the five isomeric methylfluorenes, the five methoxyfluorenes, 2-fluorofluorene, and 2-dimethylfluorene. These lithium salts exhibit in their spectra an absorption band with three maxima between 435 and 555 mμ which may be assumed to arise from a transition between the lowest vacant and highest filled π molecular orbitals.

In contrast with the results of Waack and Doran on the allyl- and benzyllithiums discussed in Chapter II, the effect of methyl substitution on the transition energy of fluorenyllithium is qualitatively predicted by consideration of the simple inductive effect and the expression $h\,\Delta\nu = (C_{jr}^2 - C_{ir}^2)\,\delta\alpha_r$ for the $i \rightarrow j$ transition.

Except for 9-methoxyfluorenyllithium, simple Hückel molecular orbital calculations correctly predict the direction of the methoxy substituent effect, but the *magnitudes* of the calculated blue shifts are too small, especially if corrected for the methoxy inductive effect. Hammons therefore concludes that the $+I_\pi$ inductive effect postulated by Clark *et al.* [10] is a major interaction mechanism of the methoxy group. According to the $+I_\pi$ effect any transition in which π electron density is increased at atom r will be shifted to higher energy by an electron lone pair substituent at r because of lone pair electron–electron repulsion. Conversely, if electron density is decreased at r in the transition, the $+I_\pi$ effect results in a lower transition energy. The use of the $+I_\pi$ effect concept correctly predicts the methoxy substituent effect on the transition energies of the methoxyfluorenyllithiums in *all* cases. The dimethylamino group is a much better π electron donor than methoxy, yet produces almost the same spectral shift as methoxy in the 2-substituted fluorenyllithium.

Absorption maxima are as follows: fluorenyl substituent, λ_{max}, mμ (ϵ); H, 480 (1220); 1-CH_3, 472 (1500); 1-CH_3O, 461 (1620); 2-CH_3, 476 (982); 2-CH_3O, 461 (1030); 2-F, 470 (973); 2 $(CH_3)_2N$, 462 (937); 3-CH_3, 489 (1210); 3-CH_3O, 497 (1260); 4-CH_3, 465 (1480); 4-CH_3O, 448 (1780); 9-CH_3, 514 (1080); 9-CH_3O, 513 (1100).

New heteroatom aromatic molecules containing boron continue to appear in the literature, such as compounds **A3** [16], **A4** [17], and **A5** [31] whose spectrum resembles that of 4-hydroxypyridine. The absorption maximum of compound **A3** should be compared with those reported [33] for the phenelenium (perinaphthindenylium) ion, λ_{max} (log ϵ) 226 (4.51), 378 (sh) (4.21), 400 mμ (4.67).

Whereas it has been concluded earlier that compound **A6** behaved as a Bronsted acid, boron NMR studies reveal [13] that compound **A6** acts as a Lewis acid to form **A7**. The ultraviolet spectrum of compound **A7** very closely resembles that of 2-hydroxybiphenyl in a strong base, which would

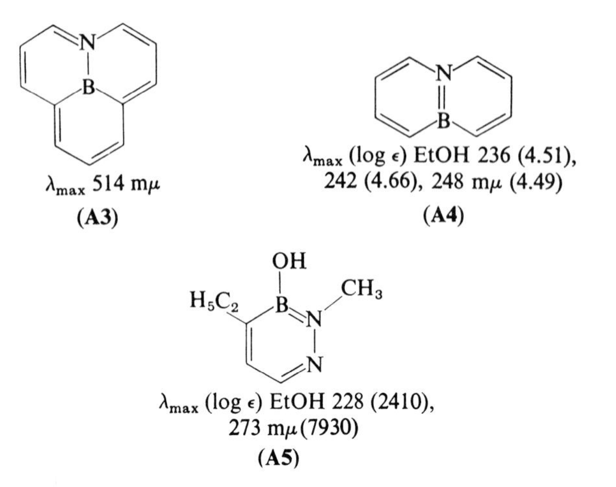

seem to imply that bonding to the —B—$(OH)_2$ group does not greatly change the electronegativity of —$O^{(-)}$. The 6-NO_2 and 8-NO_2 derivatives of the borazarophenanthene (**A8**), where R is OH or CH_3, react similarly with OH^- in EtOH to give compound **A9**, with a large red shift in the absorption maximum of the solution. Their use as indicators is suggested. For example, where R is CH_3, 6-NO_2-**8** in EtOH [λ_{max} 345 mμ (log ε 4.02)] with excess base forms, when R is CH_3, 6-NO_2-**9** [λ_{max} 457 mμ (log ε 4.25)].

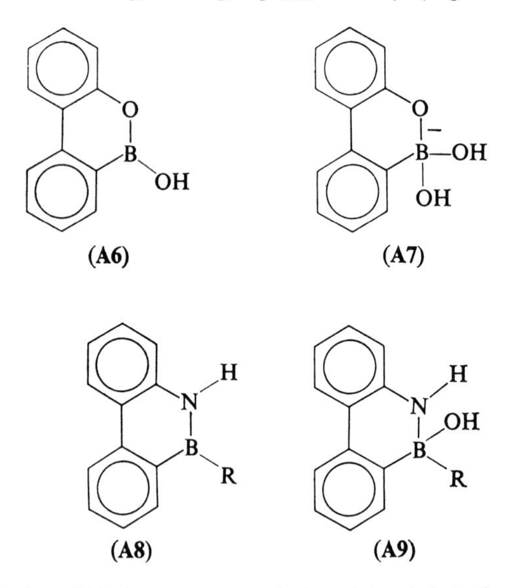

Soulie and Cadiot [41] have reported the ultraviolet absorption maxima of 21 pyridine complexes of $(C_6H_5)(Ar)BC{\equiv}CR$ and $ArB(C{\equiv}CR)_2$, where Ar is C_6H_5, p-$CH_3C_6H_4$, or p-$CH_3OC_6H_4$ and R is C_6H_5, —CH_3, —CH=

CH_2, 1-cyclopentenyl, 1-cyclohexenyl, 1-cyclopentadienyl, 1-cyclohexadienyl, etc. The ultraviolet spectra of a similar variety of acetylenic boron amines and esters have also been reported by Soulie and Cadiot [42]. These possess the general structure $RC{\equiv}CB(O\text{-alkyl})_2$ or $RC{\equiv}CB[N(\text{alkyl})_2]_2$, where R is $CH_2{=}CH(CH_3)$—, C_6H_5, 1-cyclohexenyl, and $RR'BN(\text{alkyl})_2$, where R is $CH_3C{\equiv}C$— or $CH_2{=}C(CH_3)C{\equiv}C$— and R′ is R or $n\text{-}C_4H_9$. Some representative results are indicated below. Note that the greater tendency for N than O to back bond to boron ($B^-{=}X^+$), results in a blue shift of the spectrum of compound **A13** relative to **A11**.

$CH_2{=}C(CH_3){-}C{\equiv}C{-}B(OC_4H_9)_2$
λ_{max} (ϵ) in CH
225 (1.0×10^4), 235 (1.6×10^4), 245 mμ (1.2×10^4)
(A10)

$C_6H_5C{\equiv}C{-}B(OCH_3)_2$
λ_{max} (ϵ) in CH
240 (1.4×10^4), 255 (1.7×10^4), 260 mμ (1.3×10^4)
(A11)

$CH_2{=}C(CH_3)C{\equiv}CB[N(C_2H_5)_2]_2$
λ_{max} in CH
230, 245 mμ (ϵ 2.2×10^4)
(A12)

$C_6H_5C{\equiv}CB[N(CH_3)_2]_2$
λ_{max} (ϵ) in CH
235 (1.5×10^4), 245 (1.7×10^4), 255 mμ (1×10^4)
(A13)

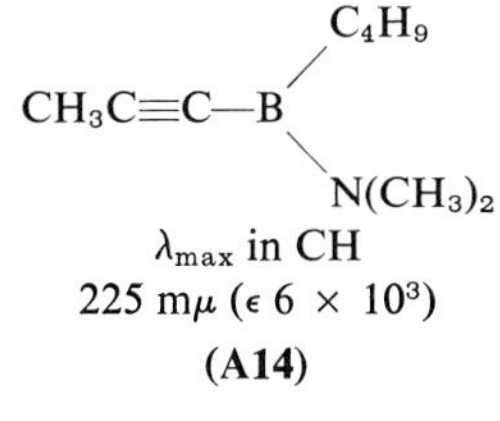

λ_{max} in CH
225 mμ (ϵ 6×10^3)
(A14)

$(CH_3C{\equiv}C)_2{-}BN(C_2H_5)_2$
λ_{max} (ϵ) in CH
220 (2×10^4), 245 mμ (1.7×10^4)
(A15)

$CH_3C{\equiv}CH$
λ_{max} 180 mμ (ϵ 4×10^3)
(A16)

$CH_3CH{=}CH{-}C{\equiv}CH$
λ_{max} 215 mμ (ϵ 8×10^3)
(A17)

Self-consistent field molecular orbital calculations including all valence shell electrons (Pople-Segal-Santry CNDO) calculations have been carried out on $(CH_2{=}CH)_nB(CH_3)_{3-n}$ by Armstrong and Perkins [2] with the conclusion that the first observed weak transition in these molecules is $\sigma_{B-C} \rightarrow \pi^*$, in agreement with the earlier assignment by Ramsey for $(CH_3)_3B$. The agreement between observed and calculated transition energy is best for $(CH_3)_3B$ (obs. 5.4–5.6 eV, calc. 5.4 eV) and worst for $(C_2H_3)_3B$ (obs. 7.40 eV, calc. 5.5 eV).

In order to evaluate the possibility of 1,3 *d–d* interactions, Pitt and coworkers [35] have measured the ultraviolet absorption spectra and the ionization potentials of a variety of geminal carbodisilanes (Si—C—Si).

Clearly the transition energy of $(CH_3)_3Si[Si(CH_3)_2]Si(CH_3)_3$ (λ_{max} 216.3 mμ) is not significantly different from that of $(CH_3)_3Si[Si(CH_3)_2]_2CH_2Si(CH_3)_3$ (λ_{max} 217.7 mμ). Although relative to the first absorption maximum of $(CH_3)_4Si$ at 167 mμ, the spectra of the cyclic carbosilanes **A18** (λ_{max} 198 mμ, ϵ 5430) and **A19** (λ_{max} 212 mμ, ϵ 7750) exhibit a large red shift, the relative ionization potentials for $(CH_3)_4Si$ (9.8 eV), **A18** (8.6 eV), and **A19** (7.8 eV) suggest that the lower transition energy is the result of ring strain in the ground state, not 1,3 *d–d* interactions. In support of this, one finds that even dimethylsilacyclobutane (ionization potential 8.8 eV) shows significant end absorption in the region of 195 mμ. A comparison of the spectra of phenyl derivatives such as **A20** (λ_{max} 220.0, 260.3 mμ) and **A21** (λ_{max} 219.7, 260.3 mμ) leads to a similar conclusion that ring strain and Si orbital rehybridization, not 1,3 *d–d* interaction, are responsible for the red shift relative to $C_6H_5Si(CH_3)_3$ transition energies. Finally, trimethylsilylmethyl methyl sulfide exhibits a transition near 202.5 mμ, indicating that any 1,3 *d* orbital interaction between Si and S is more than outweighed by 1,3 *p–d* σ bonding with the sulfur lone pair.

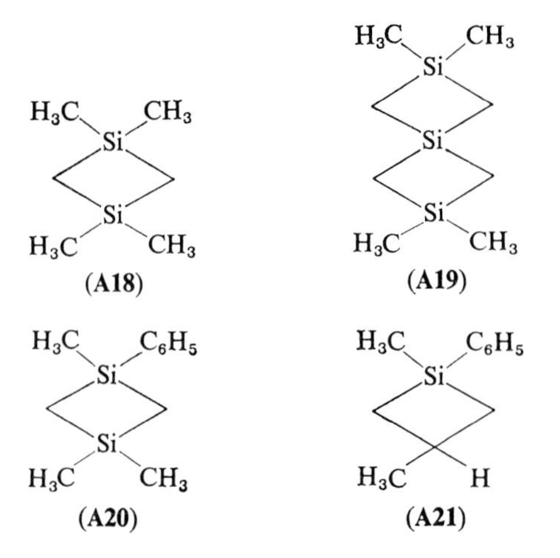

Recent calculations of the electronic transition energies of cyclopentadiene have been made by Del Bene and Jaffe [15] using a modified CNDO method. Their results indicate that the lowest energy transition is $^1A_1 \rightarrow {}^1B_2$ in which the excited state is 98% a single configuration of electron promotion from the highest filled π orbital to the lowest vacant π^*. The lowest vacant π orbital of cyclopentadiene is nodal at carbon atom 1, i.e., CH_2. If the Bene and Jaffe description of the first cyclopentadiene $\pi \rightarrow \pi^*$ transition is correct, the absence of a red shift in the spectrum of 1-trimethylsilylcyclopentadiene is easily understandable on this basis without the previous speculations involving the role of Si 3*d* orbitals in allylic silanes.

The $n \rightarrow \pi^*$ transition of $(CH_3)_3SiN{=}NSi(CH_3)_3$ (λ_{max} 786 mμ in ether) [44] represents the largest red shift relative to that of azoalkanes reported.

The very excellent work of Bock and Seidl continues to offer more definitive answers to the question of how important is the intereaction of Si d orbitals with π systems. The smaller ionization potentials [5], for example, of $(CH_3)_3CCH{=}CH_2$ (9.62 eV) and $(CH_3)_3CCH{=}CHC(CH_3)_3$ (8.99 eV), relative to $(CH_3)_3SiCH{=}CH_2$ (9.82 eV) and $(CH_3)_3SiCH{=}CHSi(CH_3)_3$ (9.32 eV), clearly suggest *d–p* π bonding, or its equivalent, in the ground state of silyl alkenes which lowers the energy of the π electrons, especially when one considers that the greater inductive effect of $(CH_3)_3Si$ should greatly decrease the ionization potentials. The carbon–carbon double bond stretching frequencies and vinyl proton chemical shifts are also in accord with significant ground state contribution by the resonance form $R_3Si^-{=}CR{-}C^+R_2$. Relevant ultraviolet spectroscopic data are given in Table AI.

TABLE AI

ABSORPTION MAXIMA OF ALKYL- AND SILYLETHYLENES IN HEXANE (R IS CH_3)

Compound	λ^1_{max}, mμ (ϵ)	λ^2_{max}, mμ (ϵ)	λ^3_{max}, mμ (ϵ)
$R_3SiCH{=}CH_2$		179 (17,000)	
$R_3CCH{=}CH_2$		178 (14,000)	
$(R_3Si)_2C{=}CH_2$	236 (300)	192 (13,000)	
trans-$R_3SiCH{=}CHSiR_3$	225 (650)	196 (20,100)	
cis-$R_3SiCH{=}CHSiR_3$	233 (920)	198 (9600)	
trans-$R_3SiCH{=}CHCR_3$		188 (12,000)	
trans-$R_3CCH{=}CHCR_3$		183 (16,050)	
trans-$R_3SiCH_2CH{=}CHCH_2SiR_3$		205 (12,150)	
$(R_3Si)_2C{=}CHSiR_3$	280 (350)	211 (17,000)	182 (14,400)
$(R_3Si)_2C{=}CHCR_3$	243 (650)	200 (13,500)	
$(R_3Si)(R_3C)C{=}CHCHR_2$		190 (11,800)	

The red shift of the olefin $\pi \rightarrow \pi^*$ transition (λ^2_{max}) with increasing R_3Si substitution is attributed to important *p–d* π interaction with the C=C π^* orbital. What is most interesting in Table AI is the very large red shift of the weak olefin mystery band (λ^1_{max}) which normally occurs in simple alkenes near 210 mμ but is found in the di- and trimethylsilylethylenes at as low as 280 mμ. This band in alkenes is most probably either a $\sigma \rightarrow \pi^*$ or $\pi \rightarrow \pi^*$ transition. Since, as the ionization potential data show, R_3Si substitution *lowers* the energy of the π orbital, and the lowest vacant σ^* orbital should not be very sensitive to the number of R_3Si substituents beyond the first, one would expect, if anything, perhaps a blue shift of a $\pi \rightarrow \sigma^*$ transition. We suggest therefore that the λ^1_{max} (Table AI) of the silylethylenes should be assigned to the $\sigma \rightarrow \pi^*$ transition. The 182 mμ transition of $(R_3Si)_2C{=}CHSiR_3$ may be either a second $\pi \rightarrow \pi^*$ or possibly $\sigma \rightarrow \pi^*$, $\pi \rightarrow \sigma^*$, or even $\sigma \rightarrow \sigma^*$ transition.

The importance of *d–p* π interactions was further confirmed in the study [6] of the $(CH_3)_3Si$-substituted butadienes where half-wave reduction potentials were more generally available. As would be expected, if *d–p* π^* interactions were important, the reduction potentials for $(CH_3)_3Si$-substituted butadienes were greater than those for the corresponding alkyl-substituted butadienes. Again, NMR and IR studies could also be interpreted on the basis of π electron donation to the R_3Si group. The ultraviolet spectra of di-1,4-, di-1,3-, di-1,2-, and di-2,3-trimethylsilylbutadiene, tetra-1,1,3,4-trimethylsilylbutadiene and tetra-1-trimethylsilyl-4-*tert*-butylbutadiene are reported [6]. Again, significant *p–d* π interaction results in a lower $\pi \rightarrow \pi^*$ transition energy for the silyl derivatives except in the case of 2,3-substitution, for example, $(CH_3)_3SiCH{=}CHCH{=}CHSi(CH_3)_3$ (λ_{max} 245 mμ) versus $(CH_3)_3CCH{=}CHCH{=}CHC(CH_3)_3$ (λ_{max} 228 mμ). The bis-2,3-trimethylsilylbutadiene absorption maximum (224 mμ) is at a shorter wavelength and lower intensity (ϵ 4670) than $CH_2{=}C(CH_3)C(CH_3){=}CH_2$ (227 mμ) (ϵ 23,000), the result presumably of a steric effect by the bulky $(CH_3)_3Si$ group which prevents coplanarity of the carbon–carbon double bonds.

The vertical ionization potential [7] of $(CH_3)_3SiC{\equiv}CH$ (10.14 eV) is slightly less than that of $(CH_3)_3CC{\equiv}CH$ (10.31 eV), but in the disubstituted acetylenes the silyl derivatives have greater ionization potentials than their carbon analogs: $(CH_3)_3SiC{\equiv}CSi(CH_3)_3$ (9.63 eV), $(CH_3)_3CC{\equiv}CC(CH_3)_3$ (9.19 eV) and $(CH_3)_3Si{-}({-}C{\equiv}C{-}){-}_2Si(CH_3)_3$ (9.23 eV), $(CH_3)_3C{-}({-}C{\equiv}C{-}){-}_2{-}C(CH_3)_3$ (8.82 eV). Triple bond IR stretching frequencies are lower in silylacetylenes than alkylacetylenes [7]. The electronic transitions in hexane of $(CH_3)_3Si{-}({-}C{\equiv}C{-}){-}_2Si(CH_3)_3$ [λ_{max} (ϵ), 180 (23,600) 181 (sh) (41,000), 195 (89,600), 202 (142,000), 225 (63), 236 (147), 238 (270), 263 (362), 278 (256), 294 mμ (10)], $(CH_3)_3Si{-}({-}C{\equiv}C{-}){-}_3Si(CH_3)_3$ [λ_{max} (ϵ), 200 (16,300), 208 (65,000), 218 (156,300), 229 (21,500), 273 (1060), 289 (1250), 309 (228), 330 mμ (147)], and $(CH_3)_3SiCH_2{-}({-}C{\equiv}C{-}){-}_2CH_2Si(CH_3)_3$ [λ_{max} (ϵ), 187 (74,000), 225, (1038), 238 (1434), 249 (1666), 211 mμ (1038)] are at 3000–4000 cm^{-1} lower in energy than those of the corresponding carbon analogs [7]. The weak first transition is assigned to $\Sigma_g{}^+ \rightarrow {}^1A_u$, and the second transition to $\Sigma_g{}^+ \rightarrow \Sigma_u{}^+$. Eastmond and Walton [21] have observed similar differences in the long wavelength transition energies of $C_6H_5(C{\equiv}C)_3Si(C_2H_5)_3$ and $C_6H_5{-}({-}C{\equiv}C{-}){-}_3H$.

The relative vertical ionization potentials [8] of $C_6H_5Si(CH_3)_3$ (8.72 eV) and $C_6H_5C(CH_3)_3$ (8.34 eV) imply again that $(CH_3)_3Si$ is π electron withdrawing, even in the ground state.

Curtis *et al.* [12] have found that the transition energy of the *p* band near 250–260 mμ in the ultraviolet spectra of biphenyls 4-$(CH_3)_3MC_6H_4$—C_6H_5 and 4,4′-$(CH_3)_3MC_6H_4$—$C_6H_4M(CH_3)_3$, where M is C, Si, or Ge, correlate well with the transition energy calculated by simple Hückel molecular orbital methods. The parameters used for the heteroatoms Si and Ge were those previously derived from the ESR spectra of the biphenyl anion radicals. Where $\alpha_M = \alpha^\circ + h_M\beta^\circ$ and $\beta_{CM} = k_{CM}\beta^\circ$, $h_{Si} = -1.20$, $h_{Ge} = -1.05$ and $k_{Si} = 0.450$, $k_{Ge} = 0.30$.

TABLE AII

ULTRAVIOLET SPECTRA OF R—$(C_6H_4)_n$—R′ IN HEPTANE

n	R	R′	λ_{max}, mμ ($\epsilon \times 10^{-3}$)	
2	H	H	247.1 (16.8)	202.0 (38.8)
2	H	$C(CH_3)_3$	252.4 (20.8)	202.0 (46.1)
2	H	$Si(CH_3)_3$	255.0 (23.4)	202.0 (45.4)
2	H	$Ge(CH_3)_3$	253.6 (22.9)	202.0 (45.4)
2	H	$Sn(CH_3)_3$	254.5 (25.1)	202.5 (49.9)
2	$(CH_3)_3C$	$(CH_3)_3C$	256.8 (24.0)	203.0 (38.1)
2	$(CH_3)_3Si$	$(CH_3)_3Si$	262.2 (30.3)	202.5 (47.5)
2	$(CH_3)_3Ge$	$(CH_3)_3Ge$	260.0 (29.5)	203.5 (38.9)
2	$(CH_3)_3Sn$	$(CH_3)_3Sn$	262.4 (32.7)	
3	$(CH_3)_3Si$	$(CH_3)_3Si$	287.0 (39.0)	
3	$(CH_3)_3Ge$	$(CH_3)_3Ge$	286.0 (44.0)	
4	$(CH_3)_3Si$	$(CH_3)_3Si$	301.0 (55.0)	
4	$(CH_3)_3Ge$	$(CH_3)_3Ge$	301.0 (57.0)	

As may be seen from Table AII the biphenyl *p* bands, which correspond to the ^1L_a benzene transition, are moderately sensitive to substitution in the 4 and 4′ positions. The *p* band is derived from a transition between the highest filled and lowest vacant molecular orbitals. The β band near 202 mμ is insensitive to 4 and 4′ substitution. The β band may be considered as the result of configuration interaction between nearly degenerate transitions from the highest filled to second lowest vacant (degenerate) and second highest filled (degenerate) to lowest vacant molecular orbitals. The second highest filled and lowest vacant biphenyl molecular orbitals are also nodal at the 4 and 4′ positions. These factors make the β band relatively insensitive to 4,4′ substitution.

Allred and Bush have also reported [1] the ultraviolet spectra for the series $(CH_3)_3M(-C_6H_4-)_nM(CH_3)_3$ where M is Si and Ge and $n = 1, 2, 3, 4$.

The absorption [1] maxima of di-1,4-$(CH_3)_3$M-naphthalene, where M is Si or Ge, occur near 298–300 mμ (ε 8000) with vibrational fine structure.

The ^1L_b transition intensities of $C_6H_5GeH_3$ and *p*-$CH_3OC_6H_4GeH_3$ indicate predominate *p–d* π electron withdrawal by GeH_3 [29b].

The ultraviolet spectra in ethanol of the 5-trimethylsilyldihydroindoles **A22** and -indoles **A23** have been reported by Belsky and co-workers [3].

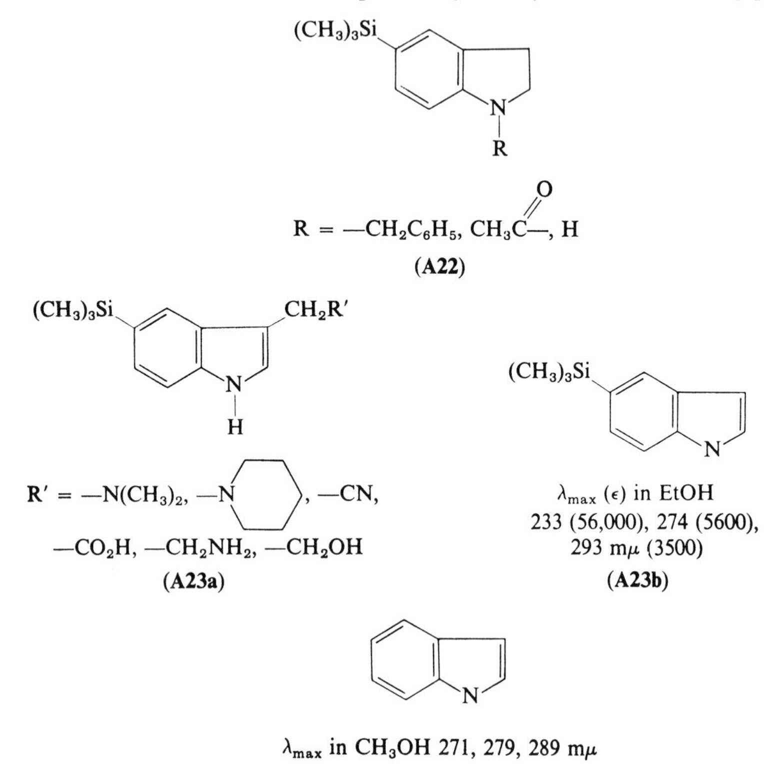

λ_{max} in CH_3OH 271, 279, 289 mμ

(A24)

The absorption maxima in cyclohexane of *cyclo*-$(C_6H_5)_8Si_4$ (205, 234, 270 mμ) and *cyclo*-$(C_6H_5)_{10}Si_5$ (205, 230 (sh), 235 mμ) have been reported by M'hirsi and Brini [30].

Simple Hückel molecular orbital calculations agree well with observed [4] charge transfer transitions in alkoxy- and siloxybenzene and -biphenyl-tetracyanoethylene complexes. This is considered to be support for postulated Si(*d*)–O(lone pair) bonding.

The $n \rightarrow \pi^*$ transition energies of $(R_3M\!-\!)_2C{=}O$, where R is alkyl or phenyl and M is Si and Ge, are suitably shifted toward the red to lower energy in comparison with $R_3MC(O)R$. The following are representative data from the work of Brook and co-workers [9]: compound, λ_{max} (ε); $[(CH_3)_3$-

$Si]_2C{=}O$, 472, 505, 537 mμ; $[(C_2H_5)_3Ge]_2C{=}O$, 464, 495, 527 mμ; $[(C_6H_5)_3Ge]_2C{=}O$ 455 (18), 485 (43), 515 mμ (66); $(C_2H_5)_3GeC(O)Si(CH_3)_3$, 535 m$\mu$.

The $n \rightarrow \pi^*$ absorption maxima in the vapor phase spectra of the silyl esters HCO_2SiH_3 (220 mμ), $CH_3CO_2SiH_3$ (214 mμ), and $CF_3CO_2SiH_3$ (217 mμ) aie found [22] at slightly longer wavelengths than the maxima of the corresponding methyl esters, HCO_2CH_3 (219 mμ), $CH_3CO_2CH_3$ (208 mμ), and $CF_3CO_2CH_3$ (215 mμ). A red shift of this magnitude might be attributable to either an inductive effect on the carbonyl oxygen lone pair electrons, or to a lowering of the π^* orbital (which must contain some contribution from the acyl oxygen) by interaction with a Si 3*d* orbital. Still a third alternative may be found in steric effects such as those suggested by Closson and Haug to account for the order of λ_{max} $n \rightarrow \pi^*$ in acetate esters: $CH_3CO_2CH_3$, 210 mμ; $CH_3CO_2C_2H_5$, 211 mμ; CH_3CO_2*i*-Pr, 212 mμ; $CH_3CO_2C(CH_3)_3$, 216 mμ. Inasmuch as the greatest red shift is between the isopropyl and *tert*-butyl esters, Closson and Haug suggested than an inductive explanation was improbable. Rather, they suggested that a direct steric interaction of the acyl R group might simultaneously increase the *n* electron energy, and at the same time, by forcing the OR group out of coplanarity with the C=O group, lower the π^* energy [11].

A moderate intensity "mystery band" often appears [34] in the ultraviolet spectra of $[(CH_3)_3Si]_2Si(CH_3)X$, where X is a lone pair electron donor: for example, X is F, 240 mμ (ϵ 2280); Cl, 233 (sh) mμ (ϵ 950); Br, 240 mμ (sh) (ϵ 370); OH, 242 mμ (ϵ 1540); OAC, 233 mμ (ϵ 3040); $OSi(CH_3)_3$, 236 mμ (ϵ 3260), all in isooctane. These mystery bands are in addition to the expected trisilane transitions which are found between 215 and 220 mμ (ϵ 5000–8000). Where X is H, CH_3, I, OC_2H_5, ONa, SH, or NH_2, the mystery band is absent.

No straightforward explanation for this 234–240 mμ absorption in the spectra of $[(CH_3)_3Si]_3Si(CH_3)X$, however, seems possible at this time, especially as any comparable transition is conspicuously absent in the spectra of the series $(CH_3)_3$ $Si[Si(CH_3)_2]_2X$; where X is F, 216.9 mμ (ϵ 6890), Cl, 217.0 mμ (ϵ 8,860), Br, 217 mμ (ϵ 7480), OH, 219.2 mμ (ϵ 6080); OAC, 215 mμ (ϵ 7250). The mystery band maximum seems remarkably independent of such factors as Si—X bond strength, or X ionization potential, which makes difficult obvious assignments to Si—$\ddot{X}$ σ or X *n* electron transitions to σ^* or $3d\pi$. Nor are matters particularly clariffed by the further observation [34] that the mystery band appears in the spectra of $[(CH_3)_3Si]_2Si(H)OCH_3$ (λ_{max} 245 mμ, ϵ 1225), $[(CH_3)_3Si]_2Si(OCH_3)_2$ (λ_{max} 234 mμ, ϵ 7420), $[(CH_3)_3Si]_2SiCl_2$ (λ_{max} 232 mμ, ϵ 3090), and $[(CH_3)_3Si]_3SiF$ (λ_{max} 250, 260, ϵ 2×10^3), but not in the spectrum of $[(CH_3)_3Si]_2SiF_2$ (λ_{max} 228, ϵ 11,480) unless it has been substantially shifted toward the blue.

Noltes [32] has kindly informed us that the transition energies of the series $C_2H_5[M(C_2H_5)_2]_nC_2H_5$, where M is Ge and Sn and $n = 2, 3, 4, 5$, and 6, also fit the simple Hückel relationship based on delocalized *d* orbitals proposed by Pitt *et al.* [35a], with the result that $\beta_{Si-Si} = \beta_{Ge-Ge} = \beta_{Sn-Sn} = -2.7$ eV. Noltes also points out that the independent systems model of Simpson *et al.* also explains qualitatively the shift to a longer wavelength with increasing chain length. The absorption maxima found are (n; λ_{max}, mμ) for Ge, *2*, 202 (sh); *3*, 217.5 (sh); *4*, 234 (sh); *5*, 248; *6*, 258 and for Sn, *2*, 322; *3*, 250 (sh); *4*, 290 (sh); *5*, 310; *6*, 325.

The absorption maxima in THF of $[(CH_3)_3Si]_3SiLi$ are reported by Gilman and Smith [23] at 236 mμ (ϵ 6000), 295 mμ (ϵ 22,000), and 370 mμ (ϵ 10,000).

The ultraviolet spectrum [36] of 1-methylphosphole (**A25**) in isooctane (λ_{max} 286 mμ, log ϵ 3.88) agrees well with that of 1-methylpyrole, and the spectrum [14] of compound **A26** resembles the spectrum of phenanthridine (**A27**) more closely than that of phenanthrene.

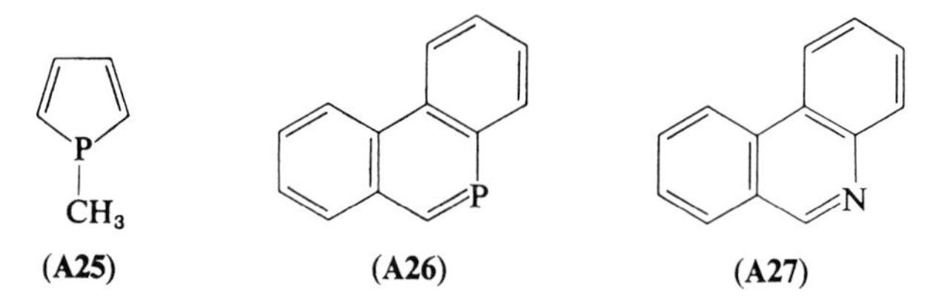

(**A25**) (**A26**) (**A27**)

The absorption maxima of compounds **A28**, **A29**, **A30**, and **A31** have been reported [19] as indicated.

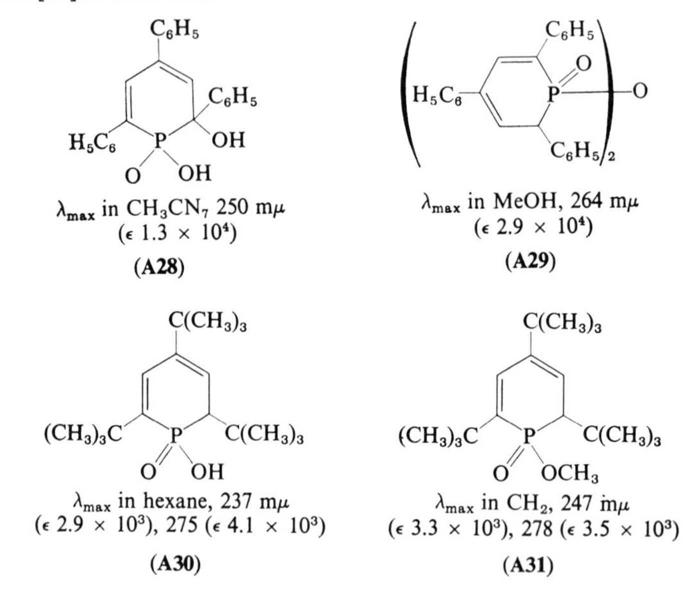

λ_{max} in CH_3CN_7 250 mμ (ϵ 1.3×10^4)

(**A28**)

λ_{max} in MeOH, 264 mμ (ϵ 2.9×10^4)

(**A29**)

λ_{max} in hexane, 237 mμ (ϵ 2.9×10^3), 275 (ϵ 4.1×10^3)

(**A30**)

λ_{max} in CH_2, 247 mμ (ϵ 3.3×10^3), 278 (ϵ 3.5×10^3)

(**A31**)

Absorption maxima have been reported by Märkl [28, 29] for phosphoranes **A32**, **A33**, and **A34**. Absorption maxima for the interesting phosphoranes **A35–38** are given by Shaw and co-workers [39].

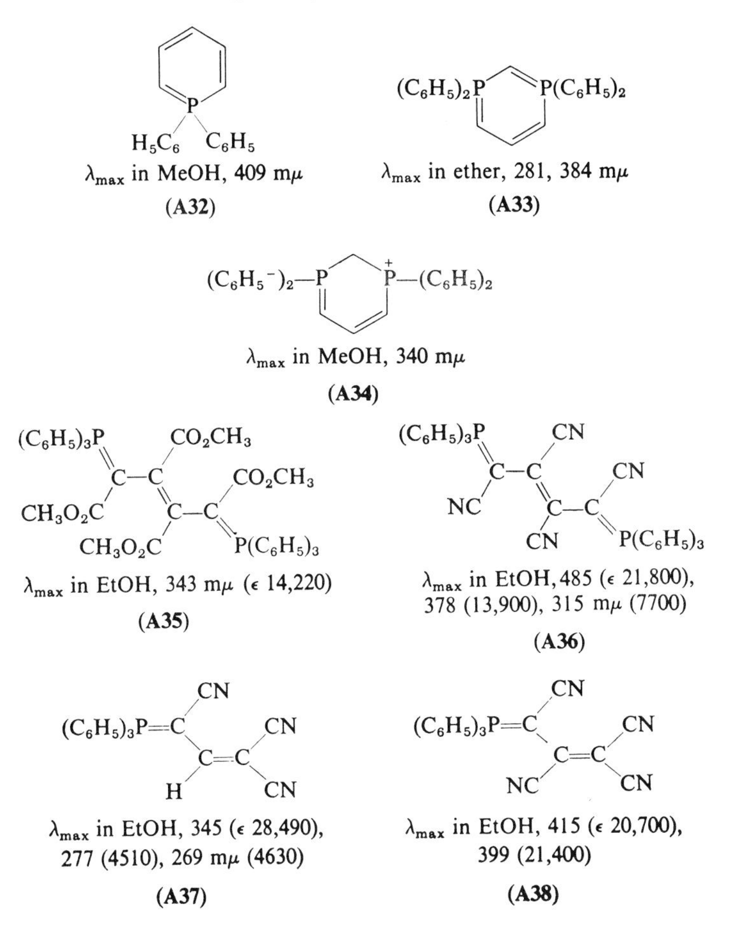

The ultraviolet spectra and measured dipole moments of $(C_6H_5)_3P$ and from the literature [25, 38] are given in Table AIII. Schindlbauer [38] explained the variation in dipole moment and ultraviolet absorption maxima of the isomeric $(CH_3C_6H_4)_3P$ on the basis of an increasing C—P—C bond angle in progression from *para*- to *meta*- to *ortho*-methyl. More recently Halpern and Mislow [25] have demonstrated that the observed dipole moments may easily be accounted for by assuming a constant 102° C—P—C bond angle but varying the angle of torsion around the phosphorus phenyl bond as a function of the methyl group position. This conclusion is strongly reinforced by the observed dipole moment of trimesitylphosphine which is

TABLE AIII. ULTRAVIOLET MAXIMA OF $(ARYL)_3P$

Aryl	Dipole moment (*D*)	λ_{max}, mμ (log ϵ)
C_6H_5	1.39–1.45	261 (4.02)
2-$CH_3C_6H_4$	0.53	275 (4.05)
3-$CH_3C_6H_4$	1.65	263 (4.04)
4-$CH_3C_6H_4$	2.14	262 (4.06)
2,4,6-$(CH_3)_2C_6H_2$	1.37	312 (4.20)

close to that of triphenylphosphine. The red shift of the tri *o*-tolylphosphine absorption maximum is attributed to either its calculated angle of twist, 36° out of the plane of the C_3 axis and C—P bond compared with 59° for triphenylphosphine, or some *ortho* inductive effect.

The latter hypothesis is held [25] more likely because of the still greater red shift observed for the trimesitylphosphine. The ^1L_a transition of mesitylene is some 0.6 eV lower in energy than the ^1L_a transition of benzene [37]. The 0.8 eV red shift of the trimesitylphosphine absorption relative to that of triphenylphosphine may be the result in large part of stronger configuration interaction between the ^1L_a excited state and a lower energy charge transfer state, as the result of better energy matching between states in the case of trimesitylphosphine.

Although the spectrum of $(C_6H_4)_4SbCl$ in EtOH is typically benzene-like λ_{max} (ϵ) 220 (sh) (40,900), 254 (sh), (1820), 258 (2210), 263.5 (2540) 270 mμ (1910), the spectra of $(C_6H_4)_4SbOR$, where R is CH_3, C_2H_5, *i*-C_3H_7, *n*-C_3H_7, or H, are without clearly defined maxima above 200 mμ [20]. The spectra of the alkoxytetraphenylantimony compounds instead exhibit a broad and intense shoulder near 200 mμ with shoulders and only inflection points in the ^1L_b region of the spectrum.

The alkoxytetraphenylantimony compounds are pentacovalent [40] and probably trigonal bipyramidal in structure. The difference between the spectra of $(C_6H_5)_4SbOR$ and $(C_6H_5)_4SbCl$ suggests that the latter is ionic with tetravalent antimony, i.e., $(C_2H_5)_4Sb^+Cl^-$.

The substantial perturbation of the benzene spectrum in $(C_6H_5)_4SbOR$ may be the result of steric crowding as found in $(C_6H_5)_4C$, or increased conjugation between phenyl rings due to a better geometry for Sb *d*–π orbital overlap.

The $n \rightarrow \pi^*$ transitions [27] of $CH_3C(O)P(CH_3)_2$ [λ_{max} (log ϵ), 347 (0.96), 359 (0.95), 371 (sh) mμ (0.80)] is shifted toward the blue in the spectra of **A39**

$$CH_3\overset{\overset{O}{\|}}{C}—{}^{+}\overset{\overset{O}{\|}}{P}(CH_3)_2 \qquad (\lambda_{max}\ 277\ m\mu,\ \log \epsilon\ 1.08)$$

(A39)

and $CH_3C(O)—^+P(CH_3)_3I^-$ (λ_{max} 325 mμ, log ϵ 1.66). Indeed the $n \rightarrow \pi^*$ transitions of $CH_3C(O)P(O)(CH_3)_2$ and $CH_3C(O)P^+(CH_3)_3$ are not much different from that of acetone at 279 mμ. The expected inductive effect of $—P^+R_3$ relative to CH_3 on the carbonyl, R—C(O)—, lone pair electrons which would lower their energy and increase the $n \rightarrow \pi^*$ transition energy is evidently counterbalanced by a decrease in π^* energy through phosphorus *d–p* π bonding. The spectra of $CH_3C(O)—P(O)(CH_3)_2$ and $CH_3C(O)—P(CH_3)_2$ have additional maxima at 224 (log ϵ 1.75) and 223 mμ (log ϵ 1.75), respectively. These were assigned [27] to $\pi \rightarrow \pi^*$ transitions, but seem too weak to this author for such an assignment. In the case of $CH_3C(O)P(CH_3)_2$ the 223 mμ transition could be assigned to a second $n \rightarrow \pi^*$ transition from the phosphorus lone pair of electrons. Replacement on phosphorus of methyl by phenyl in $CH_3C(O)P(CH_3)_2$ leads [27] to a red shift of the $n \rightarrow \pi^*$ transition, λ_{max} $CH_3C(O)P(C_6H_5)_2$ (log ϵ), 256 (2.08), 261 (2.18), 274 (2.05), 375 mμ (1.86).

Dimroth and Städe report that the spectra of compounds **A40** have two intense transitions in the region of 300 and 420 mμ, and fluoresce strongly [13a].

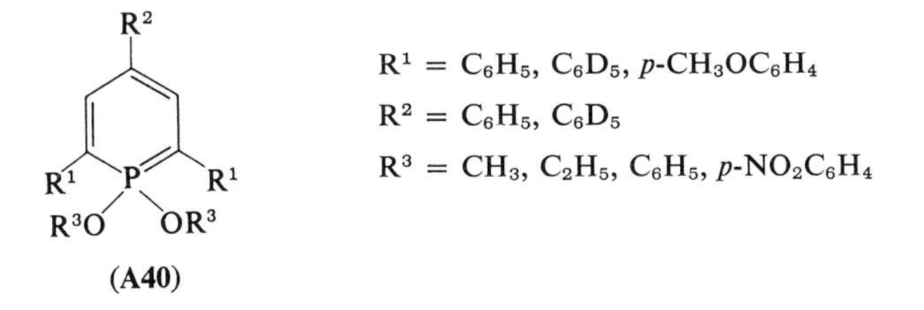

$R^1 = C_6H_5, C_6D_5$, *p*-$CH_3OC_6H_4$
$R^2 = C_6H_5, C_6D_5$
$R^3 = CH_3, C_2H_5, C_6H_5$, *p*-$NO_2C_6H_4$

(A40)

DeKoe and Bickelhaupt have reported the spectrum of 9-phenyl-10-phosphaanthracene [14a] at λ_{max} (log ϵ) 275 (4.5), 432 (3.92) mμ.

Krespan and co-workers have reported the spectrum of compound **A41** [27a] and of compounds **A42** and **A43** (27a]. These results may be compared with those of Märkl and Lieb for the compound **A44** [29a].

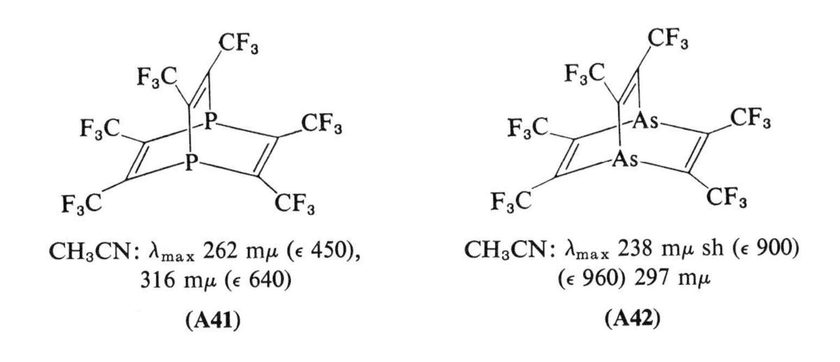

CH_3CN: λ_{max} 262 mμ (ϵ 450), 316 mμ (ϵ 640)

(A41)

CH_3CN: λ_{max} 238 mμ sh (ϵ 900) (ϵ 960) 297 mμ

(A42)

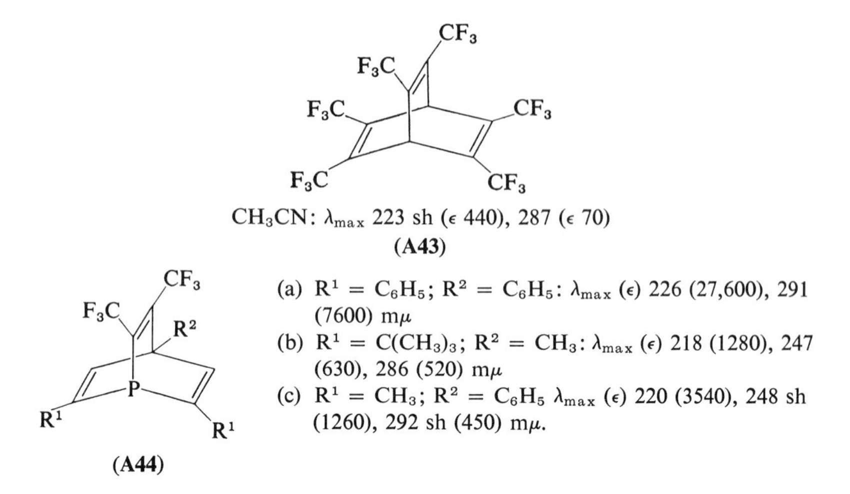

CH_3CN: λ_{max} 223 sh (ϵ 440), 287 (ϵ 70)

(A43)

(a) $R^1 = C_6H_5$; $R^2 = C_6H_5$: λ_{max} (ϵ) 226 (27,600), 291 (7600) mμ

(b) $R^1 = C(CH_3)_3$; $R^2 = CH_3$: λ_{max} (ϵ) 218 (1280), 247 (630), 286 (520) mμ

(c) $R^1 = CH_3$; $R^2 = C_6H_5$ λ_{max} (ϵ) 220 (3540), 248 sh (1260), 292 sh (450) mμ.

(A44)

Barrelene ([2.2.2]-bicyclooctatriene), which has been the subject of Pariser–Parr–Pople molecular orbital calculations by Wilcox *et al.* [45], absorbs in ethanol at λ_{max} 208 mμ (log ϵ 3.05) and 239 mμ (log ϵ 2.48), the poor agreement between $\pi \rightarrow \pi^*$ calculated and observed [46] transition energies for barrelene, suggests that the observed transitions may instead be related to the olefin $\sigma \rightarrow \pi^*$ ($\pi \rightarrow \sigma^*$) mystery band. The lone pair electrons of the compounds **A41**, **A42**, and **A44** are orthogonal to the π electron system of the olefin bonds, and we would therefore predict that in these systems the 235 mμ transition of $(C_2H_3)_3P$ or 227 mμ transition of $(C_2H_3)_3As$ would become weak and move to higher energy. This may be the origin of the 218 mμ transition of **A44b**. The long wavelength transitions of **A41**, **A42**, and **A44** possibly represent a large red shift of the olefin mystery bands of **A43** and barrelene, similar to that already noted for silylethylenes.

In addition to the 265 mμ benzene transition the simple phosphoranes $(C_6H_5)_3P{=}CHR$ (where R is H_1, CH_3, C_2H_5, n-C_3H_7), $(C_6H_5)_3P{=}C(CH_3)_2$, and $(C_6H_5)_3P{=}C(CH_2)_5$ exhibit [23a] a transition which moves from 341 mμ where R is H to 386 mμ in $(C_6H_5)_3P{=}C(CH_3)_2$. This red shift is in keeping with the expected alkyl inductive effect on p–d $\pi \rightarrow \pi^*$ transition (or $n \rightarrow 3d$). The P nmr chemical shifts correlate well with the spectroscopic changes in transition energy.

Very early work by Mohler [64, Chapter 5] reports absorption maxima for a number of chloroarsines as follows: λ_{max} (mμ) (log ϵ): $(C_6H_5)_2AsCl$ 270 (4.0); $(C_6H_5)_2AsCN$ 227 (4.4), 242 (4.0), 254 (3.4), 270 (3.0); 10-chloro-5,10-dihydrophenarsazine 198 (5.2), 220 (4.9); $C_2H_5AsCl_2$ 241 (3.25); $(ClCH{=}CH)_3As$ 207 (4.4); $(ClCH{=}CH)_2AsCl$ 209 (4.15); $(ClCH{=}CH)AsCl_2$ 214 (4.0). The difference in transition energy between $(ClCH{=}CH)_nAsCl_{3-n}$ and $(CH_2{=}CH)_nAs)C_2H_5)_{3-n}$ is 20 mμ in each case. This result suggests that the blue shift of the transition in the chlorovinylarsines is the result of a π resonance effect which raises the π^* energy and/or a $+ I_\pi$ effect of the vinyl chlorine lone pair electrons arising from increased electron repulsion with increased electron density on the β carbon in the excited state [10].

REFERENCES

1. A. L. Allred and L. W. Bush, *J. Am. Chem. Soc.* **90**, 3352 (1968).
2. D. R. Armstrong and P. G. Perkins, *Theoret. Chim. Acta* **9**, 412 (1968).
3. I. Belsky, D. Gertner, and A. Zilkha, *J. Org. Chem.* **33**, 1348 (1968).
4. H. Bock and H. Alt, *J. Organometall. Chem.* (*Amsterdam*) **13**, 103 (1968).
5. H. Bock and H. Seidl, *J. Organometall. Chem.* (*Amsterdam*) **13**, 87 (1968).
6. H. Bock and H. Seidl, *J. Am. Chem. Soc.* **90**, 5694 (1968).
7. H. Bock and H. Seidl, *J. Chem. Soc.* **1968**, 1158.
8. H. Bock and H. Seidl, *Chem. Ber.* **101**, 2815 (1968).
9. A. G. Brook, P. F. Jones, and G. J. D. Peddle, *Can. J. Chem.* **46**, 2120 (1968).
10. D. T. Clark, J. N. Murrell, and J. M. Tedder, *J. Org. Chem.* **33**, 1123 (1968).
11. W. D. Closson and P. Haug, *J. Am. Chem. Soc.* **86**, 2386 (1964).
12. M. Curtis, R. Lee and A. Allred, *J. Am. Chem. Soc.* **89**, 5150 (1967).
13. F. A. Davis and M. J. S. Dewar, *J. Org. Chem.* **33**, 3324 (1968).

13a. K. Dimroth and W. Städe, *Angew. Chem. Intern. Ed.* **7**, 881 (1968).

14. P. de Koe and F. Bickelhaupt, *Angew. Chem. Intern. Ed. Engl.* **7**, 465 (1968).

14a. P. de Koe and F. Bickelhaupt, *Angew. Chem. Intern. Ed.* **7**, 889 (1968).

15. J. Del Bene and H. H. Jaffe, *J. Chem. Phys.* **48**, 4050 (1968).
16. M. J. S. Dewar and R. Jones, *Tetrahedron Letters* **1968**, 2707.
17. M. J. S. Dewar and R. Jones, *J. Am. Chem. Soc.* **90**, 2137 (1968).
18. M. J. S. Dewar, R. Jones, and R. Logan, *J. Org. Chem.* **33**, 1353 (1968).
19. K. Dimroth, K. Vogel, W. Mach, and U. Schoeler, *Angew. Chem. Intern. Ed. Engl.* **7**, 371 (1968).
20. G. O. Doak, G. G. Land, and L. D. Freedman, *J. Organometall. Chem.* (*Amsterdam*) **12**, 443 (1968).
21. R. Eastmond and D. R. M. Walton, *Chem. Commun.* **1968**, 204.
22. E. A. V. Ebsworth and J. C. Thompson, *J. Chem. Soc. A* **1967**, 69.
23. H. Gilman and C. L. Smith, *J. Organometall. Chem.* (*Amsterdam*) **14**, 91 (1968).

23a. S. O. Grim and J. H. Ambrus, *J. Org. Chem.* **33**, 2993 (1968).

24. G. Häfelinger and A. Streitwieser, Jr., *Chem. Ber.* **101**, 2785 (1968).
25. E. J. Halpern and K. Mislow, *J. Am. Chem. Soc.* **89**, 5224 (1967).
26. J. H. Hammons, *J. Org. Chem.* **33**, 1123 (1968).
27. R. Kostyanovsky, V. Yakshin, and S. L. Zimont, *Tetrahedron* **24**, 2998 (1968).

27a. C. G. Krespan, *J. Am. Chem. Soc.* **83**, 3432 (1961).

27b. C. G. Krespan, B. C. McKusick, and T. L. Cairns, *J. Am. Chem. Soc.* **82**, 1515 (1960).

28. G. Märkl, *Angew. Chem. Intern. Ed. Engl.* **2**, 479 (1963).
29. G. Märkl, *Z. Naturforsch.* **18b**, 1136 (1963).

29a. G. Märkl and F. Lieb, *Angew. Chem. Intern. Ed.* **7**, 733 (1968).

29b. J. Meyer and A. L. Allred, *J. Phys. Chem.* **8**, 3043 (1968).

30. M. M'hirsi and M. Brini, *Bull. Soc. Chim. France* **1968**, 1509.
31. J. Namtvedt and S. Gronowitz, *Acta Chem. Scand.* **22**, 1373 (1968).
32. J. G. Noltes, Organisch Chemisch TNO Institut (Holland).

33. R. Pettit, *Chem. Ind. (London)* **1956**, 1306.
34. C. G. Pitt, private communication, Research Triangle Institute, 1968.
35. C. G. Pitt, M. S. Habercom, M. M. Bursey, and P. F. Rogerson, Meeting Am. Chem. Soc., 156th, Atlantic City, Sept. 1968.
35a. C. G. Pitt, R. Jones, and B. G. Ramsey, *J. Am. Chem. Soc.* **89**, 5471 (1967).
36. L. D. Quinn and J. G. Bryson, *J. Am. Chem. Soc.* **89**, 5985 (1967).
37. C. W. Rector, G. W. Schaeffer, and J. R. Platt, *J. Chem. Phys.* **17**, 460 (1949).
38. H. Schindlbauer, *Monatsh. Chem.* **96**, 1793 (1965).
39. M. A. Shaw, J. C. Tebby, R. S. Ward, and D. H. Williams, *J. Chem. Soc. C*, **1968**, 1609.
40. K. Shen, W. McEwen, S. LaPlaca, W. C. Hamilton, and A. Wolf, *J. Am. Chem. Soc.* **90**, 1718 (1968).
41. J. Soulie and P. Cadiot, *Bull. Soc. Chim. France* **1966**, 1981.
42. J. Soulie and P. Cadiot, *Bull. Soc. Chim. France* **1966**, 3846, 3850.
43. P. West, J. Purmont, and S. McKinley, *J. Am. Chem. Soc.* **90**, 797 (1968).
44. N. Wiberg, W. C. Joo, and W. Uhlenbrock, *Angew. Chem. Intern. Ed.* **7**, 640 (1968).
45. C. F. Wilcox, Jr., S. Winstein, and W. G. McMillan, *J. Am. Chem. Soc.* **82**, 5450 (1960).
46. H. Zimmerman and R. Pauffer, *J. Am. Chem. Soc.* **82**, 1514 (1960).

Author Index*

Numbers in parentheses are reference numbers and indicate that an author's work is referred to although his name is not cited in the text. Numbers in *italics* show the page on which the complete reference is listed.

*To indicate that a reference is cited in an appendix to a chapter, the letter of the appendix has been placed before the page number, e.g., A187. The complete references for the citations in Appendixes A, B, C, and D are located at the end of Chapters II, III, IV, and V, respectively.

G

H

I

J

K

L

M

Q

R

T

U

V

W

Y

Z

Subject Index

A

B

C

D

E

F

G

H

I

R

S

T